Exercises in PHYSICAL GEOLOGY

Seventh Edition

W. K. Hamblin
Brigham Young University
Provo, Utah

J. D. Howard
Savannah, Georgia

MACMILLAN PUBLISHING COMPANY
New York

COLLIER MACMILLAN PUBLISHERS
London

CREDITS

The authors thank the following sources for permission to reprint the maps, photographs, and images that appear in these figures:

American Association of Petroleum Geology: Figures 14.6, 15.1–15.6

American Geographical Society: Figure 18.8 (world map 1 to 5 million, sheet 14)

Department of Energy, Mines and Resources, Ottawa, Canada: Figures 8.9, 8.10, 11.2, 11.3, 12.5, 12.6, 17.20

Earth Satellite Corporation (under GEOPIC trademark), Chevy Chase, MD.: Figures 6.3B, 6.3C, 6.3E, 8.14, 9.5, 9.9, 12.7, 16.5, 17.29, 20.1

Geological Society of America: Figures 11.6, 12.2

Hubbard Scientific, Northbrook, Ill. (publisher of STEREOGRAM BOOK OF CONTOURS by H. MacMahan, Jr.): Figures 6.7–6.10

Kitt Peak Observation: Figure 19.1

United States Department of Agriculture, Agricultural Stabilization and Conservation Services, Salt Lake City, Utah: Figures 6.2H, 6.13B, 6.15, 6.17A, 6.18, 8.1, 8.2, 8.6, 8.13, 9.2, 9.3, 10.2, 10.3, 10.5, 11.5B–D, 12.3, 12.4, 13.1, 13.2, 13.4, 16.3, 16.5, 17.4, 17.6, 17.12, 17.16, 17.18, 17.22

United States Geological Survey: Figures 6.3A, 6.3G, 6.3H, 6.31, 6.12A–D, 6.13A, 7.1, 8.3–8.5, 8.7, 8.8, 8.11, 8.12, 8.15, 8.16, 9.4, 9.6, 9.8, 10.4, 10.6, 11.4, 11.5A, 12.8–12.12, 13.3, 14.1–14.4, 14.7, 16.2, 16.4, 16.6A–B, 16.7, 17.8, 17.10, 17.14, 17.23–17.28, 19.2–19.6

Woods Hole Oceanographic Institute: Figures 18.1–18.4

COVER PHOTO

Landsat mosaic of eastern Utah and western Colorado. Courtesy of Earth Satellite Corporation (under GEOPIC trademark), Chevy Chase, MD.

Printed in the United States of America

Macmillan Publishing Company
866 Third Avenue, New York, New York 10022

Collier Macmillan Canada, Inc.

ISBN 0-02-349351-8

Printing: 1 2 3 4 5 6 7 8 Year: 9 0 1 2 3 4 5 6 7 8

CONTENTS

PREFACE

This is the seventh edition of a laboratory manual that was first published in 1964. During the intervening 24 years, the advancements in geology have been greater than in any other period. The revolutionary theory of plate tectonics has been developed and is now firmly established. New technology has permitted exploration of the ocean floors. The space program has added a new perspective to the viewing of landforms and structural features of the continents, and for the first time, the surface of other planetary bodies can be compared geologically with the earth. During no other period has there been so much exploration and development of new knowledge about the earth. For this reason, we have revised this manual, in an attempt to use the new theories and discoveries.

Our objectives set forth in the first edition still stand:

1 To give students experience in examining geologic data and formulating hypotheses to explain observed facts

2 To provide an opportunity to continue laboratory-type work outside of class, so students can prepare adequately for lab sessions and review work independently

3 To give laboratory instructors maximum latitude in their instruction by providing abundant material that they can adapt to their own specific objectives.

ROCKS AND MINERALS

Rocks and minerals are the basic documents of the science of geology. In the sixth edition we prepared a new sequence of color photographs of the major rock types and minerals which continue to serve as a study guide. Most of the photographs are the actual size of the specimens, which were selected with care to show texture and physical properties. When used in the laboratory *in conjunction with the study of hand specimens*, these photographs can be an important reference and learning aid.

MAPS, AERIAL PHOTOGRAPHS, AND REMOTE SENSING IMAGES

The advancements in geology during the last several decades have brought about profound changes in the making of maps. As a result of the space program, we now have sophisticated satellite imagery of the earth's surface and radar images that can "see through" clouds. We can observe what was once unseen, and we can view the surface features of the earth from new perspectives. Landsat images of the earth can be enhanced by the computer—enlarged, manipulated in tone and color, and even reconstructed to produce stereoscopic images. In addition, most of the United States has been photographed with high-altitude infrared photography, and radar images have been made of large areas of the Appalachian region. These exciting new images are the basic data for many of the exercises in the seventh edition.

In the 1960s, the U.S. Geologic Survey published one basic series of topographic maps, the 1:24,000 scale. Now the entire nation has been mapped at a scale of 1:250,000, more than half at a scale of 1:100,000, and large areas at a scale of 1:62,500. Orthophotomaps, both in color and in black and white, have been developed to show details of the earth's surface that map symbols cannot depict easily. Perhaps more significant than producing maps at different scales is the fact that the cartographic data traditionally shown only on maps can now be digitized and manipulated by computer to display the earth's surface features in various new ways. Since graphic representations of the earth's surface features are fundamental to studying and understanding geology, we introduce students to these new maps and images, and involve them in the interpretation of the geologic processes revealed by each image.

STRUCTURAL GEOLOGY

A new series of diagrams illustrating the major structural features and their outcrop patterns is included in the section on structural geology. Portions of the geologic map of the United States have been retained from previous editions, and new maps, radar images, and computer-enhanced Landsat images have been added.

PLATE TECTONICS

The theory of plate tectonics has influenced every aspect of geology and has focused our attention on the global aspects of the science. To give students experience in analyzing geologic features on a global scale, we include a large physiographic map of the earth that serves as a basis for exercises in plate tectonics, major structural features of the continents, and geology of the ocean floor.

SEISMOLOGY AND THE EARTH'S INTERIOR

We have retained the exercise in seismology which first appeared in the sixth edition and introduces students to the way geologists study the earth's interior. Problems in seismic stratigraphy and the study of shallow geologic structures give students a chance to work with seismic records. This exercise also includes studies of *P* and *S* wave shadow zones so students can see how scientists determine the nature of the deep internal structure of the earth.

PLANETARY GEOLOGY

The exploration of the planets has added yet another dimension to the study of the earth, because it permits us to compare and contrast the geologic systems of other planetary bodies with those of our planet. We have revised this section to emphasize detailed examples from various planetary bodies that serve as contrasts to the geologic systems on the earth.

SUPPLEMENTARY EXERCISES

A major supplementary exercise at the end of the book provides students with a unique opportunity to study regional features in stereoscopic view. A color mosaic of Landsat images of the Paradox Basin in Utah and Colorado has been restructured by computer so a stereoscopic image can be obtained. (The images must be removed from the manual in order to be used as a stereoscopic pair.)

ACKNOWLEDGMENTS

We are grateful to our colleagues for their many constructive criticisms, comments, and suggestions. The wide diversity in scope and content of the laboratory experience in physical geology is apparent. We have altered exercises when the consensus was clear to do so, but have relied on our own judgment when opinions varied.

We also thank Wards Natural History Establishment, Rochester, New York, for letting us use many specimens to make photographs of rocks and minerals, and the Photo Mechanix Studios, Bloomington, Minnesota, for their excellent work in photographing rock and mineral specimens.

We are especially grateful to GeoScience Resources for the excellent photomicrographs used in the rocks and minerals section. Extra effort by the GeoScience Resources staff to obtain the best specimens available combined with their high quality photography provides students with a unique visual aid to the understanding of rock and mineral composition.

The U.S. Geological Survey topographic maps, which appear in blue, brown, and black in this book, are made from plates obtained from the U.S. Geological Survey maps originally published in several colors. These maps are specially edited and designed for teaching purposes and are not intended to duplicate the complete maps. We are grateful to the U.S. Geological Survey for permission to reproduce these topographic maps, sections of geologic maps, and aerial photographs.

We also thank the Agricultural Stabilization and Conservation Services of the U.S. Department of Agriculture for the use of aerial photographs from their files, and the Canadian Geological Survey and the Mapping Branch of the Department of Energy, Mines and Resources of Canada for permission to publish their aerial photographs.

We sincerely thank Dale Claflin, who spent many hours preparing illustrations.

Exercise 1

CRYSTAL GROWTH

OBJECTIVE

To gain an understanding of the crystalline structure of minerals by observing the process of crystal growth.

MAIN CONCEPT

Minerals grow by crystallization, a process by which atoms are arranged in a specific geometric pattern.

SUPPORTING IDEAS

1 Crystal form is an expression of the mineral's atomic structure.

2 All crystals of a given mineral possess the same properties of symmetry, even though the crystals may differ in size and shape.

3 Similar pairs of crystal faces of a given mineral always meet at the same angle (law of constancy of interfacial angles).

DISCUSSION

A mineral is composed of elements or groups of elements that unite in nature to form an inorganic crystalline solid. Every crystalline substance has a definite internal structure in which the atoms occur in specific proportions and are arranged in an orderly geometric pattern. This systematic arrangement of atoms is one of the most significant aspects of a mineral. It exists throughout the entire specimen, and if crystallization occurs under ideal conditions, the arrangement will be expressed in perfect crystal faces. For example, moisture in the air may freeze and develop into ice crystals. The form of the solid ice crystals is an external expression of the orderly arrangement of the water molecules. Although each individual crystal of ice is different in size and shape, all possess the same properties of symmetry. Similarly, a solution rich in sodium (Na) and chlorine (Cl) ions will, on evaporation, develop into crystals of halite (NaCl), or common salt. All crystals of halite will have the same internal structure and the same properties of symmetry in their crystal form.

Not all mineral crystals have perfect faces or perfect crystal form because of restrictions in the environment in which they crystallize. Minerals with poor or imperfect crystal faces do, however, still possess a systematic internal structure, as indicated by properties such as cleavage and the symmetrical patterns produced from X-ray studies. In contrast, the atoms in a liquid or gas have little or no systematic structure and no definite form. Substances that become solid without crystallizing are *amorphous,* that is, without a definite internal structure. Glass, for example, is rigid and solid, but the atoms of glass are not arranged in a systematic pattern, so physicists regard glass as an extremely viscous liquid.

A significant feature that expresses the internal structure of crystals is the constancy of their interfacial angles. The angles between similar crystal faces of a specific mineral will be the same, even though the size of the crystals may vary. This unique feature is of prime importance in differentiating minerals that are similar in other respects. Numerous precise measurements of the angles between crystal faces have demonstrated repeatedly that although crystals of the same substance may exhibit different overall shapes, the angles between corresponding crystal faces are identical. This phenomenon is known as the *law of constancy of interfacial angles.*

GROWTH OF CRYSTALS FROM SOLUTION

Many minerals are precipitated from aqueous solutions by evaporation at atmospheric temperature and pressure. Crystal growth of such minerals can be observed in the laboratory by evaporating prepared concentrated solutions.

PROCEDURE

Your instructor will provide a concentrated solution of sodium chloride, sodium nitrate, potassium aluminum sulfate (alum), and copper acetate.

Place a drop of each solution on a slide. Label each solution on a piece of tape placed at the end of the slide. As the water evaporates, crystals of each compound will appear, and as evaporation continues, the crystals will grow larger. Observe the slide through a microscope or magnifying glass at several different times during the process of crystal growth. (While you are waiting for sufficient evaporation to occur to form crystals, you may begin the next part of this exercise concerning the growth of crystals from a melt.)

PROBLEMS

1 Sketch and label two crystals from each compound.

2 Which compound has a cubic (box-like) form?

3 Why aren't all of the crystals perfect cubes?

4 Do the crystals of each compound satisfy the law of constancy of interfacial angles?

5 Does each compound have a unique crystal form?

6 Can minerals be identified from crystal form alone?

GROWTH OF CRYSTALS FROM A MELT

Most minerals that originate from a melt (or magma) crystallize at temperatures well above the boiling point of water, so observing their crystal growth directly is difficult. Thymol, an organic chemical, crystallizes near room temperature, and although it differs from a magma, certain principles of crystallization demonstrated in a thymol melt pertain generally to a magma. Thymol is not poisonous but should be handled with forceps, for it can irritate the skin and eyes.

In performing the experiment, remember to do the following:

1 Observe crystal growth and distinguish a single crystal from an aggregate of crystals.

2 Study the law of constancy of interfacial angles by observing crystals growing from a melt solution.

3 Discover for yourself some of the factors that govern crystal growth.

SLOW COOLING WITHOUT "SEED" CRYSTALS

Place a petri dish containing a small amount of crystalline thymol on a hot plate, adjusted to low heat, until all of the crystals are melted. Let the melt continue to heat for 1 or 2 minutes more, then remove the dish from the hot plate and set it aside to cool slowly where it will not be disturbed. You will examine this melt near the end of the lab.

SLOW COOLING WITH "SEED" CRYSTALS

Repeat the procedure just described, but transfer the petri dish to the stage of the microscope *as soon as the thymol is melted.* Add four or five "seed" crystals to the melt, and observe the melt under the microscope or with a magnifying lens. If the seed crystals melt, wait for a few moments and add seed crystals again. As the melt cools, observe the crystals beginning to grow.

PROBLEMS

1 Sketch the shape of one or two single crystals and describe briefly the manner of crystal growth. (Consider such factors as internal expansion versus external accretion, rate of growth, direction of growth, angles between crystal faces, the effect of limited space on crystal shape, the difference between a single crystal and an aggregate of crystals, and the texture produced by interlocking crystals.)

2 How do the growth lines of a crystal compare with the outline of a single crystal in a solid aggregate?

3 What role do the "seed" crystals play in initiating crystal growth?

4 When less than half of the melt remains, you may notice the formation of some spontaneous nucleating crystals. Study these as they grow. Make a sketch of the final crystalline aggregate. Individual crystals may be recognized by characteristic growth lines and reflection of light. The aggregate of crystals produced in this experiment is similar to that found in many common igneous rocks. Both are composed of many interlocking crystals originating from a melt.

5 Examine several specimens of granite supplied by your instructor. Discriminate between the individual crystals, and make a sketch of the texture of the granite. How does the texture of an aggregate of thymol crystals compare with the texture of the minerals in a granite?

RAPID COOLING

Place a petri dish containing a small amount of thymol on a hot plate until only a few crystals remain. Transfer the dish to the top of an ice cube, and let the melt cool there for a fraction of a minute. As the melt cools rapidly, observe the crystals beginning to grow. Almost immediately, several independent, spontaneous nucleating centers of crystallization will develop. Quickly transfer the petri dish to the stage of a microscope and observe the nature of the crystal growth. Repeat the experiment as often as time permits, so you are sure of your general observations and conclusions.

COMPARISON OF CRYSTALS PRODUCED BY SLOW COOLING AND RAPID COOLING

Examine the thymol in the dish prepared for the first part of this exercise concerned with slow cooling without "seed" crystals.

PROBLEM

1 What effect does the rate of cooling have on crystal size?

Exercise 2

MINERALS

OBJECTIVES

To become familiar with the common physical properties of minerals and to be able to use these properties in mineral identification.

MAIN CONCEPT

Each mineral species is characterized by a specific (atomic) internal structure and by a specific chemical composition that varies only within certain limits. As a result, all specimens of a given mineral will possess certain diagnostic physical properties.

SUPPORTING IDEAS

1 The more important physical properties of minerals are: (a) crystal form, (b) cleavage, (c) hardness, (d) specific gravity, (e) streak, and (f) luster.

2 Other physical properties such as color, tenacity, and magnetism are important diagnostic properties of some minerals.

3 The physical properties of a mineral can be used to identify mineral specimens.

4 The most important rock-forming minerals are quartz, feldspar, olivine, pyroxene, amphibole, mica, calcite, and clay minerals.

DISCUSSION

PHYSICAL PROPERTIES OF MINERALS

CRYSTAL FORM. When a crystal is allowed to grow in an unrestricted environment, it will develop natural crystal faces that produce a perfect geometric pattern. The shape of such a crystal is a reflection of the internal structure and can be used to identify many mineral species. Remember, however, that two or more different minerals may have essentially the same internal structure and may develop similar crystals. Some of the common minerals in which crystal form is especially diagnostic are quartz, halite, garnet, fluorite, pyrite, galena, amphibole, and pyroxene.

CLEAVAGE. The forces that hold the atoms together in a crystalline structure are not necessarily equal in all directions. If definite planes of weakness exist, the mineral will cleave, or break, along the planes of weakness much more easily than in other directions. The surface along which the break develops is referred to as the *cleavage plane,* and the orientation of the plane is the *cleavage direction.* Crystals may have one, two, three, four, or six cleavage directions, as illustrated in Figures 2.1A–2.1G. Some minerals, such as gypsum, cleave in multiple directions but possess better cleavage in one direction than in another.

Perfect cleavage is recognized easily, for it develops a characteristically smooth, even surface that reflects light like a mirror (Figure 2.2A). Cleavage planes can occur in a small, steplike manner, however, and may appear at first to be an irregular fracture. If the specimen is rotated in front of a light, these small, parallel cleavage planes will reflect light just as a large, smooth cleavage surface does (Figure 2.2B). In contrast, an uneven fracture will not concentrate light in any particular direction (Figure 2.2C).

Keep in mind the distinction between cleavage planes and crystal faces. Cleavage results from planes of weakness within the crystal structure along which the crystal breaks. Crystal faces reflect the geometry of the atomic structure. You should also remember that both are properties of single crystals. In a crystalline aggregate, the individual crystals will break along their own cleavage planes, if any exist, but the aggregate does not possess cleavage.

In some minerals, the crystal structure is so well knit that there is no tendency to break along one plane in preference to another. Such minerals do not possess cleavage, but break or fracture in an irregular manner. Common fractures are conchoidal, irregular, and fibrous.

HARDNESS. Hardness is a measure of the ability of a mineral to resist abrasion. Like cleavage and fracture, this property is also related to crystal structure and to interatomic bond strength.

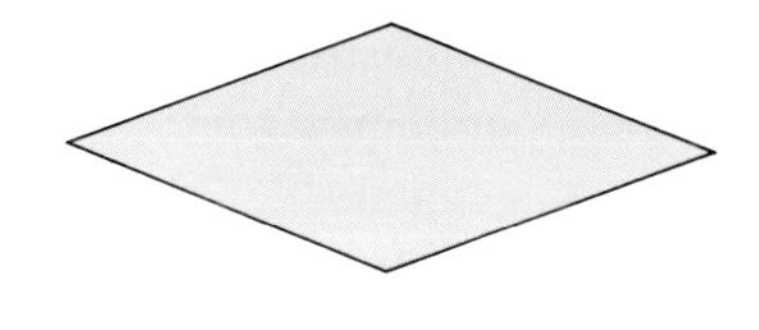

A. CLEAVAGE IN ONE DIRECTION. Example: muscovite.

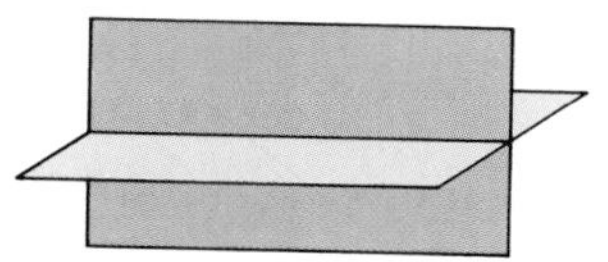

B. CLEAVAGE IN TWO DIRECTIONS AT RIGHT ANGLES. Example: feldspar.

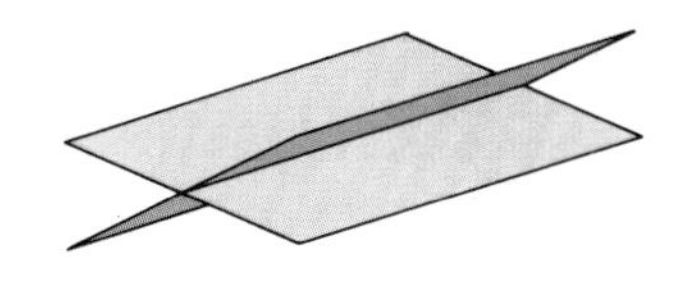

C. CLEAVAGE IN TWO DIRECTIONS NOT AT RIGHT ANGLES. Example: amphibole.

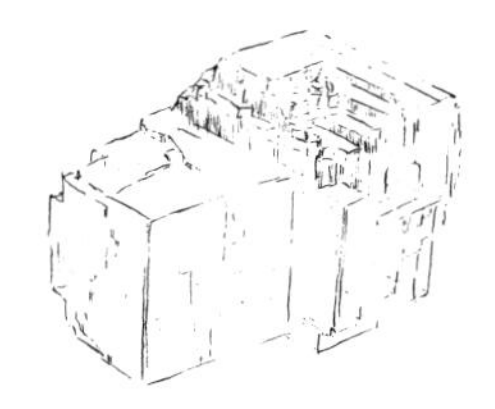
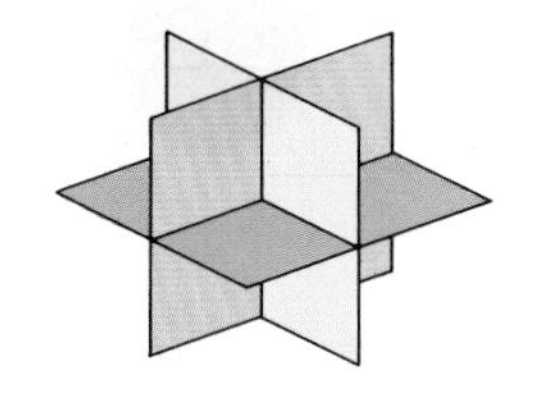

D. CLEAVAGE IN THREE DIRECTIONS AT RIGHT ANGLES. Example: halite.

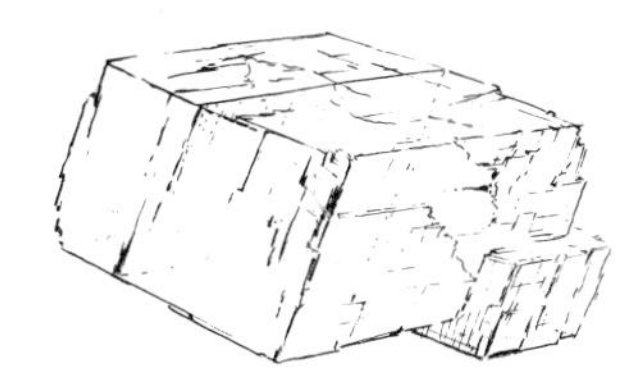

E. CLEAVAGE IN THREE DIRECTIONS NOT AT RIGHT ANGLES. Example: calcite.

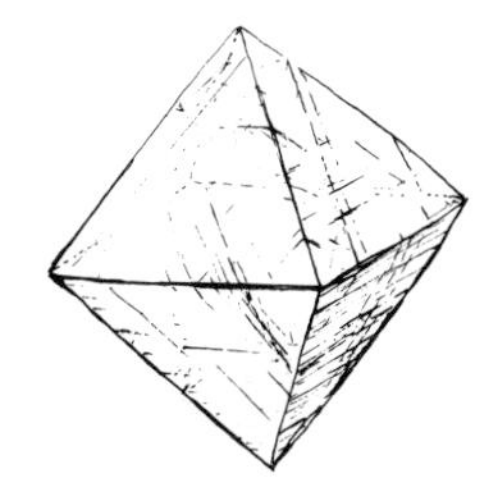
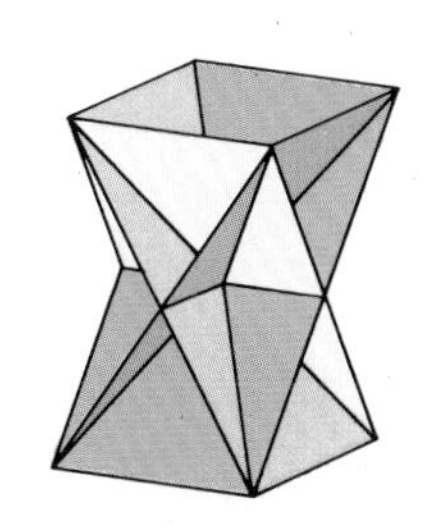

F. CLEAVAGE IN FOUR DIRECTIONS. Example: fluorite.

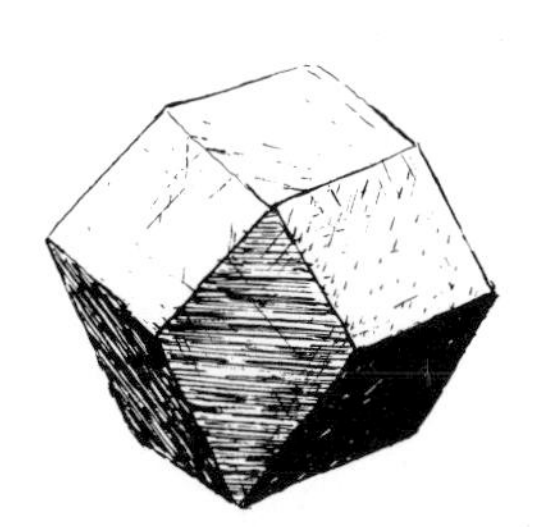
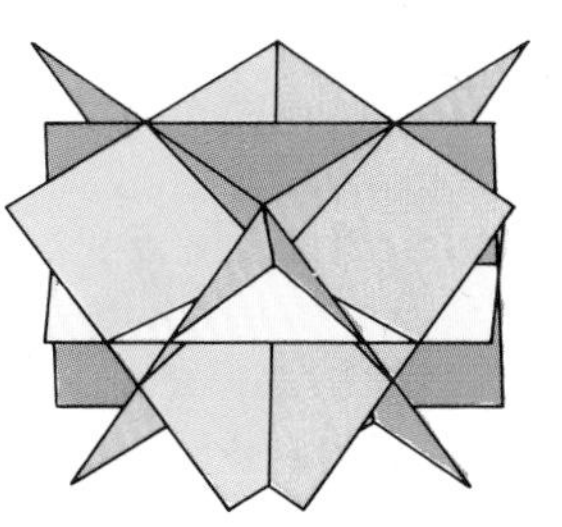

G. CLEAVAGE IN SIX DIRECTIONS. Example: sphalerite.

FIGURE 2.1 **CRYSTAL CLEAVAGE.**

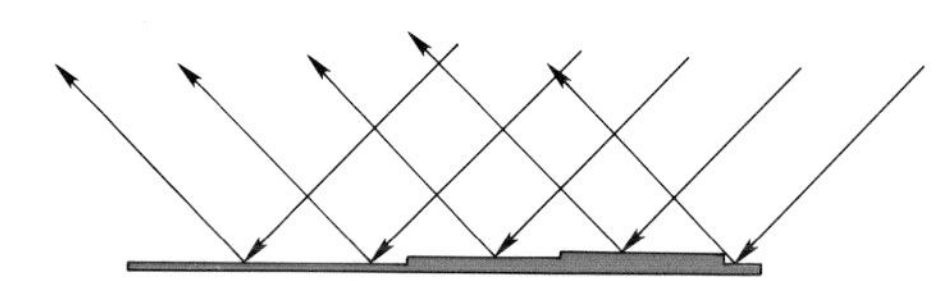

A. REFLECTED LIGHT FROM SMOOTH CLEAVAGE.

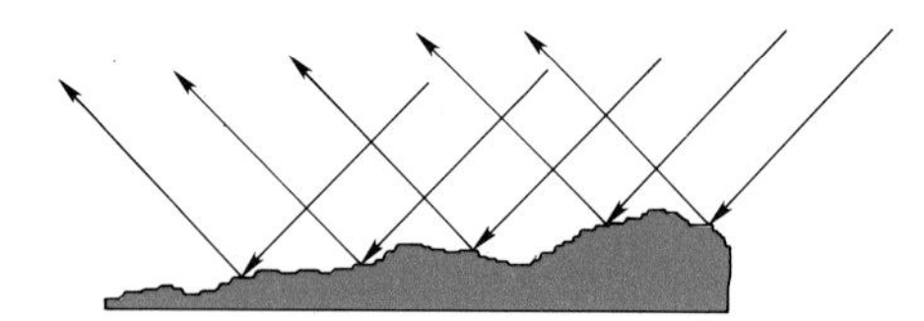

B. REFLECTED LIGHT FROM STEPPED CLEAVAGE.

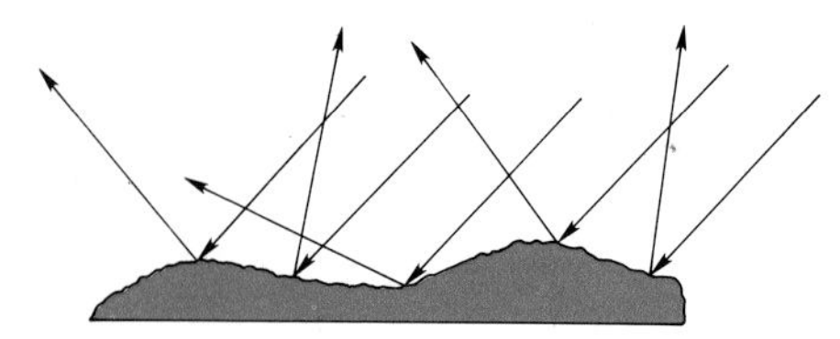

C. REFLECTED LIGHT FROM FRACTURE.

FIGURE 2.2

REFLECTION OF LIGHT FROM CLEAVAGE AND FRACTURE SURFACES.

Hardness is a relative property and was one of the first properties used to determine unknown minerals in the field. If two minerals are rubbed together, the harder one scratches the softer. Over a century ago, the German mineralogist Friedrich Mohs assigned an arbitrary relative number to ten common minerals. Diamond, the hardest mineral known, was placed at the top of his hardness scale and assigned the number 10. Softer minerals were ranked in descending order, perhaps without a realization of the unequal degree of hardness between the various ranks. More scientific hardness tests have been devised since in which the absolute hardness values can be measured. A comparison of Mohs scale and an absolute hardness scale is shown in Figure 2.3A. Among the softer minerals, there is little difference in quantitative hardness, but rather large steps occur between quartz, topaz, and corundum, and above all between corundum and diamond.

The Mohs scale of hardness and the hardness of some common everyday objects (Figure 2.3B) can be used effectively to test the hardness of unknown minerals. For example, calcite will scratch a fingernail, gypsum, and talc, but it will not scratch a knife blade, fluorite, or quartz. The Mohs scale is not exact, but the convenience of the scale more than makes up for its lack of precision. It is used often as a simple preliminary test of a mineral's hardness property.

Hardness is generally a reliable diagnostic physical property of a mineral, but variations in composition may render some mineral specimens harder or softer than normal. Moreover, because weathering can affect hardness, making tests on fresh surfaces is important

SPECIFIC GRAVITY. Specific gravity is the ratio between the mass of a mineral and the mass of an equal volume of water. It is one of the most constant physical properties of a mineral. You can determine a mineral's specific gravity accurately by dividing its weight in air by the weight of an equal volume of water. For purposes of general laboratory work and field work, however, you can estimate specific gravity with surprising accuracy simply by lifting a mineral specimen in your hand and making an estimate. Galena, with a specific gravity of approximately 7.5, and pyrite with a specific gravity of about 5, are typical of minerals with high specific gravity. They are distinguished easily from common rock-forming minerals such as quartz, feldspar, and calcite, which have a specific gravity between 2.6 and 2.8.

COLOR. Color is one of the most obvious physical properties, and for some minerals, such as galena (gray), azurite (blue), and olivine (green), it is diagnostic. Other minerals are found in differing hues, which may be caused by variations in composition or inclusion of impurities. Quartz, for example, ranges through a spectrum of colorless clear crystals to purple, red, white, and jet black. In brief, color may be diagnostic for a few minerals, but in others it is almost without diagnostic significance. *Color should be considered in mineral identification but should never be used as the major identifying characteristic.*

STREAK. When a mineral is powdered, it usually exhibits a much more diagnostic color than when it occurs in large pieces. The color of the powdered min-

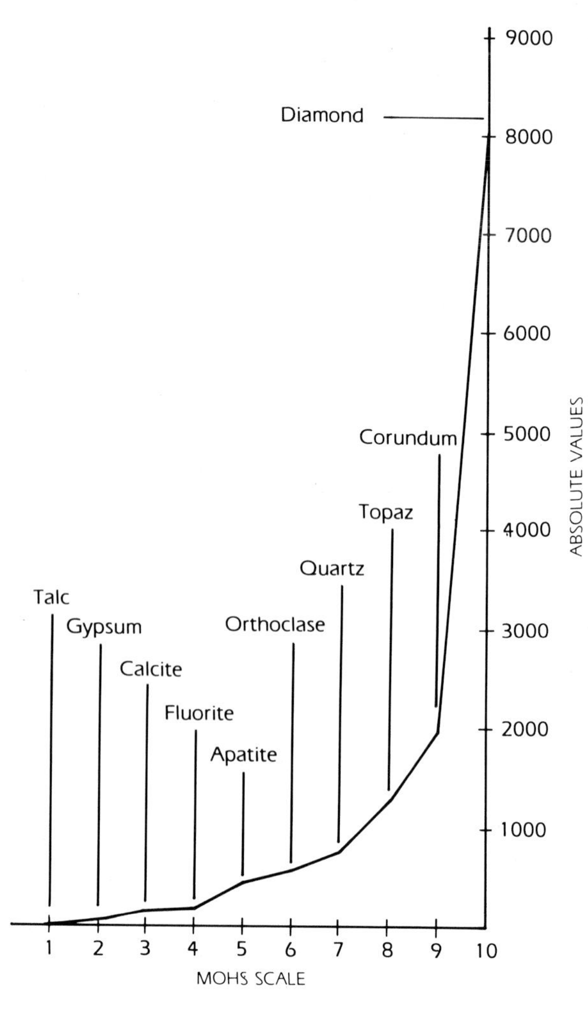

A. MOHS HARDNESS SCALE COMPARED WITH AN ABSOLUTE HARDNESS SCALE.

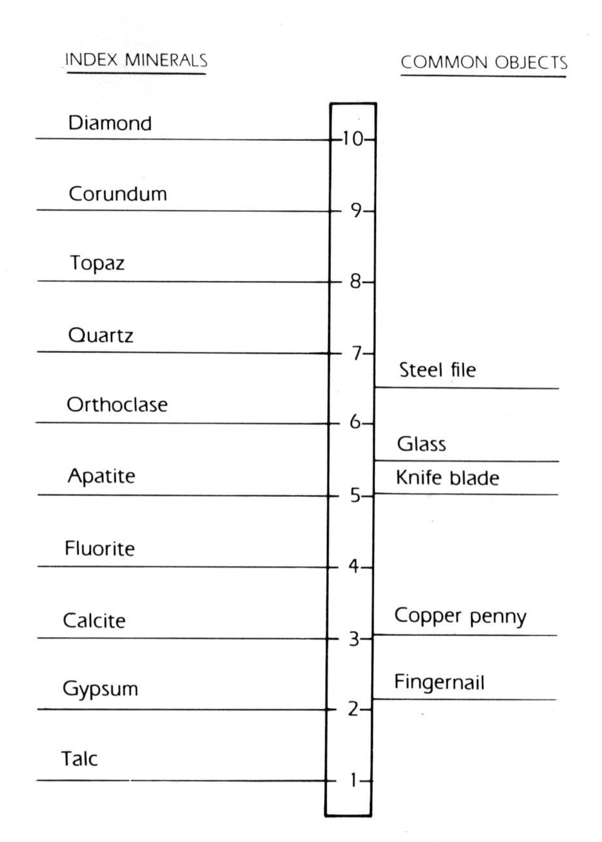

B. MOHS HARDNESS SCALE COMPARED WITH THE HARDNESS OF SOME COMMON EVERYDAY OBJECTS.

FIGURE 2.3 **HARDNESS SCALES.**

eral is referred to as *streak.* In the laboratory, streak is obtained by rubbing the mineral vigorously on an unglazed porcelain plate. Commonly, the streak of a mineral will be different from the color seen in the hand specimen, so do not anticipate the streak color by simple visual examination of a mineral fragment.

Most minerals with a nonmetallic luster have a white or pastel streak. For this reason, streak is not very useful in distinguishing nonmetallic minerals. Minerals with a hardness greater than that of porcelain will scratch the plate and will not produce streak.

LUSTER. The terms "metallic" and "nonmetallic" describe broadly the basic types of *luster,* or the appearance of light reflected from a mineral. Pyrite and galena have a metallic luster on unweathered surfaces. A variety of nonmetallic lusters can be distinguished, the most important of which are described as vitreous, pearly, resinous, silky, and greasy. To become acquainted with these, study examples of each luster type from specimens provided by your instructor. Some minerals, because of their characteristically weathered or porous nature, have a distinctive dull or earthy luster.

TENACITY. The manner in which a mineral resists breakage is called *tenacity.* This property is described by the following terms:

1 Brittle—crushes to angular fragments. Example: quartz.

2 Malleable—can be modified in shape without breaking and can be flattened to a thin sheet. Example: native copper.

3 Sectile—can be cut with a knife into thin shavings. Example: talc.

4 Flexible—will bend but does not regain its original shape when force is released. Example: gypsum.

5 Elastic—will bend and regain its original shape when force is released. Example: muscovite or biotite.

OPTICAL PROPERTIES. Minerals have a number of important optical properties, most of which require a polarizing microscope for effective study. The general way in which a mineral transmits light can, however, be observed in hand specimens and is useful in identifying some minerals. Optical properties are described as follows:

1 Transparent—objects are visible when viewed through the mineral. Example: some species of quartz, calcite, or biotite.

2 Translucent—light, but not an image, is transmitted through the mineral. Example: some varieties of gypsum.

3 Opaque—no light is transmitted, even on the thinnest specimen edges. Example: magnetite or pyrite.

REACTION TO HYDROCHLORIC ACID. Calcite ($CaCO_3$) is one of the most common minerals on the earth's surface, and when treated with dilute hydrochloric acid (HCl), it will bubble or effervesce vigorously. This simple chemical test is very diagnostic and can be used to distinguish calcite from most of the other common minerals. Dolomite ($CaMg(CO_3)_2$), a mineral similar to calcite, will react with cold, dilute hydrochloric acid, but only if the specimen is powdered.

DOUBLE REFRACTION. This is a property of the transparent variety of calcite whereby light passing through the crystal or cleavage fragment is split into two rays. An object viewed through the mineral therefore shows a double image.

MAGNETISM. Magnetite is one of the few minerals to show obvious magnetic attraction. Some other minerals do have distinctive magnetic properties that can, with proper equipment, be used for mineral separation, but magnetite is the only common mineral that is actually attracted by a small magnet.

TASTE. The salty taste of halite is a definite and unmistakable property of that mineral. Other evaporite minerals usually have a bitter taste but are not exceedingly common.

FIGURE 2.4

MINERAL CLASSIFICATION CHART.

METALLIC LUSTER

Streak	Properties	Mineral
Gray streak	**Perfect cubic cleavage;** heavy, Sp. Gr. = 7.6; H = 2.5; **silver gray color**	GALENA PbS
Black streak	**Magnetic;** black to dark gray; Sp. Gr. = 5.2; H = 6; commonly occurs in granular masses; single crystals are octahedral	MAGNETITE Fe_3O_4
Gray to black streak	**Steel gray;** soft, smudges fingers and marks paper, greasy feel; H = 1; Sp. Gr. = 2; luster may be dull	GRAPHITE C
Greenish black streak	**Golden yellow color;** may tarnish purple; H = 4; Sp. Gr. = 4.3	CHALCOPYRITE $CuFeS_2$
	Brass yellow; cubic crystals; common in granular aggregates; H = 6–6.5; Sp. Gr. = 5; uneven fracture	PYRITE FeS_2
Reddish brown streak	**Steel gray; red to red-brown streak;** black to dark brown; granular, fibrous, or micaceous; single crystals are thick plates; H = 5–6; Sp. Gr. = 5; uneven fracture	HEMATITE Fe_2O_3
Yellow brown streak	**Yellow, brown, or black;** hard structureless or radial fibrous masses; H = 5–5.5; Sp. Gr. = 3.5–4; yellow brown streak.	LIMONITE $Fe_2O_3 \cdot H_2O$

non met.- dark color, Bright green, H. 3.5-4, fibrous crystals or granular masses. sol in HCl with Effervescence — malachite $CuCO_3\ Cu(OH)_2$

Azure to lt. blue, tabular crystals, granular crusts, etc. H. 3.5-4, lighter blue streak. — Azurite $2CuCO_3\ Cu(OH)_2$

NONMETALLIC LUSTER—DARK COLOR

Hardness	Cleavage	Properties	Mineral
Harder than glass	Cleavage prominent	**Cleavage—2 directions nearly at 90°;** dark green to black; short prismatic 8-sided crystals; H = 6; Sp. Gr. = 3.5	PYROXENE GROUP Complex Ca, Mg, Fe, Al silicates
		Cleavage—2 directions at approximately 60° and 120°; dark green to black or brown; long prismatic 6-sided crystals; H = 6; Sp. Gr. = 3.35	AMPHIBOLE GROUP Complex Na, Ca, Mg, Fe, Al silicates
		Gray to blue-gray; good cleavage in two directions at approximately 90°; striations on cleavage planes	PLAGIOCLASE FELDSPAR $NaAlSi_3O_8$ to $CaAl_2Si_2O_8$
	Cleavage absent	**Various shades of green; sometimes yellowish; commonly occurs in aggregates of small glassy grains;** transparent to translucent; glassy luster; H = 6.5–7; Sp. Gr. = 3.5–4.5	OLIVINE $(Fe, Mg)_2SiO_4$
		Red, brown, or yellow; glassy luster; conchoidal fracture resembles poor cleavage; commonly occurs in well-formed 12-sided crystals; H = 7–7.5; Sp. Gr. = 3.5–4.5	GARNET GROUP Fe, Mg, Ca, Al silicates
		Conchoidal fracture; H = 7; gray to gray-black; vitreous luster	QUARTZ SiO_2
Softer than glass	Cleavage prominent	**Brown to black; 1 perfect cleavage;** thin, flexible, and elastic when in thin sheets; H = 2.5–3; Sp. Gr. = 3–3.5	BIOTITE $K(Mg, Fe)_3AlSi_3O_{10}(OH)_2$
		Green to very dark green; 1 cleavage direction; commonly occurs in foliated or scaly masses; nonelastic plates; H = 2–2.5; Sp. Gr. = 2.5–3.5	CHLORITE Hydrous Mg, Fe, Al silicate
		Yellowish brown; resinous luster; cleavage in 6 directions; yellowish brown or nearly white streaks; H = 3.5–4; Sp. Gr. = 4	SPHALERITE ZnS
		Four perfect cleavage directions; H = 4; Sp. Gr. = 3; green through deep purple; transparent to translucent; cubic crystals	FLUORITE CaF_2
	Cleavage absent	**Red, earthy appearance;** red streak; H = 1.5	HEMATITE Fe_2O_3 (earth variety)
		Yellowish brown streak; yellowish brown to dark brown; commonly in compacted earth masses; H = 1.5	LIMONITE $Fe_2O_3 \cdot H_2O$

NOTE: The most diagnostic properties for each mineral are indicated by bold type.

NONMETALLIC LUSTER—LIGHT COLOR

Hardness	Cleavage	Properties	Mineral
Harder than glass	**Cleavage prominent**	**Good cleavage in 2 directions at approximately 90°; commonly flesh colored to dark pink,** pearly to vitreous luster; H = 6–6.5; Sp. Gr. = 2.5	**Potassium feldspars:** $KAlSi_3O_8$
		Good cleavage in 2 directions at approximately 90°; white to gray; striations on some cleavage planes	**Plagioclase feldspars:** $NaAlSi_3O_8$ to $CaAl_2Si_2O_8$
	Cleavage absent	**Conchoidal fracture;** H = 7; Sp. Gr. = 2.65; transparent to translucent; vitreous luster; 6-sided prismatic crystals terminated by 6-sided triangular faces in well-developed crystals; vitreous to waxy; colors range from milky white, rose pink, violet, to smoky gray	QUARTZ SiO_2 (silica) **Varieties:** Milky; Smoky; Rose; Amethyst
		Conchoidal fracture; H = 7; variable color; translucent to opaque; dull or clouded luster; colors range widely from white, gray, red, to black	CRYPTOCRYSTALLINE QUARTZ SiO_2 **Varieties:** Agate; Flint; Chert; Jasper; Opal
Softer than glass	**Cleavage prominent**	**Perfect cubic cleavage; salty taste;** colorless to white; soluble in water; H = 2–2.5; Sp. Gr. = 2	HALITE $NaCl$
		Perfect cleavage in 1 direction; poor in 2 others; H = 2; white; transparent; nonelastic; Sp. Gr. = 2.3 **Varieties:** Selenite: colorless, transparent; Alabaster: aggregates of small crystals; Satin spar: fibrous, silky luster	GYPSUM $CaSO_4 \cdot 2H_2O$
		Perfect cleavage in 3 directions at approximately 75°; effervesces in HCl; H = 3; colorless, white, or pale yellow, rarely gray or blue; transparent to opaque; Sp. Gr. = 2.7	CALCITE $CaCO_3$ (fine-grained crystalline aggregates form limestone and marble)
		Three directions of cleavage as in calcite; effervesces in HCl only if powdered; H = 3.5–4; Sp. Gr. = 2.8; color variable but commonly white or pink; rhomb-faced crystals	DOLOMITE $CaMg(CO_3)_2$
		Good cleavage in 4 directions; H = 4; Sp. Gr. = 3; colorless, yellow, blue, green, or violet; transparent to translucent; cubic crystals	FLUORITE CaF_2
		Perfect cleavage in 1 direction, producing thin, elastic sheets; H = 2–3; Sp. Gr. = 2.8; transparent and colorless in thin sheets	MUSCOVITE $KAl_3AlSi_3O_{10}(OH)_2$
		Green to white; soapy feel; pearly luster; H = 1; Sp. Gr. = 2.8; foliated or compact masses; one direction of cleavage forms thin scales and shreds	TALC $Mg_3Si_4O_{10}(OH)_2$
	Cleavage absent	**White to red; earthy masses;** crystals so small no cleavage visible; soft, H = 1.2; becomes plastic when moistened; earthy odor	KAOLINITE $Al_4Si_4O_{10}(OH)_8$

Non Met luster - light color
Yellow, reb-brown, crystals tabular or needle-like or in granular masses and crusts. Soft. H = 1.5 - 2.5. luster resinous Sulfur odor when rubbed or heated

DESCRIPTION OF COMMON ROCK-FORMING MINERALS

QUARTZ

Quartz is one of the most common minerals that you are likely to encounter because it is a major constituent in many igneous, sedimentary, and metamorphic rocks. If allowed to grow in an unrestricted environment, quartz will form well-developed hexagonal (six-sided) crystals that terminate in a pyramid. The crystal form of quartz is thus a distinctive property. In most rocks, however, the crystal form of quartz grains is not expressed. In igneous rocks and in veins, quartz is one of the last minerals to form, so it fills the available space between other minerals, and the crystals thus have irregular shapes. In sedimentary rocks, quartz grains are rounded by the abrasion they have undergone during transportation by wind and water. In metamorphic rocks, quartz grains and other mineral grains are generally deformed and recrystallized. The quartz may therefore appear in a variety of sizes and shapes.

The color, crystal size, and general appearance of quartz vary greatly, so a large number of varieties of quartz have been named. These can be classified in two major groups: (1) macrocrystalline quartz and (2) cryptocrystalline quartz. Further subdivision in each group can be made on the basis of color or some other special feature. However, regardless of such features as color, crystal size, crystal shape, and mode of origin, all quartz is characterized by the following: (1) hardness of 7 (it will scratch glass and steel), (2) conchoidal fracture, and (3) glassy luster.

Macrocrystalline quartz is composed of individual crystals that can be seen with the naked eye or with low-power magnification. Varieties are distinguished mainly on the basis of color, the major varieties being clear, rose, milky, smoky, and amethyst. Cryptocrystalline quartz is composed of aggregates of innumerable submicroscopic crystals. Color variations give rise to many varieties such as chert, flint, agate, jasper, and opal (see Figures 2.14–2.19).

FIGURE 2.5

QUARTZ CRYSTALS often develop in groups of slender, six-sided prisms that terminate in a pyramid. Striations are common across the prism faces. Crystals are usually milky, colorless, or transparent.

FIGURE 2.6

QUARTZ CRYSTALS grow in a variety of sizes, shapes, and colors. In every quartz crystal, regardless of where, when, or how it grew, the angles between corresponding crystal faces are identical.

FIGURE 2.7

AMETHYST is a transparent quartz crystal in purple or violet hues that usually develops as a secondary mineral in veins and cavities. The color is due to small inclusions of ferric iron.

FIGURE 2.8

MILKY QUARTZ occurs in veins and is commonly massive, without well-developed crystal faces. Minute fluid inclusions produce the white color.

FIGURE 2.9

SMOKY QUARTZ is a brown-, gray-, or black-colored quartz characteristic of intrusive igneous rocks. It is transparent to translucent in luster. Radiation from radioactive minerals will develop a smoky appearance in colorless quartz and may be responsible for much smoky quartz in nature.

FIGURE 2.10

ROSE QUARTZ—The color is due to small amounts of titanium. Rose quartz is generally coarsely crystalline, transparent to translucent, and rarely forms well-developed crystals.

FIGURE 2.11

QUARTZ IN GRANITE—Quartz is one of the last minerals to form in granite, and thus fills the spaces between crystals of other minerals formed during earlier periods of crystallization. Quartz in granite is characterized by glassy luster and conchoidal fracture, and is usually smoky.

FIGURE 2.12

QUARTZ IN SANDSTONE—Most sand grains are quartz crystals that have been rounded and abraded during transportation by wind or water. The surface of some grains may be pitted or frosted. Broken grains show glassy luster and conchoidal fracture.

FIGURE 2.13

QUARTZ IN QUARTZITE—Heat and pressure may fuse sand grains to form a dense, massive rock called quartzite. Quartzite is resistant to weathering and abrasion, so quartzite pebbles are common constituents of many gravels.

FIGURE 2.14

CHERT is a light-colored variety of cryptocrystalline quartz. It occurs commonly as nodules in limestone.

FIGURE 2.15

FLINT is similar to chert but is characterized by its dark color. Both break with a conchoidal fracture.

FIGURE 2.16

AGATE is a variety of cryptocrystalline quartz in which colorful banding occurs due to successive periods of deposition.

FIGURE 2.17

JASPER is a variety of cryptocrystalline quartz with a distinctive red color produced by minute inclusions of hematite.

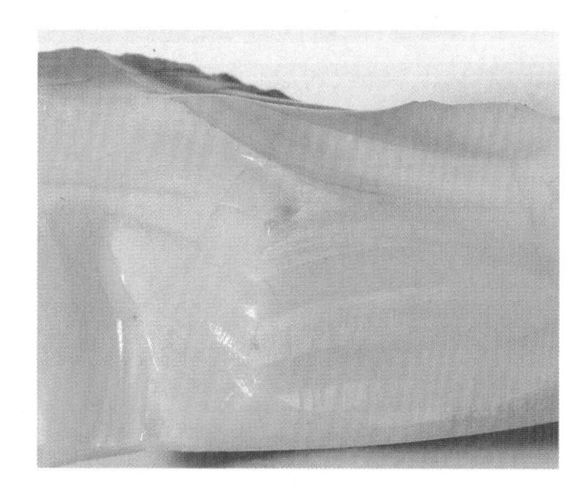

FIGURE 2.18

OPAL is a variety of cryptocrystalline quartz that contains various amounts of water. It is distinguished from other varieties by its waxy luster.

FIGURE 2.19

PETRIFIED WOOD—Cryptocrystalline quartz commonly replaces organic material in the process of fossilization. In petrified wood, many delicate structures are preserved because of the small crystal size.

THE FELDSPAR GROUP

Feldspar is the most abundant mineral in the earth's crust. Like quartz, it is a basic constituent of many igneous, sedimentary, and metamorphic rocks. All feldspars are aluminum silicates of either potassium, sodium, or calcium, and all are related closely in form and physical properties. Two main subgroups are recognized: (1) the plagioclase feldspars, which range in composition from $CaAl_2Si_2O_8$ to $NaAlSi_3O_8$, and (2) the potassium feldspars, which have a composition of $KAlSi_3O_8$. Varieties of potassium feldspar occur because of variations in the arrangement of Al and Si ions in the crystal structure. The varieties of feldspar can best be distinguished by chemical analysis, or by optical or X-ray measurements of the crystal structure. Color in feldspars is quite variable and is due to small amounts of iron or magnesium, and to certain aspects of the crystal structure. Color is *not* diagnostic of a particular variety of feldspar. The most important physical properties of the feldspars are: (1) two-directional cleavage at approximately right angles, (2) hardness of 6, and (3) pearly luster.

FIGURE 2.20

SODIUM PLAGIOCLASE (ALBITE)—Sodium plagioclase feldspar is commonly light colored and has a porcelain luster. The two-directional cleavage, characteristic of all feldspars, is well expressed in most specimens.

FIGURE 2.21

CALCIUM PLAGIOCLASE (LABRADORITE)—Striations on certain cleavage surfaces are diagnostic of calcium plagioclase and distinguish this group from the potassium feldspars. Ca-plagioclase is dark colored and shows iridescence.

FIGURE 2.22

POTASSIUM FELDSPAR—Although color is generally not diagnostic, most pink and green feldspars belong to the potassium subgroup. Well-developed cleavage in two directions is similar to that in plagioclase feldspar.

FIGURE 2.23

FELDSPARS IN GRANITE—Potassium feldspar is a major constituent of granite and is commonly expressed by rectangular pink crystals. Sodium feldspar in granite is characteristically white. Both may occur in the same rock specimen.

FIGURE 2.24

FELDSPAR IN SANDSTONE—Weathering and erosion of a granite will disaggregate the feldspar and quartz. Feldspars break down rapidly by weathering to form clay minerals and are therefore not a major constituent of most sediments. Angular grains of feldspar are, however, found in some sandstones deposited adjacent to a granite source.

FIGURE 2.25

PHOTOMICROGRAPH OF FELDSPAR IN BASALT (length of field = 8 mm)—Many fine-grained, dark-colored igneous rocks (basalts) contain large amounts of plagioclase in crystals too small to be seen in hand specimens. Under the microscope, however, the plagioclase feldspars appear as elongated rectangular crystals in a felt-like fabric.

CALCITE AND DOLOMITE

Calcite is a major rock-forming mineral and appears in a great variety of forms. It occurs as the major mineral constituent in rock formations of limestone, chalk, and marble. It is also commonly deposited in caves, in hot and cold springs, and in veins. Regardless of form, the following three physical properties distinguish calcite from other minerals: (1) perfect rhombohedral cleavage, (2) hardness of 3, and (3) reaction with dilute hydrochloric acid. Calcite is thus one of the easiest minerals to identify. The major problem for beginning students is to become familiar with the variety of forms in which it occurs.

FIGURE 2.26

CRYSTALS OF CALCITE are extremely variable in form. They may be tabular, hexagonal, pyramidal, or rhombohedral. Large crystals may develop in caves, veins, and other voids, but calcite in most limestone usually occurs as compact aggregates of very small crystals.

FIGURE 2.27

RHOMBOHEDRON OF CLEAR CALCITE—The perfect rhombohedral cleavage of calcite is one of its most distinctive properties. Regardless of crystal form or mode of occurrence, all calcite cleaves perfectly in three directions not at right angles. Most calcite is white, although various impurities may tint it almost any color or even black. Colorless, clear calcite shows strong double refraction.

FIGURE 2.28

CRYSTALLINE LIMESTONE—Calcite is the dominant mineral in limestone, and in some formations is the only mineral present. It may occur as interlocking crystals that range in size from submicroscopic to over an inch long. Additional forms of calcite in limestone are discussed and illustrated on page 44.

FIGURE 2.29

PHOTOMICROGRAPH OF CALCITE (length of field = 8 mm)—The physical properties of a mineral are commonly well expressed on a microscopic scale. In this limestone specimen, perfect rhombohedral cleavage is clearly visible in the three interlocking grains.

FIGURE 2.30

FRACTURE FILLED WITH CALCITE—Calcite is one of the most soluble of the common minerals. It is dissolved readily by both surface and ground water, and may precipitate subsequently from solution to fill cavities and fractures. In this specimen, both the rock (limestone) and the fracture fill are composed of calcite.

FIGURE 2.31

DOLOMITE ($CaMg(CO_3)_2$) is a carbonate mineral similar in many respects to calcite. Crystals are commonly rhombohedral and occur in compact aggregates. Pure dolomite is usually pink, but may be colorless, white, gray, green, or black. Dolomite is distinguished from calcite by its effervescence in HCl in powdered form only.

OTHER ROCK-FORMING MINERALS

Specimens of other important rock-forming minerals are shown in Figures 2.32–2.43. These include the ferromagnesian minerals, evaporite minerals, and clay minerals.

The ferromagnesian minerals (Figures 2.32–2.40) are complex silicates that contain appreciable amounts of iron and magnesium. They are generally dark green to black in color and have a high specific gravity. The most important minerals in this group are olivine, pyroxenes, amphiboles, micas, chlorite, garnet, talc, and serpentine.

Halite and gypsum (Figures 2.41 and 2.42) are the two most common minerals formed by the evaporation of sea water or saline lake water and constitute important formations in some sequences of sedimentary rocks.

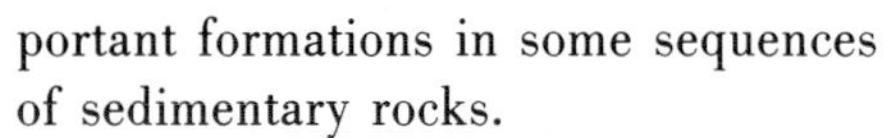

The clay minerals constitute the major part of shales and soils. They result from the alteration of aluminum silicates, especially feldspars. More than a dozen clay minerals can be distinguished; kaolinite (Figure 2.43) is a common variety.

FIGURE 2.32

OLIVINE ranges in composition from Mg_2SiO_4 to Fe_2SiO_4. A complete solid solution of Mg and Fe exists in the crystal structure. The most common varieties of olivine are rich in Mg and are characterized by an olive green color and by equidimensional glassy grains that lack cleavage. Olivine usually occurs in aggregates.

FIGURE 2.33

PYROXENES are complex silicates that contain substantial amounts of Ca, Mg, and Fe. They range in color from green to black or brown. Crystals are short and stubby, and when seen in cross section, they appear nearly square. Two cleavage planes intersect at right angles. Both the cleavage and crystal forms are generally poorly expressed in hand specimens but are well defined microscopically. Pyroxenes are essential constituents of dark-colored igneous rock rich in iron and are common in very high-grade metamorphic rock.

FIGURE 2.34

AMPHIBOLES are complex silicates with a composition similar to that of pyroxene, but the amphibole crystal structure contains H_2O. Many physical properties of the two mineral groups are similar. Amphiboles are distinguished by long, columnar crystals and by visible cleavage in two directions intersecting at 56 and 124 degrees. Amphiboles (in particular the variety known as hornblende) are common as phenocrysts in many volcanic rocks.

FIGURE 2.35

MUSCOVITE is a variety of mica, a distinctive mineral group with cleavage so perfect in one direction that most specimens can be separated into sheets thinner than a piece of paper. The two major varieties are (1) muscovite, or white mica, and (2) biotite, or black mica.

FIGURE 2.36

BIOTITE—In addition to perfect cleavage, mica is characterized by a pearly to vitreous luster. Cleavage sheets are elastic and flexible, and the mineral is notably soft. Micas are common in light-colored granitic rocks, and to a lesser extent in light-colored volcanics. They are also widespread in a variety of low- to intermediate-grade metamorphic rocks and in impure sandstones.

FIGURE 2.37

CHLORITE is a common metamorphic mineral that results from the alteration of silicates such as pyroxene, amphibole, biotite, or garnet. It is characteristically dark green in color and occurs in aggregates of minute crystals or in finely disseminated particles. Individual crystals have perfect cleavage in one direction, which produces small, scaly, nonelastic plates.

FIGURE 2.38

GARNET—The garnet mineral group includes a series of silicate minerals that characteristically occur in well-formed, nearly spherical, 12-sided crystals. They have a hardness of 7 to 7.5, and are commonly red, although brown, yellow, green, and black varieties do occur, depending on composition. Garnets have no cleavage, but may fracture along subparallel surfaces. The dominant fracture, however, is conchoidal. Most garnets are translucent to transparent and have a vitreous luster. They are common minerals in metamorphic rocks.

FIGURE 2.39

TALC is a secondary mineral formed from the alteration of magnesium silicates such as olivine, pyroxene, and amphiboles. It is distinguished from other minerals by its extreme softness (1 on the Mohs hardness scale), greasy feel, and perfect cleavage in one direction. Talc usually occurs as white to green foliated masses. Cleavage flakes are nonelastic.

FIGURE 2.40

SERPENTINE is usually compact and massive, and often multicolored. A fibrous variety (asbestos) consists of delicate, fine, parallel fibers that can be separated easily. It is usually found in veins.

FIGURE 2.41

HALITE is common salt (NaCl). It is characterized by cubic crystals that are usually clear and transparent, but impurities may impart reddish hues. It has perfect cleavage in three directions at right angles, and a hardness between 2 and 2.5. Halite is by far the most common water-soluble mineral. It is precipitated from sea water or from saline lakes by evaporation, and accumulates as salt beds interstratified with gypsum, silt, or shale. Salt layers may flow under pressure and may invade a weak area in an adjacent bedding plane to form pluglike bodies of solid salt. (The salt domes of the Gulf Coast are an example of this.)

FIGURE 2.42

GYPSUM is a common sedimentary mineral formed by evaporation of sea water or saline lakes. It is generally colorless to white and has a glassy to silky luster. Crystals are tabular. Gypsum is further characterized by its softness (2 on the Mohs scale), and by three unequal cleavage planes, one of which is perfect. Cleavage sheets are nonelastic. Varieties of gypsum include satin spar (fibrous gypsum with silky luster), alabaster (aggregates of fine crystals), and selenite (large, colorless, transparent crystals).

FIGURE 2.43

KAOLINITE is one of several clay minerals that form from the decomposition of aluminum silicates, especially feldspar. It occurs in earthy aggregates of nearly submicroscopic, platy crystals. In hand specimens, it appears claylike and is usually chalk white, although it may be stained red, brown, or black. It has a dull, earthy luster and disaggregates easily. Masses of kaolinite can be cut or shaped with a knife.

FIGURE 2.44

PYRITE is a yellow, metallic mineral composed of FeS_2. It commonly crystallizes in cubes, has a hardness between 6 and 6.5, and breaks with a conchoidal fracture.

FIGURE 2.45

CHALCOPYRITE is a metallic mineral composed of copper, iron, and sulfur. It is characterized by a brass yellow color, which often appears tarnished, and is slightly iridescent.

FIGURE 2.46

GALENA (lead sulfide) is the most important lead mineral. It commonly crystallizes in cubes and is recognized easily by its perfect cubic cleavage and high specific gravity (7.5).

FIGURE 2.47

GRAPHITE is composed of carbon atoms united in a sheetlike structure. It commonly occurs as foliated masses and is distinguished by extreme softness, perfect cleavage in one direction, and greasy feel.

FIGURE 2.48

SPHALERITE is the most important zinc mineral. It is difficult to recognize because of its extremely variable color (yellow, red, or brown to black). The most diagnostic properties are its resinous luster and perfect cleavage in six directions. Note how the various cleavage planes reflect light in this specimen.

FIGURE 2.49

NATIVE COPPER usually crystallizes in poorly formed cubes, thus producing an irregular mass, twisted and wirelike in form. It has a copper color and metallic luster, and is highly ductile and malleable.

FIGURE 2.50

MAGNETITE is a common iron ore and is the only mineral that is strongly magnetic. The magnetic property of this specimen is shown by the manner in which it attracts hairlike iron filings. Note how the alignment of the filings conforms to the north and south poles of the specimen.

FIGURE 2.51

HEMATITE is an iron mineral that can vary greatly in appearance. Three varieties are common: (2) specular hematite (metallic luster) (3) earthy hematite, and (1) oolitic hematite. All are characterized by a deep red streak.

FIGURE 2.52

LIMONITE is a yellowish brown, hydrous iron oxide that results from the alteration or weathering of iron minerals. It is amorphous and occurs commonly in both earthy and metallic varieties. The most diagnostic feature of limonite is its yellowish brown streak.

PROBLEMS

Examine each mineral specimen provided for this exercise by your instructor and briefly describe its physical properties as accurately as possible.

PROCEDURE

It is important to study a mineral specimen systematically and to record all properties observed, even though these may seem completely obvious. An effective approach is as follows:

1 Determine the type of luster, the color, and the streak.

2 Determine the type of crystal or the texture of crystal aggregates.

3 Test for hardness, cleavage or fracture, and other physical properties.

Do not break the specimens without permission from your instructor. After you have recorded all physical properties observed, identify the mineral by referring to the mineral identification chart on pages 12–13. Note that Figure 2.4 is arranged in a systematic manner. The minerals are first grouped in three categories on the basis of luster and color. A second grouping is made on the basis of hardness or streak, and a third on the basis of cleavage or fracture. Once a specific mineral is categorized according to these properties, it can then be identified from the brief description of other characteristics. For example, if you observe that a mineral is light colored, has a nonmetallic luster, is softer than steel or glass but will scratch a penny, and has good cleavage in four directions, you would identify the mineral by referring to the Nonmetallic Luster—Light Color section of the chart. Next, you would look under the list of minerals that are softer than glass and select the one that has four good cleavage directions. The only mineral possessing all of these properties is fluorite.

Colored photographs and descriptions of some of the common minerals are shown in Figures 2.5–2.52. When you have identified a mineral specimen, compare it with its picture in this manual. Note the similarities and determine which are significant physical properties. *The colored photographs are not a key to mineral identification and should not be used as such.* They are intended to serve as a study aid and reference only.

PROBLEMS

1 Why would misidentification be likely if you attempted to identify unknown samples simply by comparing them with the photographs on pages 14–20, instead of testing the physical properties and using the mineral identification chart?

2 Explain the law of constancy of interfacial angles.

3 List the important types of physical properties of minerals.

4 Give a complete and accurate definition of a mineral.

5 What is a solid solution?

6 List the rock-forming minerals in which solid solution is important.

7 What is the difference between a crystal face and a cleavage plane?

8 Give a brief definition of the term *hardness.*

9 What common mineral will react with dilute HCl?

10 What is the typical crystal form of quartz?

11 Are hand specimens of microcrystalline quartz single crystals or aggregates of crystals?

12 What is the most common occurrence of chert?

13 Is there a fundamental difference between milky quartz, smoky quartz, and rose quartz?

14 What are the diagnostic physical properties of the feldspars?

15 What are the two main subgroups of the feldspars? How do they differ?

16 How would you distinguish between Ca-plagioclase and K-feldspar?

17 What are the important physical properties of calcite?

18 How can microcrystalline calcite be recognized in a hand specimen?

19 What are the diagnostic properties of olivine?

20 In what rocks is olivine an important constituent?

21 How can you distinguish between amphibole and pyroxene?

22 What are the diagnostic properties of halite?

23 What are the diagnostic properties of gypsum?

24 Is a hand specimen of kaolinite an aggregate of crystals or a single crystal?

25 List the important rock-forming minerals.

Exercise 3

IGNEOUS ROCKS

OBJECTIVE

To recognize the major types of igneous rocks and to understand the genetic significance of their texture and composition.

MAIN CONCEPT

Igneous rocks can be classified on the basis of composition and texture. The composition of a rock provides information about the magma from which the rock formed and about the tectonic setting where it originated. The texture of a rock gives important insight into the cooling history of the magma.

SUPPORTING IDEAS

1 Igneous rocks are composed of silicate minerals, the most important of which are (a) plagioclase, (b) K-feldspar, (c) quartz, (d) mica, (e) amphibole, (f) pyroxene, and (g) olivine.

2 The major textures in igneous rocks are (a) phaneritic, (b) porphyritic-phaneritic, (c) aphanitic, (d) porphyritic-aphanitic, (e) glassy, and (f) fragmental (pyroclastic).

3 Basaltic magmas originate by partial melting of the upper mantle at divergent plate margins.

4 Granitic magmas originate at subduction zones by partial melting of the oceanic crust.

DISCUSSION

COMPOSITION OF IGNEOUS ROCKS

Approximately 99% of the total bulk of most igneous rocks is made up of only eight elements—oxygen, silicon, aluminum, iron, calcium, sodium, potassium, and magnesium. Most of these elements occur in the crystal structures of the rock-forming silicate minerals to produce feldspars, olivine, pyroxenes, amphiboles, quartz, and mica. These six minerals constitute over 95% of the volume of all common igneous rocks and are therefore of paramount importance in a study of their classification and origin. Magmas rich in silica and aluminum are referred to as *sialic,* and tend to produce more quartz, potassium feldspar, and sodium plagioclase. They generally form light-colored rocks. Magmas rich in iron, magnesium, and calcium are referred to as *mafic,* and produce greater quantities of olivine, pyroxene, amphibole, and calcium plagioclase. The resulting rocks are dark colored because of an abundance of the dark ferromagnesian minerals. (There are, however, many exceptions to such generalizations, and the mineral composition of igneous rocks can only be roughly approximated by an observation of color.)

Crystallization of minerals from a magma occurs between 1200 and 600 °C. Those minerals with the highest freezing point crystallize first and thus develop well-formed crystal faces. Minerals that crystallize at lower temperatures are forced to grow in the spaces between the earlier-formed crystals, and are commonly irregular in shape with few well-developed crystal faces. From laboratory studies of artificially produced magmas, and from petrographic studies of igneous rocks, a general order of crystallization has been established. This sequence is summarized in Figure 3.1 and is fundamental to the study of igneous rocks.

It is apparent from Figure 3.1 that in a mafic magma, olivine and Ca-plagioclase are the first minerals to form, followed by pyroxenes, amphiboles, and Na-plagioclase. Such a magma crystallizes between 900 and 1200 °C and produces rocks of the gabbro-basalt family. In magmas rich in silica and aluminum, biotite and K-feldspar form first, followed by quartz and muscovite. Rocks of the granite-rhyolite family develop from such a magma at temperatures below 900 °C.

TEXTURE OF IGNEOUS ROCKS

Texture refers to the size, shape, and boundary relationships of adjacent minerals in a rock mass. In igneous rocks, texture develops primarily in response to composition and rate of cooling of

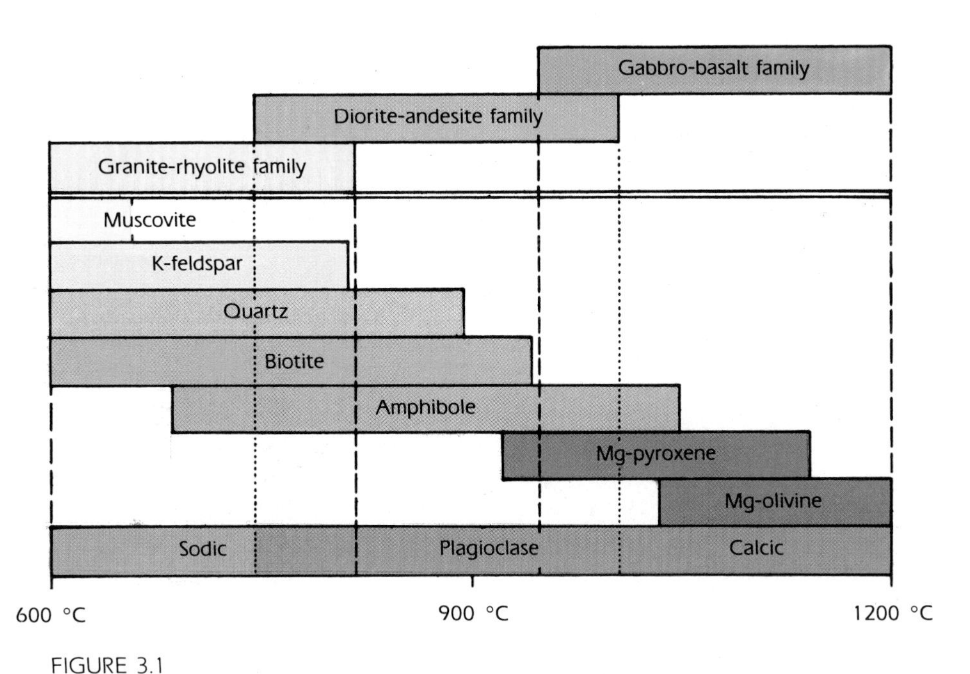

FIGURE 3.1

ORDER OF CRYSTALLIZATION OF COMMON ROCK-FORMING MINERALS.

the magma. Magmas located deep in the earth's crust cool slowly. Individual crystals grow to a more or less uniform size and may be more than an inch in diameter. In contrast, a lava extruded at the earth's surface cools rapidly, so the mineral crystals have only a short time in which to grow. The crystals from such a magma are typically so small that they cannot be seen without the aid of a microscope, and the resultant rock appears massive and structureless. Regardless of crystal size, the texture of most igneous rocks is distinguished by a network of interlocking crystals.

Igneous rock textures are divided into the following types: (1) phaneritic, (2) porphyritic-phaneritic, (3) aphanitic, (4) porphyritic-aphanitic, (5) glassy, and (6) fragmental.

PHANERITIC TEXTURE. In phaneritic textures, the individual crystals are large enough to be plainly visible to the naked eye (Figure 3.2). The grains in each specimen are approximately equal in size and form an interlocking mosaic. The size of crystals in a phaneritic texture can range from those barely visible to crystals of more than an inch in length. Phaneritic texture develops from magmas that cool slowly, and commonly develops in intrusive igneous bodies such as batholiths or stocks. Very coarse phaneritic rocks, in which the crystals are several feet long, almost invariably are found in large veins.

PORPHYRITIC-PHANERITIC TEXTURE. A porphyritic-phaneritic texture is characterized by two distinct crystal sizes, both of which can be seen with the naked eye. The smaller crystals constitute a matrix, or groundmass, that surrounds the larger crystals, called *phenocrysts* (Figure 3.3).

APHANITIC TEXTURE. In aphanitic texture, individual crystals are so small that they cannot be detected without the aid of a microscope (Figure 3.4). Rocks with this texture therefore appear to be massive and structureless. When a thin section of an aphanitic rock is viewed under a microscope, however, the crystalline structure is readily apparent. The rock is seen to be composed of numerous small crystals and, usually, some glass. Examples of aphanitic texture viewed under a microscope are shown in the photomicrographs in Figures 3.12 and 3.15.

PORPHYRITIC-APHANITIC TEXTURE. A porphyritic-aphanitic texture is defined as a rock with an aphanitic matrix in which embedded phenocrysts make up more than 10% of the total rock volume (Figure 3.5). The phenocrysts are visible to the unaided eye. When phenocrysts are abundant, the rock may at first glance appear to be phaneritic. Careful study of the area between the phenocrysts will indicate whether the matrix is aphanitic or phaneritic.

GLASSY TEXTURE. This texture is similar to that of ordinary glass. It may occur in massive units (Figure 3.6A), or in a threadlike mesh similar to spun glass (Figure 3.6B). Crystals cannot be discerned in a glassy texture, even when the specimen is viewed under high magnification.

FRAGMENTAL TEXTURE. Fragmental textures consist of broken, angular fragments of rock material (Figure 3.7). In pyroclastic rocks, the fragmental material is composed of pumice, glass, and broken crystals. Some sorting and stratification are generally present. Material finer than 4 mm is called *tuff*. Material larger than 4 mm is referred to as *volcanic breccia*. If the fragments are exceptionally hot when deposited, they may fuse or weld together to form a dense mass.

A. PHANERITIC TEXTURE in this specimen is produced by relatively large grains of quartz (glassy grains) and feldspar (porcelainlike grains). The grains are about 0.5 in. across and are intergrown, so well-developed crystal faces have not formed. Black grains are biotite.

B. FINE-GRAINED PHANERITIC TEXTURE in this specimen is produced by small crystals of glassy quartz, porcelainlike feldspar, and black mica.

FIGURE 3.2 **EXAMPLES OF PHANERITIC TEXTURE** (actual size).

A. PORPHYRITIC-PHANERITIC TEXTURE in this specimen consists of phenocrysts of large, rectangular crystals of pink K-feldspar. The matrix is composed of much smaller grains of white porcelainlike plagioclase, glassy-appearing smoky quartz, and black ferromagnesian minerals.

B. PORPHYRITIC-PHANERITIC TEXTURE in this specimen consists of phenocrysts of well-formed amphibole crystals set in a fine-grained phaneritic matrix.

FIGURE 3.3 **EXAMPLES OF PORPHYRITIC-PHANERITIC TEXTURE** (actual size).

A. APHANITIC TEXTURE may appear dense and structureless, but under the microscope individual mineral crystals can be distinguished (see Figure 3.15). Note that some secondary crystals have grown in cavities, so a few crystal specks can be seen without a microscope.

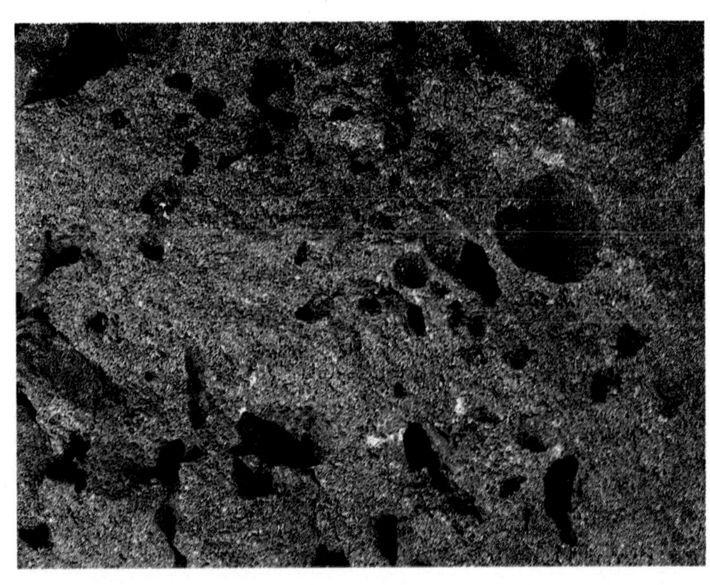

B. APHANITIC TEXTURE may contain vesicles (holes formed by gas bubbles trapped in the lava as it cools). The rock is composed of microscopic crystals and is typical of aphanitic texture.

FIGURE 3.4 **EXAMPLES OF APHANITIC TEXTURE** (actual size).

A. PORPHYRITIC-APHANITIC TEXTURE in this specimen consists of large phenocrysts of white feldspar set in a dark matrix. Smaller, black phenocrysts of amphibole are also apparent. Close study of the dark material indicates aphanitic texture.

B. PORPHYRITIC-APHANITIC TEXTURE in this specimen has small, white, rectangular plagioclase crystals set in a dense aphanitic matrix.

FIGURE 3.5 **EXAMPLES OF PORPHYRITIC-APHANITIC TEXTURE** (actual size).

A. GLASSY TEXTURE can be dense and massive, as in this specimen. Note the conchoidal fracture and knife-sharp edges.

B. GLASSY TEXTURE in this specimen occurs as tangled, threadlike filaments. Pore spaces are produced by gases that escape as the lava cools.

FIGURE 3.6 **EXAMPLES OF GLASSY TEXTURE** (actual size).

A. PYROCLASTIC TEXTURE in this specimen consists of white, fine-grained volcanic ash with larger fragments of brown pumice and some angular rock fragments. An ash fall produced this texture.

B. PYROCLASTIC TEXTURE produced by an ash flow is commonly dense and massive. Fragmental material in this specimen was very hot when extruded and flowed en masse close to the ground. As the flow cooled, the hot particles fused together. Black lenses were originally frothy pumice fragments like those in specimen A, but were so hot that the cellular structure collapsed as the mass became welded together.

FIGURE 3.7 **EXAMPLES OF FRAGMENTAL, OR PYROCLASTIC, TEXTURE** (actual size).

CLASSIFICATION AND IDENTIFICATION OF IGNEOUS ROCKS

The most useful and significant classification system for igneous rocks is based on two criteria: composition and texture. These criteria are important not only in describing the rock so that it can be distinguished from other rock types, but also in drawing important implications about the rock's origin. A chart (Figure 3.8) in which variations in composition are shown horizontally and variations in texture are shown vertically provides an effective framework for classifying and naming igneous rocks.

Three major mineralogical criteria are used to classify igneous rocks.

1 Presence or absence of quartz—Quartz is an essential mineral in sialic rocks, and an accessory mineral in intermediate[1] or mafic rocks. (In the classification of igneous rocks, an "essential mineral" is a mineral on which the rock classification is based. An accessory mineral occurs in minor amounts and is not a factor in classification.)

2 Composition of the feldspars—Potassium feldspars and sodium plagioclase are essential minerals in sialic rocks but are rare or absent in intermediate and mafic rocks. Calcium plagioclase is characteristic of mafic rocks.

3 Proportion and kinds of ferromagnesian minerals—As a general rule, mafic rocks are rich in ferromagnesian minerals, while sialic rocks are rich in quartz. Olivine is generally restricted to mafic rocks. Pyroxenes and amphiboles are present in mafic to intermediate rocks. Biotite is common in intermediate and sialic rocks.

The graph at the top of Figure 3.8 may appear at first to be complex but is actually quite simple. It is designed to show the range in composition of the major igneous rock types. Each color zone represents a rock family (yellow—the granite-rhyolite family, yellow green—the diorite-andesite family, green—the gabbro-basalt family, and darkest green—the peridotite family). The area on the graph allotted to each mineral represents the percentage of the rock composed of that mineral. For example, peridotite may be composed entirely of olivine, but pyroxene can occur in amounts up to 30%, and Ca-plagioclase in amounts up to 15%. In the granite-rhyolite family, K-feldspar may form over 75% of the rock (the extreme left side of the graph). Note how the percentage of K-feldspar decreases as quartz increases, and at the right margin of the granite-rhyolite zone, both quartz and K-feldspar decrease and plagioclase content increases.

1. The term *intermediate* is used for rocks with a composition "intermediate" between sialic and mafic.

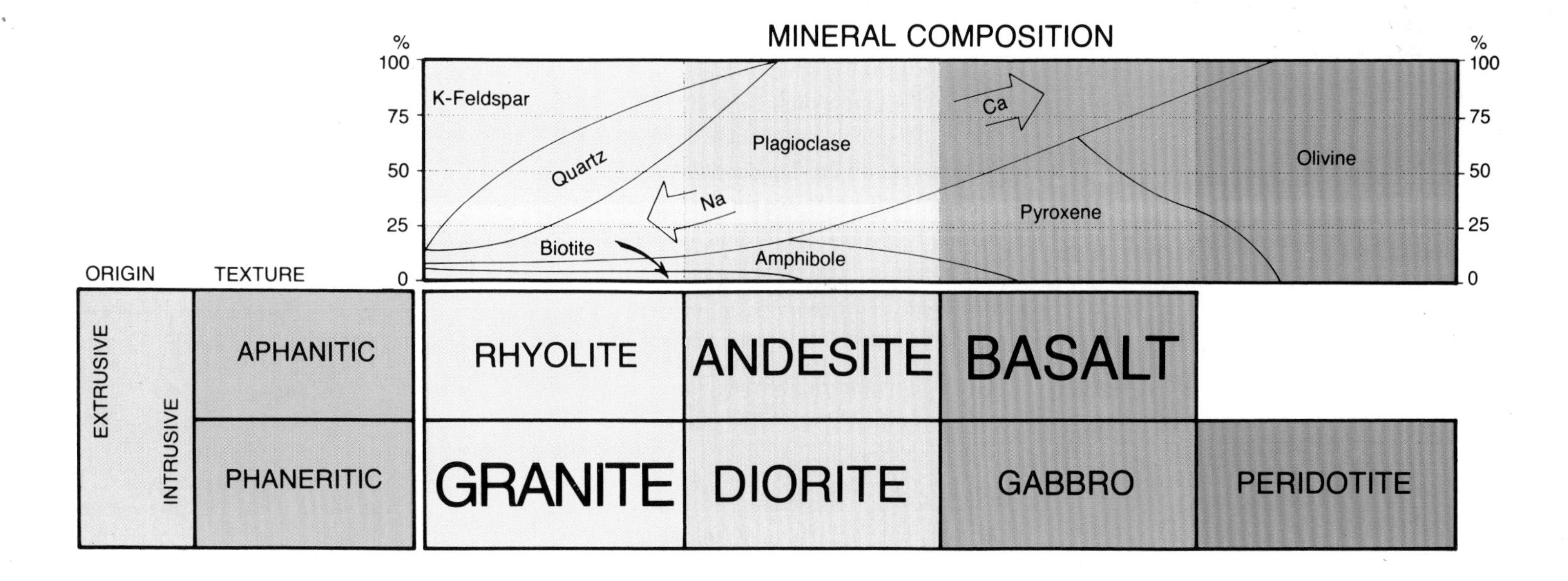

FIGURE 3.8

CLASSIFICATION OF IGNEOUS ROCKS.

From Figure 3.8, we can make the following summary of the composition of the igneous rock families.

THE GRANITE-RHYOLITE FAMILY

The granite-rhyolite family is characterized by the following mineral composition:

Quartz	10–40%
Potassium feldspar	30–60%
Plagioclase	0–33%
Biotite and amphibole	10–33%

The magmas that produce these rocks are high in potassium, silicon, and sodium, and are low in iron, magnesium, and calcium. Granites and rhyolites are therefore characteristically light colored.

THE DIORITE-ANDESITE FAMILY

The diorite-andesite family is intermediate in composition between the granite-rhyolite and gabbro-basalt families. It is characterized by the following composition:

Plagioclase	55–70%
Amphibole and biotite	25–40%

Plagioclase is approximately 50% albite and 50% anorthite. Potassium feldspar and quartz are present in minor amounts only. The diorite-andesite family is therefore characteristically gray in color.

THE GABBRO-BASALT FAMILY

The gabbro-basalt family has the following composition:

Plagioclase (mostly Ca)	45–70%
Ferromagnesian minerals (olivine, pyroxene, and amphibole)	25–50%

These rocks crystallize from magmas that are relatively high in iron, magnesium, and calcium, but deficient in silica. Rock coloration is characteristically black or dark green.

THE PERIDOTITE FAMILY

The peridotite family is characterized by the following mineral composition:

Olivine	85–100%
Pyroxene	0–10%
Ca-plagioclase	0–5%
Ore minerals (i.e., magnetite, ilmenite, chromite)	0–10%

HINTS ON MINERAL IDENTIFICATION IN HAND SPECIMENS

Identification of the major rock types in the earth's crust may be difficult at first for the beginning student. The most obvious properties (i.e., color, size, shape, and form) that we use to distinguish most things in the physical world are not important in identifying rock type. Indeed, the specific color, size, and shape of a given rock have little significance. The rock properties that permit us to discriminate between the various igneous rock types are mineral composition and texture. A student of geology must learn to ignore specific color, shape, and size, and to concentrate instead on composition and texture.

QUARTZ. Occurs as irregular, *glassy* grains commonly clear to smoky in color. No cleavage.

MUSCOVITE. Brass-colored flakes associated with quartz or K-feldspar. Perfect cleavage in one direction.

K-FELDSPAR. Porcelain luster. Commonly colored pink, white, or gray. Cleavage in two directions at right angles may be detected. Cleavage planes flash light when specimen is rotated.

PLAGIOCLASE. Usually gray or white in granite; dark bluish color in gabbro. Striations common. Two cleavage directions at right angles may be detected.

BIOTITE. Small black flakes. Perfect cleavage in one direction. Reflects light.

AMPHIBOLE. Long, black crystals in a light-colored matrix. Cleavage at 60 and 120 degrees.

PYROXENE. Short, dull, greenish black minerals in darker rocks. Cleavage in two directions at 90 degrees.

OLIVINE. Glassy, light green grains.

The photographs in Figures 3.9–3.23 illustrate the characteristics of the major igneous rock types as seen in hand specimens and under the microscope. *Remember, the photographs are not a key to rock identification. Use them as a visual reference only for examples.*

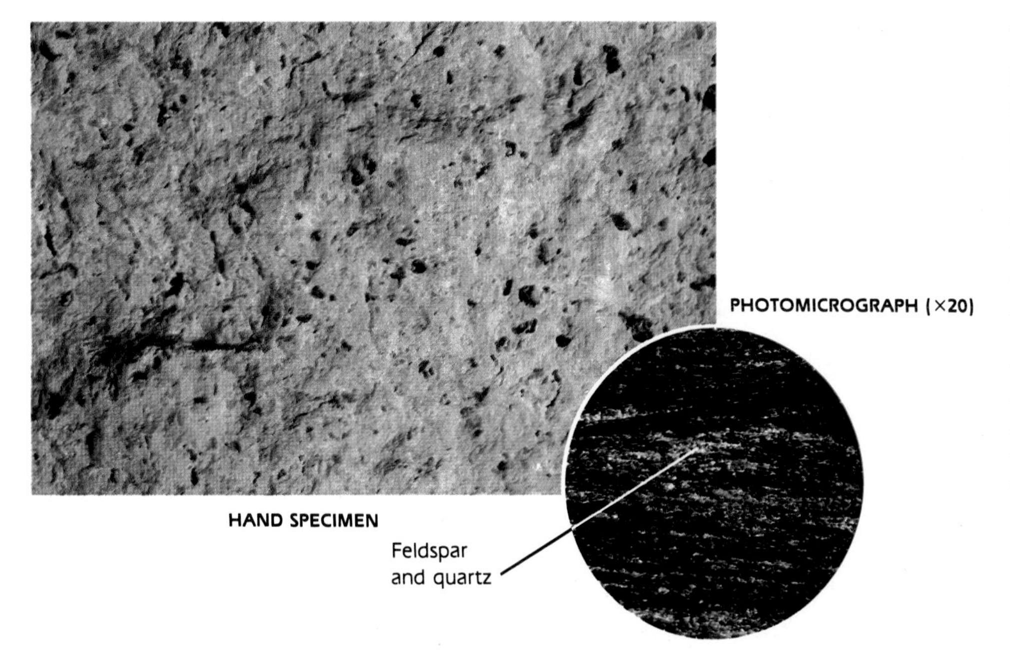

FIGURE 3.9

RHYOLITE is an aphanitic rock with the same composition as granite. It is commonly white, gray, or pink, and almost always contains a few phenocrysts of feldspar or quartz (2–10%). If phenocrysts constitute more than 10% of the volume, the rock is properly termed "porphyritic rhyolite." Because the texture is aphanitic, the only minerals that can be identified in a rhyolite hand specimen are those occurring as phenocrysts. This specimen is from a typical porphyritic rhyolite deposit in the western United States. The dark phenocrysts are smoky quartz, but most of the rock has a characteristic aphanitic texture. In thin section, flow structures are commonly apparent, and a considerable amount of glass is present between the fine-grained crystals. The specimen in Figure 3.4A is also a rhyolite.

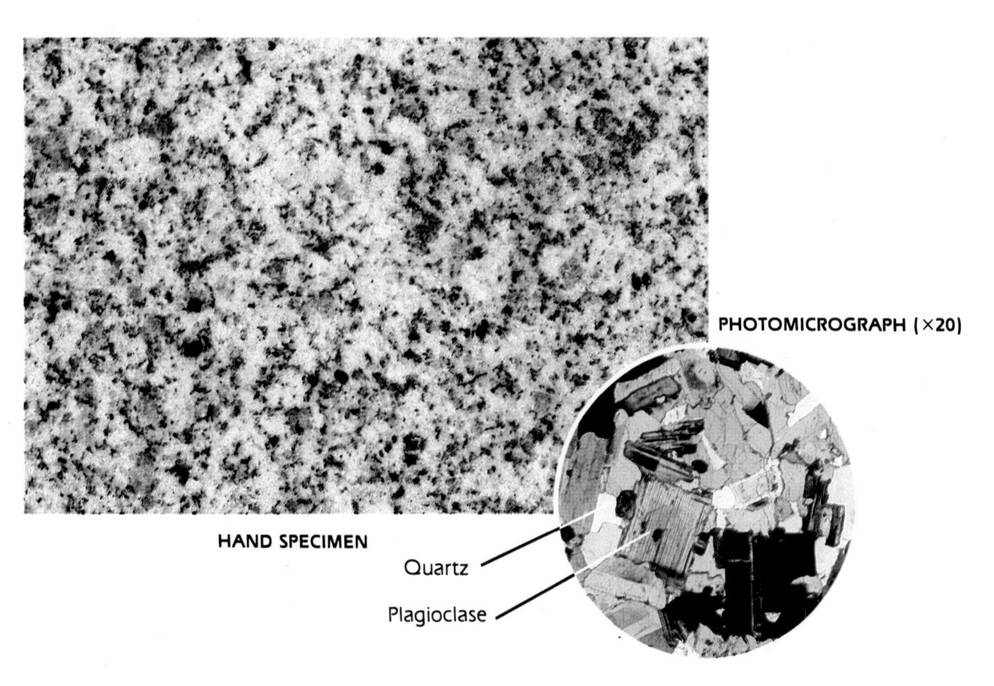

FIGURE 3.10

GRANITE is a phaneritic igneous rock composed predominantly of feldspar and quartz. Biotite, amphibole, and plagioclase (early-formed crystals) are generally euhedral (i.e., have well-developed crystal faces). In this specimen, white crystals are sodium-rich plagioclase, light crystals are quartz, and dark crystals are biotite. Many granites are gray, but if K-feldspar dominates, coloration may be pink or red. Specimens in Figures 3.2A and B are also granites. From one granitic body to another, there may be a wide range in the average crystal size.

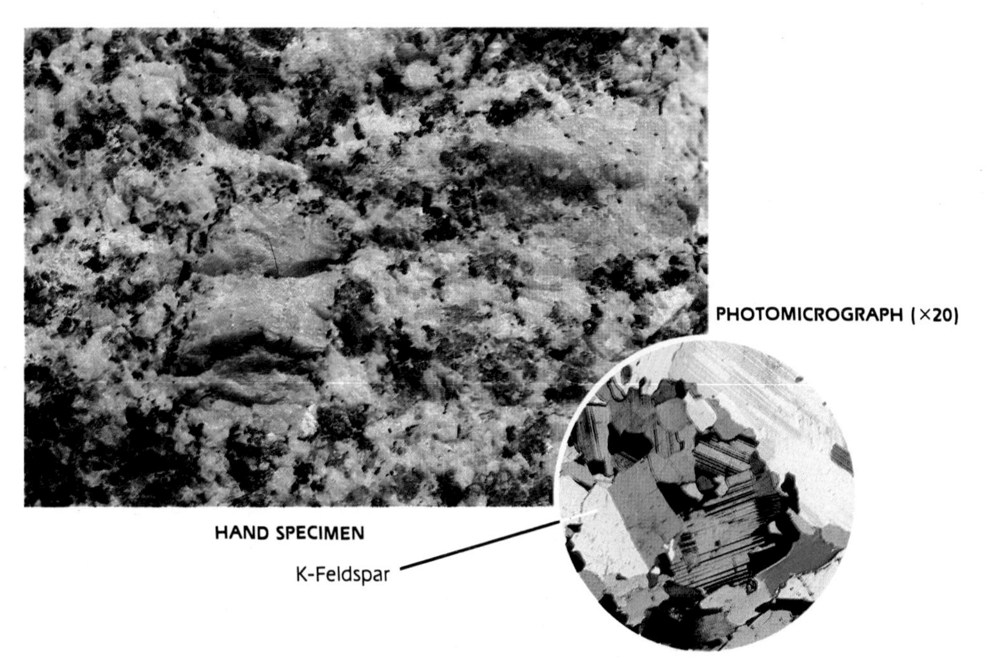

FIGURE 3.11

PORPHYRITIC GRANITE usually indicates two stages of cooling: an initial stage in which large crystals formed, followed by a period of more rapid cooling in which smaller grains developed. In this typical specimen, the phenocrysts are large crystals of pink K-feldspar. The porcelain white mineral is plagioclase, and the glassy grains are quartz. Small biotite grains are also apparent. Although grain size, color, and general specimen appearance are quite different from the specimen in Figure 3.10, composition of the two rocks is essentially the same. Both are composed mostly of quartz, K-feldspar, and calcium plagioclase. The only significant difference is in texture. The specimen in Figure 3.3A is also porphyritic granite.

FIGURE 3.12

ANDESITE is an aphanitic rock composed of Na-plagioclase, amphibole, and pyroxene, with little or no quartz. Specimens are generally dark gray, green, brown, or red. The name "andesite" comes from the Andes Mountains in South America where volcanic eruptions have produced this rock type in great abundance. Andesitic magmas originate in subduction zones by partial melting of the oceanic crust. After basalt, andesite is the most abundant volcanic rock. Completely aphanitic andesite is relatively rare, since most flows contain some phenocrysts.

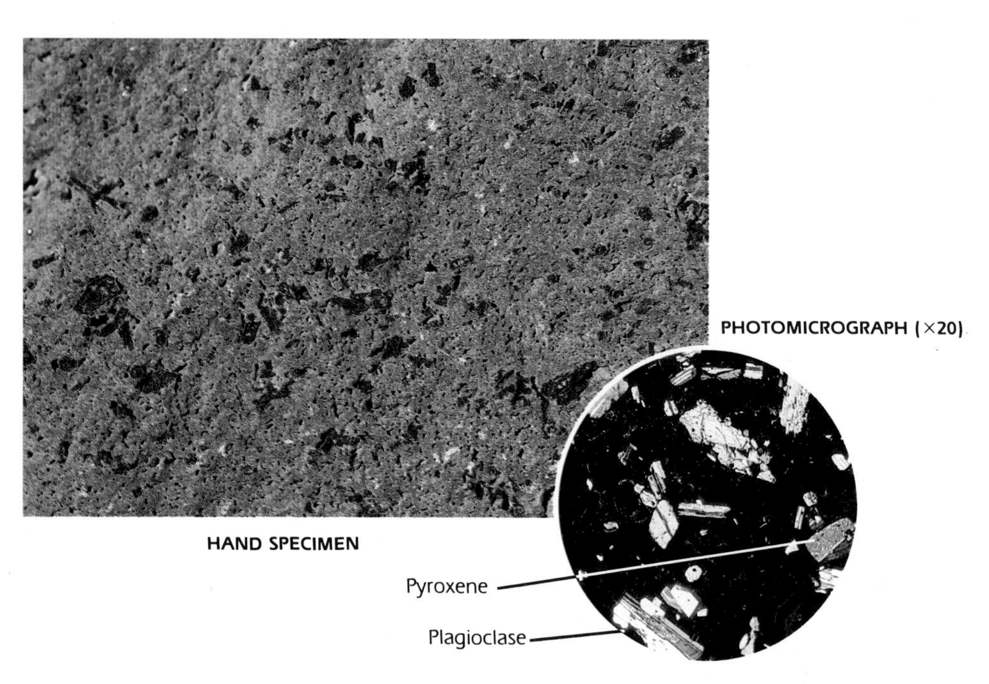

FIGURE 3.13

PORPHYRITIC ANDESITE is the most common variety of intermediate extrusive rock. Phenocrysts are composed mainly of plagioclase, amphibole, or biotite set in an aphanitic matrix of plagioclase and some glass. In this specimen, black phenocrysts are amphibole. Porphyritic andesite and basalt comprise 95% of all volcanic material. The specimen in Figure 3.5B is also porphyritic andesite.

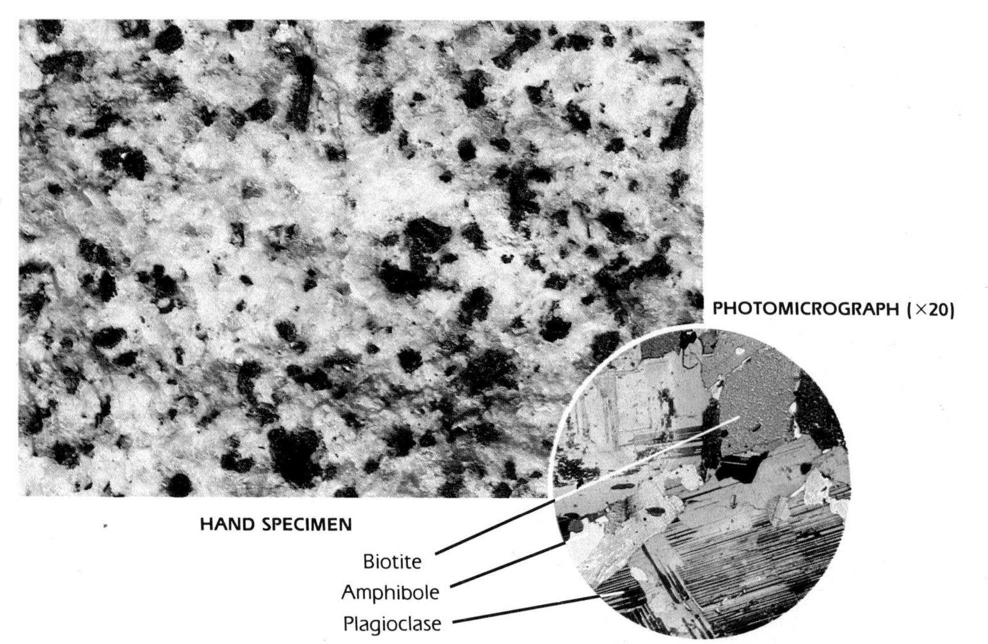

FIGURE 3.14

DIORITE has a texture essentially the same as granite. (Compare this specimen with that in Figure 3.2A.) The two differ in composition only. Essential minerals in a granite are K-feldspar, quartz, and Na-plagioclase, whereas diorite is composed predominantly of Na-Ca plagioclase and ferromagnesian minerals. In this specimen, ferromagnesian minerals (i.e., amphibole) make up most of the rock and give it a speckled appearance. In thin section, it is clear that quartz and K-feldspar are absent: plagioclase and amphibole predominate. Quartz accounts for less than 5% of the total volume.

Diorite occurs in large intrusives such as stocks or batholiths. It is found also in dikes, sills, and laccoliths.

FIGURE 3.15

BASALT is an aphanitic rock composed predominantly of calcium-rich plagioclase and pyroxene, with smaller amounts of olivine and amphibole. It is characteristically black, dense, and massive. Although individual crystals cannot be seen without magnification, basalt is generally easy to recognize. Many specimens are clinkerlike in appearance, with half of the total volume consisting of small holes, or vesicles. Vesicular texture develops as gas bubbles rise toward the top of a flow and are trapped in cooling lava. Under the microscope, crystals of individual minerals can be seen and studied. Plagioclase crystals commonly occur as a mesh of lathlike crystals surrounding crystals of pyroxene, olivine, or amphibole.

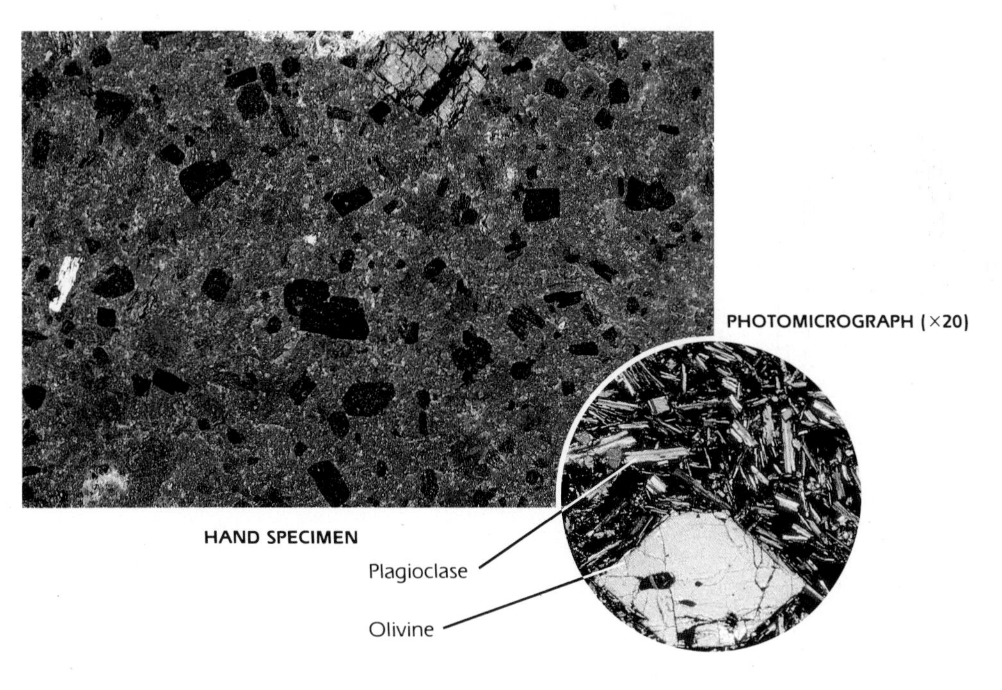

FIGURE 3.16

PORPHYRITIC BASALT is common because many basaltic lavas have some early-formed crystals of olivine, pyroxene, amphibole, or Ca-plagioclase. In this specimen, phenocrysts are amphibole. Figure 3.5A is also a porphyritic basalt, but phenocrysts are plagioclase.

Basalt is the most abundant extrusive rock. It is the bedrock for the oceanic crust, and is found also in great floods on some continents. Basaltic magma forms from partial melting of the upper mantle, and is extruded mostly along rift zones.

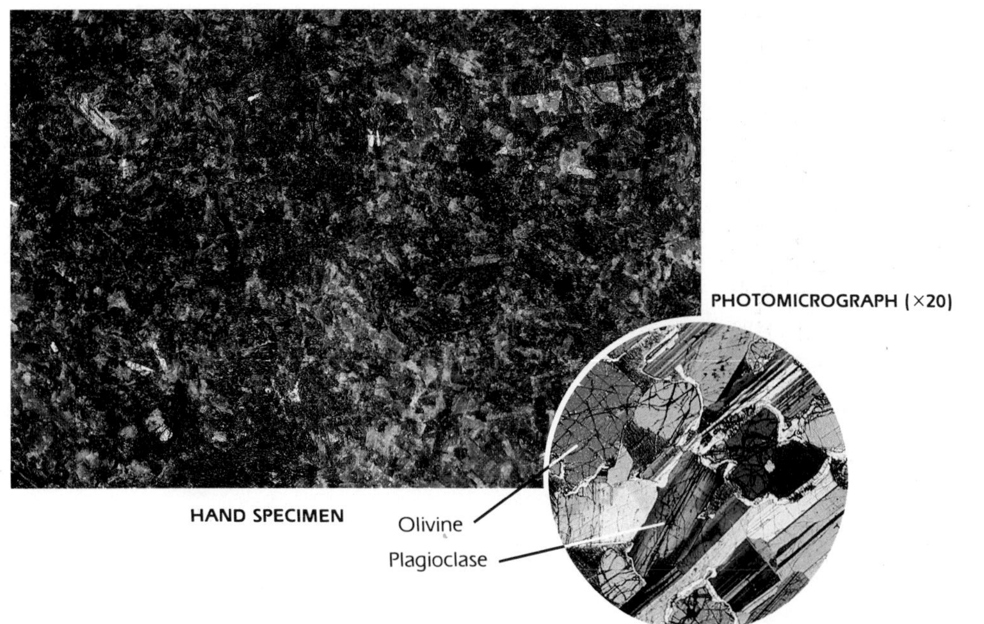

FIGURE 3.17

GABBRO has phaneritic texture similar to granite, but is composed almost entirely of pyroxene and Ca-plagioclase with minor amounts of olivine. Composition is thus the same as basalt. Calcium-rich plagioclase is the dominant feldspar and usually occurs as dark, elongated crystals like those in this specimen. Coloration is characteristically dark green, dark gray, or almost black because of the predominance of Ca-plagioclase and ferromagnesian minerals. Gabbro is not common in the continental crust, but large exposures occur north of Duluth, Minnesota, in Labrador, and in Finland. Most gabbro is produced at divergent plate margins and forms the bottom layer of the oceanic crust.

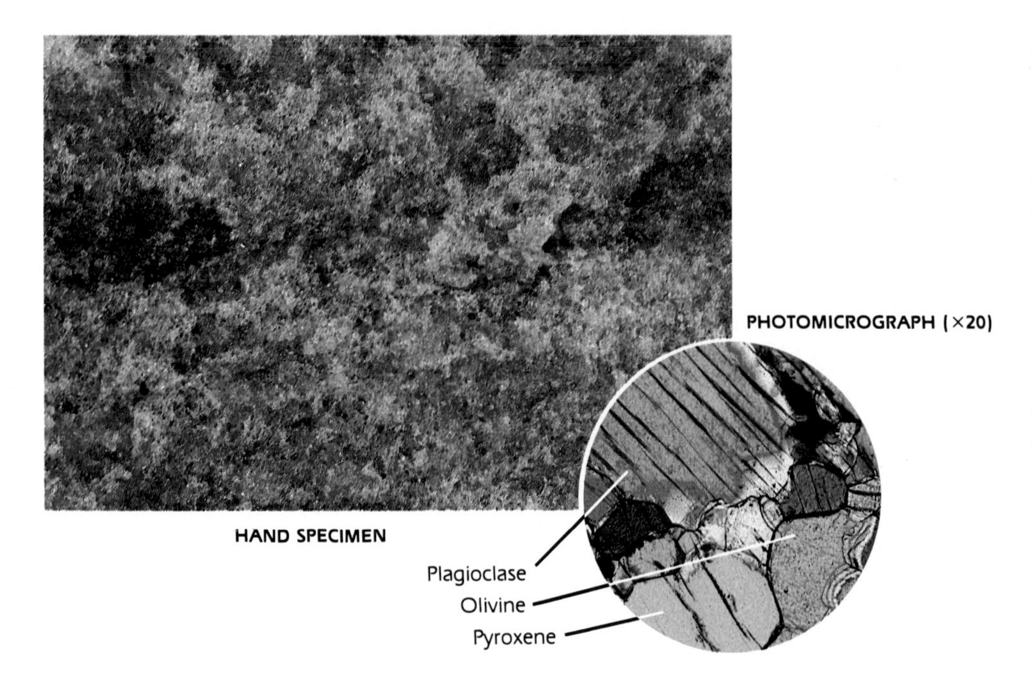

FIGURE 3.18

PERIDOTITE is relatively easy to identify because it is composed almost entirely of olivine, but significant amounts of pyroxene and lesser amounts of Ca-plagioclase may occur. This specimen is composed mostly of olivine. Olivine crystals are characteristically light green and have a distinctive glassy luster. They are typically equidimensional and are medium to coarse grained. In hand specimens, this peridotite thus resembles a green, glassy sandstone.

Rocks of the peridotite family have phaneritic texture and originate far below the surface. No extrusive rocks have equivalent composition.

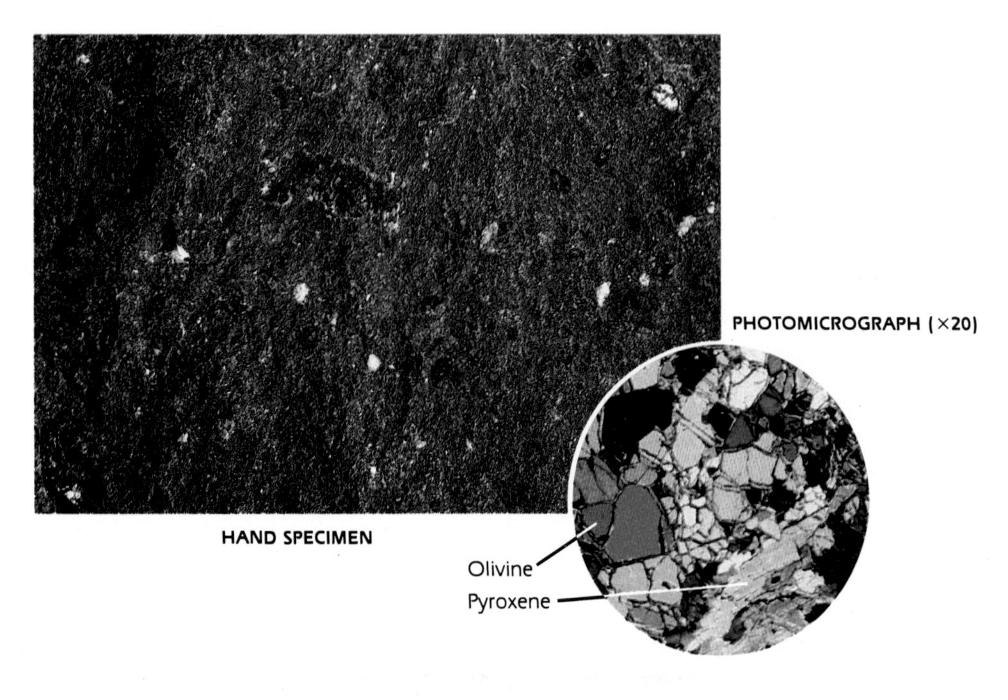

FIGURE 3.19

PERIDOTITE (KIMBERLITE) may have a relatively fine-grained texture and may lack the glassy green appearance of the specimen in Figure 3.18, especially if pyroxene, Ca-plagioclase, and some accessory minerals are present. This specimen is a kimberlite, a variety of peridotite that is host rock for diamonds.

Peridotite is not an abundant rock type in the earth's crust. It is important, however, in a study of the earth's dynamics, because it is believed that the entire upper mantle is composed almost exclusively of this type of rock. Indeed, it is probable that the melting and movement of peridotite in the asthenosphere produces the tectonic system. As the asthenosphere moves slowly in convection cells, peridotite in the upward-moving currents is subjected progressively to less and less pressure, and begins to melt. Plagioclase and pyroxene, having lower melting points than olivine, melt first to form a basaltic magma, which forms new oceanic crust in the rift zones.

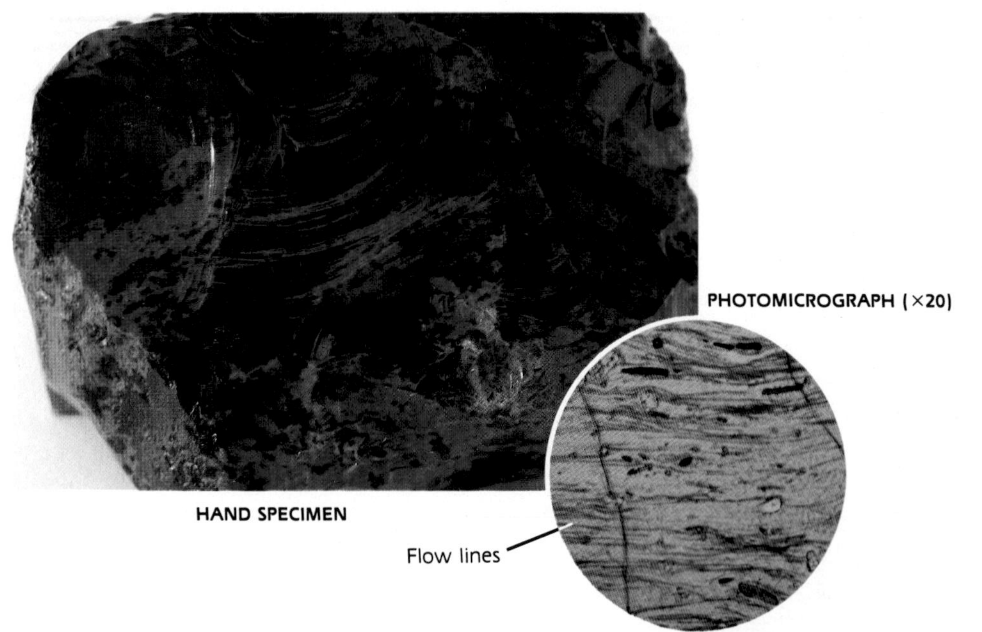

FIGURE 3.20

OBSIDIAN is a massive volcanic glass. As shown in this specimen, it typically breaks with a conchoidal fracture and has a bright, glassy luster. Despite its composition (mostly SiO_2), it is characteristically jet black due to countless dustlike particles of magnetite or ferromagnesian minerals. More rarely, yellow, red, or brown hues are produced by oxidized magnetite or hematite. In thin section, flow lines are common, and the rock appears totally black under polarized light. Obsidian is not crystalline, but does contain skeletal crystal embryos called crystallites. Many aphanitic rocks contain appreciable quantities of glass, which fills the space between crystals. The specimen in Figure 3.6A is also obsidian.

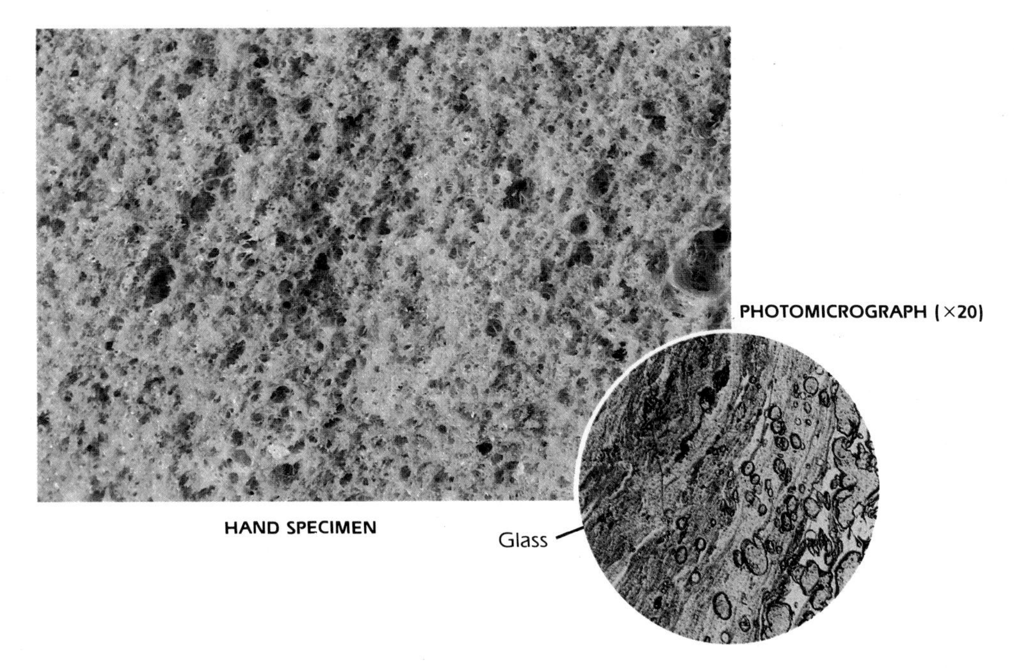

FIGURE 3.21

PUMICE is a very porous volcanic glass. Its texture (both in hand specimen and thin section) consists of subparallel, silky glass fibers that are tangled together. It originates when a release of pressure in a volcano permits rapid expansion of gases through the upper part of an ascending column of obsidian lava. The lava froths or foams with innumerable minute bubbles. When this frothy lava cools rapidly, a spun-glass texture is produced. The specimen in Figure 3.6B is also pumice.

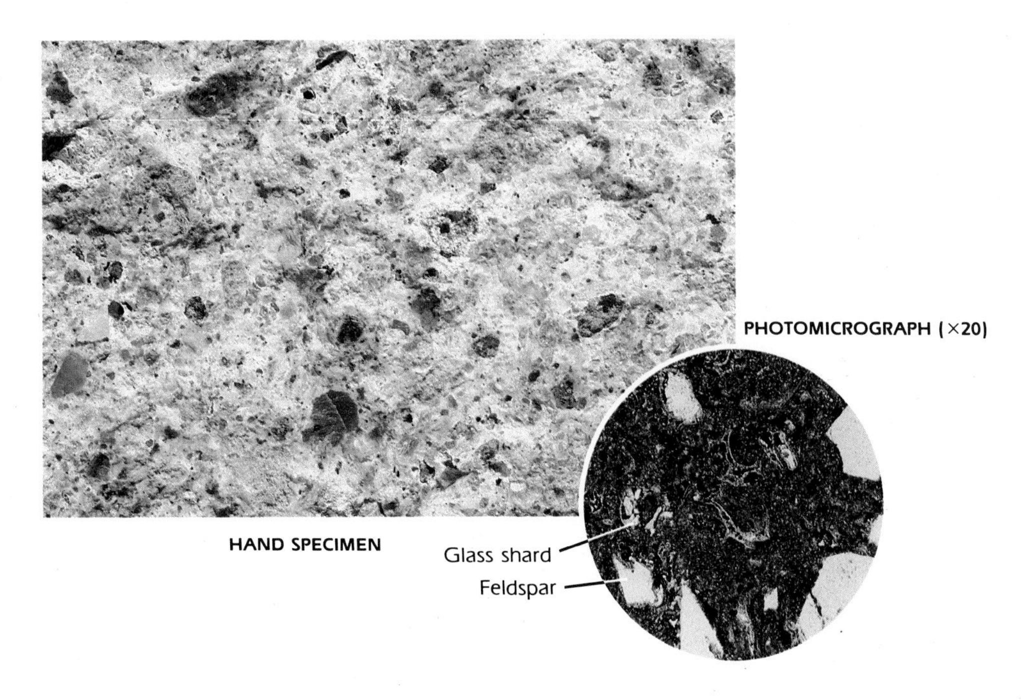

FIGURE 3.22

TUFF is composed of volcanic ash, which produces a fragmental texture more characteristic of sedimentary rocks. It forms from fragmental material and droplets of lava that erupt from volcanic vents and are transported through the air before settling back to earth. This hot fragmental material is referred to as "pyroclastic." Rocks formed from the consolidation of such material are classified by grain size as follows: (1) tuff (fine ash and dust less than .25 in. in diameter), (2) volcanic breccia (coarse ash and angular blocks .25 to 2 in. in diameter), and (3) agglomerates (volcanic bombs and blocks greater than 2 in. in diameter).

This specimen is composed of coarse fragments of pumice and fine-grained ash. Coloration commonly ranges from yellow to gray, pink, light brown, or dark grayish brown. Tuff is characteristically lightweight and poorly indurated. Volcanic breccia and agglomerate are commonly more dense and darker in color than tuff, and contain fragments of pumice, obsidian, aphanitic rock, and some fragments of the surrounding rock (see also Figure 3.7A).

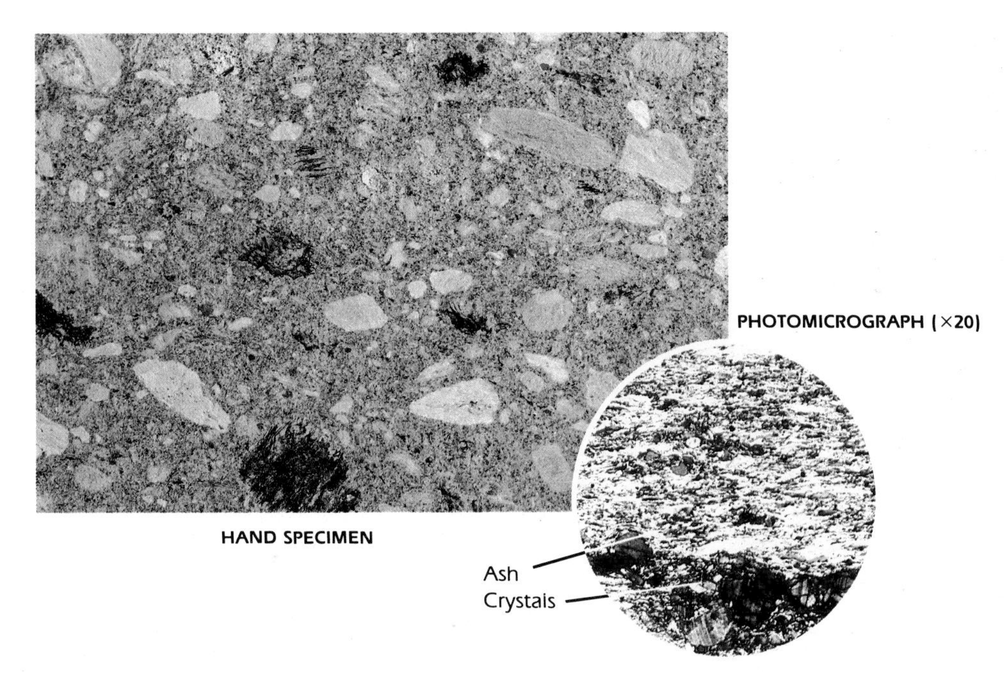

FIGURE 3.23

ASH-FLOW TUFF is composed of volcanic ash (i.e., fragments of volcanic glass, broken crystals, droplets of lava, and fragments of country rock fused together in a tight, coherent mass). It differs from ash-fall tuff (Figure 3.22) in that the fragments were very hot when extruded. As a result, the fragments fused together, flattened, or were bent out of shape. This unique texture forms from a cloud of hot, incandescent ash that moves very rapidly, like a flow. The specimen in Figure 3.7B is also an ash-flow tuff.

THE ORIGIN OF MAGMA

The origin of magma is not known from direct observations, but during the last few decades, our understanding of the chemistry and physics of liquid rock material has increased greatly. This knowledge is based on the observation of synthetic magmas made in the laboratory, and on the observation of volcanic products and studies of geophysical properties of the earth.

We know from seismic evidence that the earth is solid to a depth of 1800 mi, and that only the outer core of the earth is liquid. The liquid core, however, cannot be the source of magma. The core's density ranges from 9 to 10 gm/cm^3, a density much greater than that of any magma or igneous rock in the crust. (Basalt, one of the densest igneous rocks, has a density of only 3 gm/cm^3.) Magma must therefore originate from the localized melting of solid rock in the upper mantle and lower crust, at depths ranging from only 20 to 100 mi below the surface.

To understand the origin of magma, you must remember that a rock does *not* have a specific melting point. Each mineral within the rock begins to melt at a different temperature. As shown in the diagram in Figure 3.1, muscovite, K-feldspar, and quartz not only crystallize in sequence at temperatures ranging from 600 °C to a little over 800 °C, but they melt in this temperature range as well. Amphibole melts at a higher temperature, followed by the melting of pyroxene and olivine. Sodium-rich plagioclase begins to melt at about 600 °C, whereas calcium-rich plagioclase may not melt until it reaches a temperature of 1200 °C. In reality, melting and crystallization are complex physical and chemical processes. It is, however, well established that some minerals melt at lower temperatures than others, and this fact is important to an understanding of the general way in which a magma originates.

In addition to temperature, remember that other important factors, such as pressure, amount of water present, and overall composition of the rock also influence melting. For example, melting can be initiated by either an increase in temperature or a decrease in pressure, and is enhanced by the amount of water present in the pore spaces of the rock.

The diagrams in Figure 3.24 summarize our present understanding of how and where a magma originates. Global studies of the distribution of flood basalts, volcanoes, batholiths, and associated mountain belts indicate that basaltic magma is generated at divergent plate margins and that granitic-andesitic magma is generated in subduction zones. In both areas, variations in temperature and pressure occur as a result of plate movement, and magma is produced by partial melting of the lower crust or upper mantle.

GENERATION OF BASALTIC MAGMA

Basaltic magma is believed to be generated at divergent plate margins by partial melting of the asthenosphere (Figure 3.24A). The asthenosphere, as well as the lower mantle, is almost certainly composed of peridotite, a rock made up largely of the minerals olivine, pyroxene, and minor amounts of plagioclase. The balance between temperature and pressure in the asthenosphere is just about right for peridotite to begin to melt. Below the asthenosphere, the pressure is too great for melting to occur, and at more shallow depths above the asthenosphere, the temperature is too low.

The general process of how basaltic magma originates is as follows: As material in the asthenosphere moves slowly upward and outward, the lithosphere splits and moves apart. This reduces pressure above a section of the asthenosphere, and melting begins. Plagioclase melts first, followed by pyroxene, and ultimately by the melting of olivine. If only part of the mantle rock melts, much of the olivine remains solid, so the resulting magma is richer in those elements that compose plagioclase and pyroxene. In fact, that is the composition of basaltic magma. Scientists therefore hypothesize that basaltic magma originates from partial melting of the asthenosphere at rift zones, where pressure is reduced as a result of the splitting and spreading apart of lithospheric plates. A decrease in pressure thus plays an important role in the generation of basaltic magma. The basaltic magma, being less dense than the surrounding peridotite, rises along the spreading center and is extruded as basalt flows in the rift zone.

GENERATION OF GRANITIC MAGMA

In a subduction zone, the basaltic oceanic crust and water-saturated oceanic sediments descend into the mantle. This mixture of crust and sediment is heated by the friction between colliding lithospheric plates and by the higher temperatures at depth as the mixture is submerged (Figure 3.24B). As the basalt and oceanic sediments are heated, the silica-rich minerals begin to melt first, some at temperatures as low as 500 °C. Partial melting of the basaltic crust thus produces a magma richer in silica than the magma produced at divergent plate boundaries (see Figure 3.1). This silica-rich magma rises upward in the orogenic belt to produce granitic intrusions and andesitic-rhyolitic flows. The extreme pressure in the roots of a deformed mountain belt at converging plate margins would also increase the temperature enough to begin the melting of some of the minerals in metamorphic rocks. The liquid would rise, collect in larger bodies, and produce chambers of granitic magma.

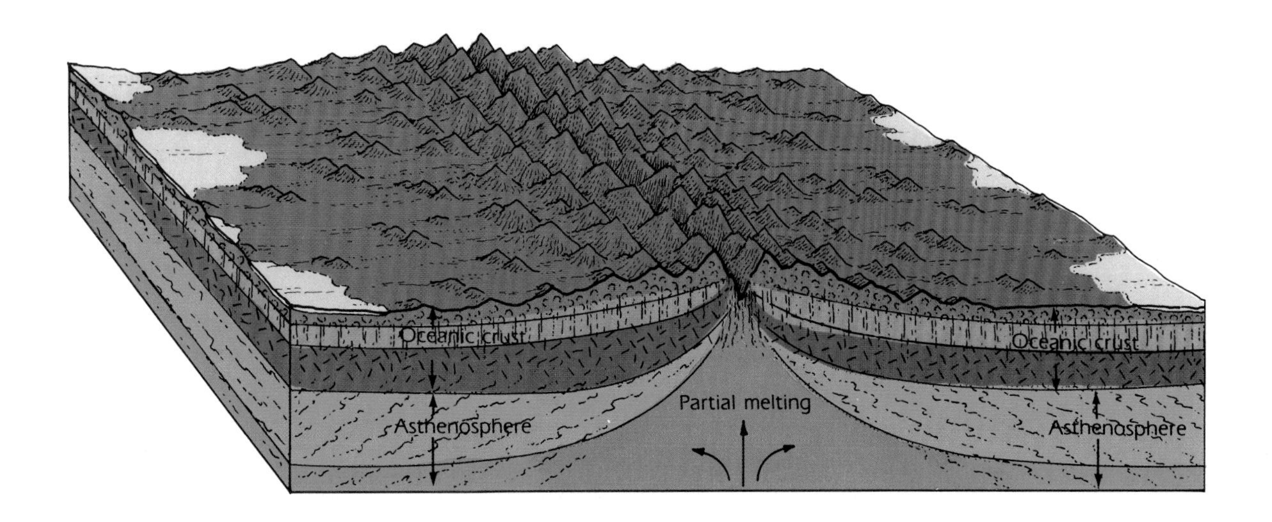

A. GENERATION OF BASALTIC MAGMA AT DIVERGENT PLATE MARGINS by partial melting of the asthenosphere.

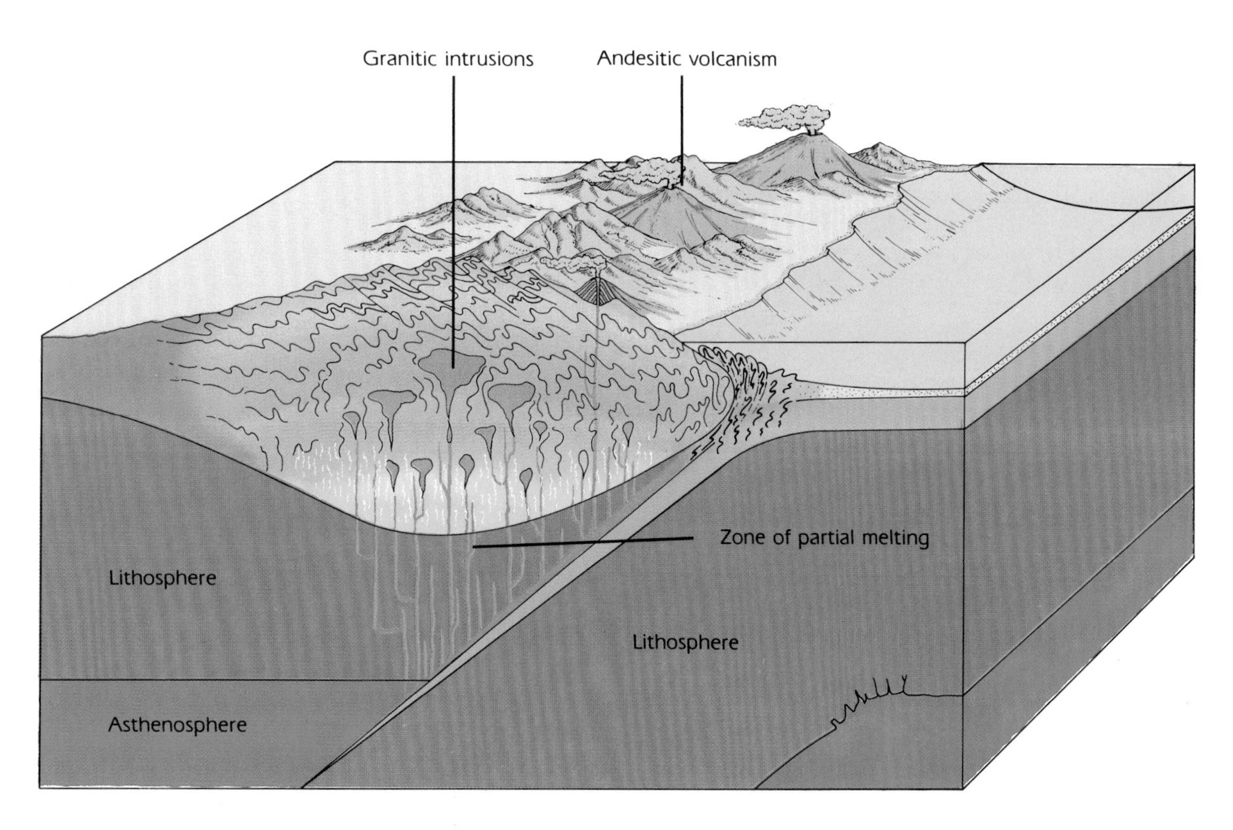

B. GENERATION OF GRANITIC MAGMA AT CONVERGENT PLATE MARGINS by partial melting of the oceanic crust.

FIGURE 3.24 **THE ORIGIN OF MAGMA.**

PROCEDURES

To identify specimens provided by your instructor, proceed as follows:

1 Examine the rock and determine the type of texture. You can then refer to the rock as (a) phaneritic, (b) porphyritic-phaneritic, (c) aphanitic, (d) porphyritic-aphanitic, (e) glassy, or (f) fragmental.

2 Determine the percentage of dark minerals in the rock. You can then refer to the rock as (a) silicic—few dark minerals, generally a light gray color; (b) intermediate—nearly 50% dark minerals, dark gray in color; or (c) mafic—over 70% dark minerals, very dark to black.

3 Determine the approximate percentage and type of feldspar. (a) Pink feldspar is almost invariably a potassium feldspar. (b) White or gray feldspar may be either potassium feldspar or plagioclase. If the feldspar has striations, it is definitely plagioclase.

4 Determine the approximate percentage of quartz. (a) 10% to 40% quartz—granite-rhyolite family; (b) less than 10% quartz—diorite-andesite family; (c) no quartz—gabbro-basalt family.

5 Use the rock chart (Figure 3.1) and determine the rock name.

Example:

a A phaneritic rock composed of 25% olivine, 50% pyroxene, and 25% plagioclase would be a gabbro. (See graph of mineral composition at top of Figure 3.1.)

b An aphanitic rock containing phenocrysts of amphibole in a light-colored matrix, presumably plagioclase, would probably be a porphyritic andesite. An aphanitic rock containing phenocrysts of quartz would be a porphyritic rhyolite.

c A phaneritic rock containing 15% amphibole, no quartz, and 85% plagioclase would be diorite.

Aphanitic rocks have few distinguishable crystals and are difficult to identify with certainty from a hand specimen alone.

The following general guidelines will be useful in the identification of igneous rocks:

1 If quartz phenocrysts are present, the rock is a rhyolite.

2 If potassium feldspar phenocrysts are present, the rock is a rhyolite.

3 If amphibole phenocrysts are present, the rock is an andesite.

4 If pyroxene or olivine phenocrysts are present, the rock is a basalt.

If phenocrysts are conspicuous, the texture is porphyritic, and the adjective "porphyritic" is added to the rock name (i.e., porphyritic granite or porphyritic basalt).

Note that some types of igneous rock do not fit well in the classification system used in Figure 3.8. The most important exceptions are described here.

OBSIDIAN. A massive volcanic glass, usually jet black due to the presence of dustlike particles of magnetite and ferromagnesian minerals. Most obsidian is rich in silica and has a chemical composition similar to that of granite and rhyolite.

PUMICE. A porous volcanic glass with a texture consisting of subparallel glass fibers tangled together.

TUFF AND BRECCIA. A volcanic rock of fragmental texture formed from consolidated volcanic ash. Tuff is fine grained, similar to fine sand or mud. Breccia is coarse grained, like a gravel. The composition of fragmental rocks is variable.

ASH-FLOW TUFF. If the grains of ash, pumice, and crystal fragments are fused together, the term *ash-flow tuff* is used.

PROBLEMS

1 Use the procedures described to identify specimens provided by your instructor.

2 Compare your specimens with the illustrations and descriptions on pages 29–33. What are the distinguishing characteristics of each rock type?

3 How does your specimen differ from the same rock type illustrated in this manual? Is this difference a fundamental difference in texture and composition, or a minor difference in a less diagnostic feature such as color or specimen shape?

4 In Hawaii, many basaltic flows are only 1 to 2 ft thick. On the Columbia Plateau, similar flows are 20 to 40 ft thick. What factor is important in determining the thickness of basalt flows?

5 Could phenocrysts be found in an obsidian? Explain your answer briefly.

6 What is the difference between granite and diorite?

7 What is the origin of vesicles in a basalt?

8 What is the difference between basalt and andesite?

9 Which rock type would be most likely to form a sill? a laccolith? Why?

10 What minerals are likely to form phenocrysts in a basalt? Why?

11 A granite may be pink, red, or various shades of gray, and may be coarse, medium, or fine grained. What characteristics are typical of all granites and distinguish granites from other rock types?

12 What geologic events are implied if a granite body is exposed at the surface?

13 What rock type is most abundant in the oceanic crust? What is the origin of magma on the ocean floors?

14 What igneous rock types form in a mountain belt? Why are these rocks different from those formed at the oceanic ridge?

Exercise 4

SEDIMENTARY ROCKS

OBJECTIVE

To recognize the major types of sedimentary rock and to understand the genetic significance of texture and composition.

MAIN CONCEPT

Sedimentary rocks are classified on the basis of texture and composition. Two main groups are recognized: (1) clastic (formed from fragments of other rocks), and (2) nonclastic (formed by chemical or organic processes).

SUPPORTING IDEAS

1 Sedimentary rocks are composed of material that has been weathered, transported, and deposited by processes operating at the earth's surface (i.e., running water, wind, glaciation).

2 Most minerals that form sedimentary rocks are relatively stable at surface temperatures and pressures.

3 The most important constituents in sedimentary rocks are (a) quartz, (b) calcite, (c) clay minerals, (d) rock fragments, and (e) feldspars.

4 The most important environments of sedimentation are (a) deltas, (b) shallow-marine environments, (c) beaches and bars, (d) deep-marine environments, and (e) reefs. Other important settings in which sedimentation occurs are (a) fluvial environments, (b) alluvial fans, (c) deserts, and (d) glaciers.

DISCUSSION

COMPOSITION OF SEDIMENTARY ROCKS

Because sediment can be derived from any preexisting source rock, one might expect the composition of sedimentary rocks to be extremely variable and complex. This is indeed true if the sediment is deposited close to the source area. If weathering and erosion are prolonged, however, sedimentary processes concentrate materials that are similar in size, shape, and composition in separate, distinct deposits. Most sedimentary rocks are thus composed of the materials that are abundant in other rocks and that are stable under the conditions of surface temperature and pressure. The great bulk of most sedimentary rocks is composed of only four constituents: (1) quartz, (2) carbonates, (3) clay, and (4) rock fragments.

QUARTZ. Quartz is one of the most abundant clastic minerals in sedimentary rocks. This is because quartz is one of the most abundant minerals in the granitic continental crust and because it is extremely hard, resistant, and chemically stable. Weathering processes decompose and disintegrate less stable minerals such as feldspars, olivine, pyroxene, amphibole, and mica. Quartz, however, remains unaltered, and is transported by running water and commonly concentrated as deposits of sand in fluvial and beach environments. Silica in solution or in particles of colloidal size is also a product of the weathering of igneous rocks, and is commonly precipitated as a cement in certain coarse-grained sediments.

CALCITE. Calcite is the major constituent of limestone and is the most common cementing material for sands and shales. Calcium is derived from igneous rocks that are rich in calcium-bearing minerals such as calcic plagioclase. The carbonate is derived from water and carbon dioxide. Calcium is precipitated as $CaCO_3$, or is extracted from sea water by organisms and concentrated as shell material. When the organisms die, the shell fragments commonly accumulate as clastic particles that ultimately form a variety of limestones.

CLAY. Clay minerals develop from the weathering of silicates, particularly the feldspars. They are very fine grained, and are concentrated in mud and shale. The abundance of feldspar in the earth's crust, and the fact that it decomposes readily under atmospheric conditions, account for the large amount of clay minerals in sedimentary rocks.

ROCK FRAGMENTS. Fragments of the parent rock, in which constituent minerals have not disaggregated, are the major constituents of coarse-grained, clastic deposits such as gravel. Rock fragments are also present as grains in some sandstones; the parent rock is usually a basalt, slate, or some other fine-grained rock.

OTHER MINERALS IN SEDIMENTARY ROCKS. Deposits of quartz, calcite, and clay, either alone or in various combinations, form most sedimentary rocks. Other minerals are, however, sometimes abundant enough to form distinct strata. Dolomite, $CaMg(CO_3)_2$, may replace calcite in limestone. Feldspars and mica may be concentrated in some sandstones if weathering, erosion, and deposition occur rapidly. Halite and gypsum are precipitated by the evaporation of salt lakes or sea water, and in certain environments may accumulate in thick layers. Decayed plant material may accumulate in swamps and may become thick beds of coal.

TEXTURE OF SEDIMENTARY ROCKS

Texture in sedimentary rocks is significant because it provides important clues concerning the distance that the sediment has been transported and the environment in which it was deposited. Two basic texture types are recognized: (1) clastic texture (mineral fragments and rock debris) and (2) nonclastic texture (mostly crystals that have grown from solutions, or organic material).

CLASTIC TEXTURE

Examples of clastic texture are shown in Figure 4.1. The size of the clastic particles in sedimentary rock ranges from large blocks, many feet in diameter, to fine dust. Particles are referred to as coarse grained (over 2 mm in diameter), medium grained (1/16 to 2 mm in diameter), and fine grained (less than 1/16 mm in diameter).

Rock fragments and mineral particles may be rounded (Figure 4.1A) or angular (Figure 4.1B), depending on the amount of abrasion they have undergone. Sediment moved by ice or by the direct action of gravity is commonly angular, whereas particles carried by wind or water are rounded by continual abrasion. In general, grain size and shape are a rough measure of the distance of transport. For example, large, angular boulders indicate a nearby source, because significant transport by streams rapidly rounds off the rock corners and wears down the boulder.

Sorting refers to the separation of particles by size. It is a very important textural characteristic because it can provide clues concerning the history of transportation and the environment in which the sediment accumulated. A layer of well-sorted material (Figures 4.1C and D) is composed of grains of one dominant size, and usually of one type of composition. Such layering results only after considerable transport, during which particles similar in size and density are concentrated by prolonged current action. Poorly sorted material (Figure 4.1A) contains grains of several different sizes. Glaciers do not sort material, but deposit coarse and fine particles together. Mudflows also do not produce sorting. Wind and water are the best sorting agents, concentrating materials of one dominant grain size in dunes, beaches, bars, and mud flats.

After a sediment is deposited, it may be compacted and cemented to form solid rock. The dominant cementing materials are calcite, quartz, iron, and chert. These may be carried by ground water from a foreign source or may be derived by solution activity from the site of sedimentation. The extent of cementation is an important textural characteristic in clastic rocks.

NONCLASTIC TEXTURE

Examples of nonclastic texture are shown in Figure 4.2.

CRYSTALLINE TEXTURE. Crystalline texture is in marked contrast to the clastic texture just described. The minerals precipitated from sea water, ground water, or lakes form this texture, which consists of a network of interlocking crystals similar to the texture in some igneous rocks. Crystalline textures are described as coarse (greater than 2 mm), medium (1/16 to 2 mm, Figure 4.2A), and fine or microcrystalline (less than 1/16 mm, Figure 4.2B). Deposits from springs and deposits formed in caves are commonly microcrystalline with a banded appearance that results from chemical variations and impurities during deposition.

SKELETAL TEXTURE. The skeletal texture shown in Figure 4.2C was formed by the accumulation of skeletal parts of invertebrate marine life. Calcium carbonate is removed from sea water by marine organisms to make their shells and other hard parts. When the organisms die, their shell material settles to the sea floor and may be concentrated as shell fragments on a beach or near a reef. The texture of the resulting rock is similar to a clastic texture, but the material is unique in that it consists of the skeletal fragments of organisms. Skeletal texture predominates in many limestones.

OOLITIC TEXTURE. Oolitic texture resembles sand, but the individual grains have concentric layers (Figure 4.2D). This layering is produced when calcium carbonate, precipitated on the sea floor, is agitated by wave or current action and accumulates around a tiny shell fragment or grain of silt. As the particle moves to and fro, thin spherical layers are built up by accretion. Close examination of an oolitic texture reveals a concentric structure around each nucleus, and generally minor amounts of associated shell debris.

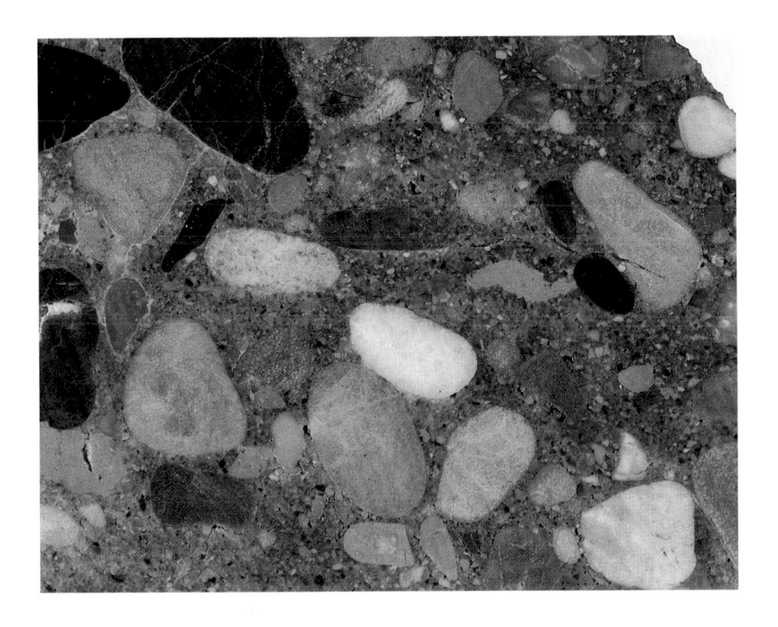

A. COARSE-GRAINED CLASTIC TEXTURE consists of particles larger than 2 mm in diameter. In this specimen, the particles are well-rounded but poorly sorted.

B. COARSE-GRAINED ANGULAR PEBBLES are an important textural characteristic in this specimen. The angular particles indicate that little abrasion occurred and suggest a short distance of transport.

C. MEDIUM-GRAINED TEXTURE consists of particles from $\frac{1}{16}$ to 2 mm in diameter—the size of common sand. In this specimen, the particles are well rounded and well sorted.

D. FINE-GRAINED TEXTURE consists of particles less than $\frac{1}{16}$ mm in diameter, the grain size of mud. The individual particles are too small to be seen without high magnification. Many fine-grained clastic rocks are stratified with fine laminations like the specimen shown here.

FIGURE 4.1 **EXAMPLES OF CLASTIC TEXTURE** (actual size).

A. CRYSTALLINE TEXTURE in sedimentary rocks consists of a network of interlocking crystals, and is similar in many respects to igneous rock textures.

B. MICROCRYSTALLINE TEXTURE is similar to the texture of aphanitic igneous rocks. The grains are so small that they can be seen only with high magnification.

C. SKELETAL TEXTURE consists of fragments of shells. Many different varieties of this texture occur, depending on the type and size of shell debris, and the nature of the cementing material. Note that the small sand-size particles are not quartz, but shell fragments.

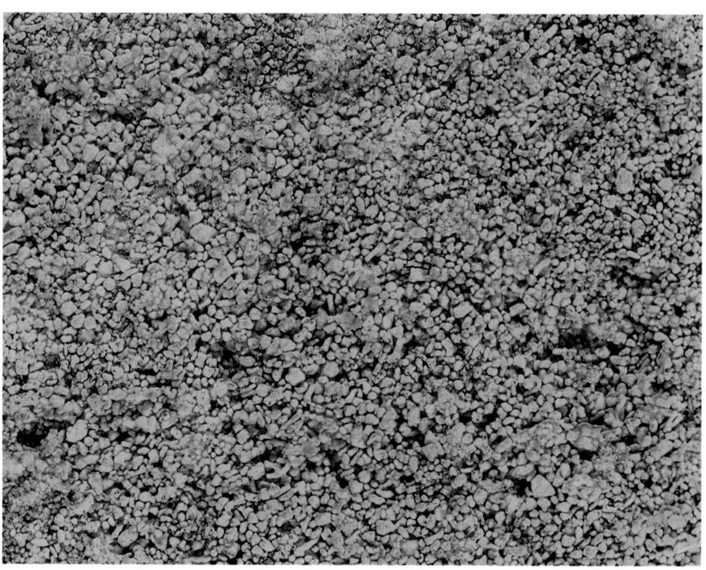

D. OOLITIC TEXTURE, shown in this specimen, resembles medium-grained clastic rock texture. The grains are unique, however, in that they consist of thin concentric layers developed on a small silt particle or shell fragment. These grain nuclei were shifted to and fro by waves, and picked up newly precipitated calcium carbonate, just as a rolling snowball picks up layers of new snow as it moves.

FIGURE 4.2 **EXAMPLES OF NONCLASTIC TEXTURE** (actual size).

CLASSIFICATION OF SEDIMENTARY ROCKS

Although sedimentary processes produce a wide variety of rocks that differ in general appearance, most sedimentary rocks fit into three major categories: (1) clastic, (2) chemical, or (3) organic (Figure 4.3).

CLASTIC ROCKS. Clastic rocks consist predominantly of fragments and debris of other rocks. These materials are familiar as gravel, sand, and mud, and when consolidated, form conglomerate, sandstone, and shale, respectively (see Figures 4.4–4.9). Clastic rocks are thus classified according to grain size with subdivisions based on composition. The grain size range, composition types, and nomenclature are shown on the classification chart in Figure 4.3.

CHEMICAL ROCKS. Chemical rocks are precipitated directly from water, usually as a result of evaporation or changes in the chemistry of the water. Limestone and dolomite are the most abundant types of chemically formed rocks. Limestone contains more than 50% calcite. Other materials present in limestone may include clay, quartz, rock fragments, or iron oxide. The calcite that is precipitated chemically may form crystalline limestone, microcrystalline limestone, or oolitic limestone (Figures 4.10–4.12). Gypsum and halite also form important evaporite deposits, and chert is common as a component in many limestone layers (Figures 4.17–4.19).

ORGANIC ROCKS. Most organic rocks are composed of fragments of calcite shells of invertebrate animals. They thus form varieties of limestone such as coquina, skeletal limestone, and chalk (Figures 4.13–4.15). Peat and coal (Figures 4.20 and 4.21) are sedimentary rocks formed from alteration of plant debris.

FIGURE 4.3 **CLASSIFICATION OF SEDIMENTARY ROCKS.**

A. CLASTIC ROCKS

	Texture	Composition	Rock Name
Clastic	Coarse grained	Rounded fragments of any rock type—quartz, quartzite, chert dominant	CONGLOMERATE
		Angular fragments of any rock type—quartz, quartzite, chert dominant	BRECCIA
	Coarse to fine grained	Poorly sorted, nonstratified, angular fragments of any rock type	TILLITE
	Medium grained	Quartz and rock fragments	SANDSTONE
	Fine grained	Quartz and clay minerals	SILTSTONE
	Very fine grained	Quartz and clay minerals	SHALE
B. CHEMICAL PRECIPITATES			
Chemical or Organic	Medium to coarse grained	Calcite ($CaCO_3$)	CRYSTALLINE LIMESTONE
	Microcrystalline, conchoidal fracture		MICRITE
	Aggregates of oolites		OOLITIC LIMESTONE
	Fossils and fossil fragments loosely cemented		COQUINA
	Abundant fossils in calcareous matrix		FOSSILIFEROUS LIMESTONE
	Shells of microscopic organisms, clay—soft		CHALK
	Banded calcite		TRAVERTINE
	Textural varieties similar to limestone	Dolomite ($CaMg(CO_3)_2$)	DOLOMITE
	Cryptocrystalline, dense, conchoidal fracture	Chalcedony (SiO_2)	CHERT
	Fine to coarse crystalline	Gypsum ($CaSO_4 \cdot 2H_2O$)	GYPSUM
	Fine to coarse crystalline	Halite (NaCl)	ROCK SALT

FIGURE 4.4

CONGLOMERATE consists of coarse fragments (greater than 2 mm in diameter) held together by a matrix of sand, clay, and cement. The individual pebbles are usually well rounded and moderately well sorted. Cobbles and pebbles in conglomerate may consist of any mineral or rock, but resistant materials such as quartz, quartzite, and chert are especially common. Fragments of limestone, granite, or other rock types may predominate, however, in some deposits.

The only agents capable of transporting the large fragments that make up a conglomerate are swiftly moving water, glaciers, and the direct action of gravity. The most important environments for conglomerate deposition are alluvial fans, river channels, and beaches.

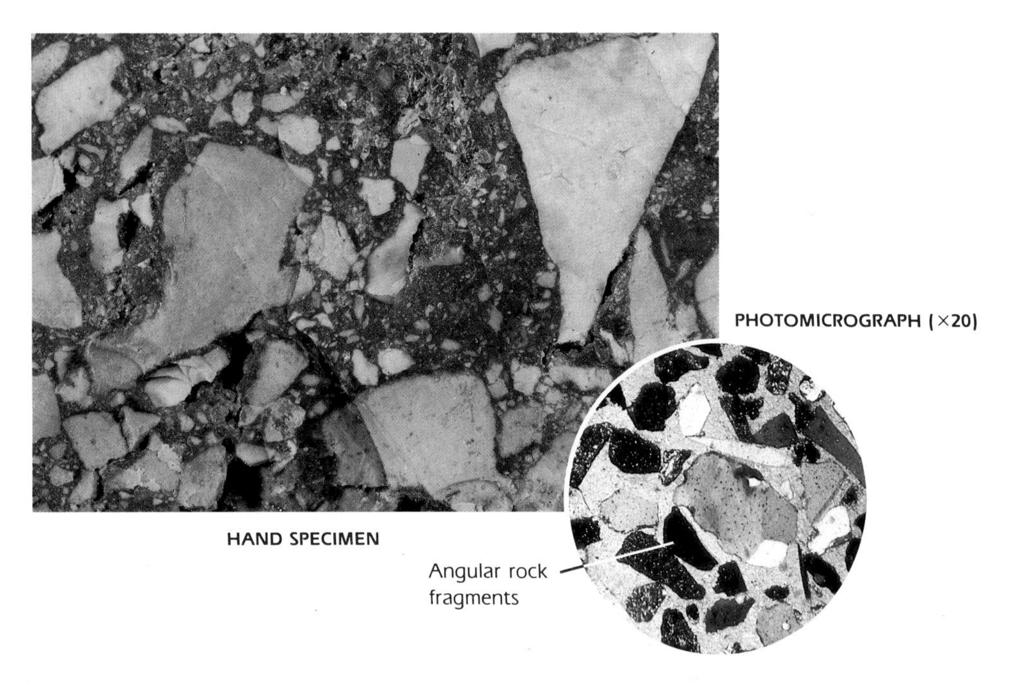

FIGURE 4.5

BRECCIA is a coarse-grained clastic rock in which fragments are angular and show little evidence of abrasion. Material is commonly poorly sorted, with a fine-grained matrix.

The most significant breccia deposits result from meteorite impact, landslides, and other types of mass movement. Movement along a fault also produces a breccia zone, as does a collapsed cave.

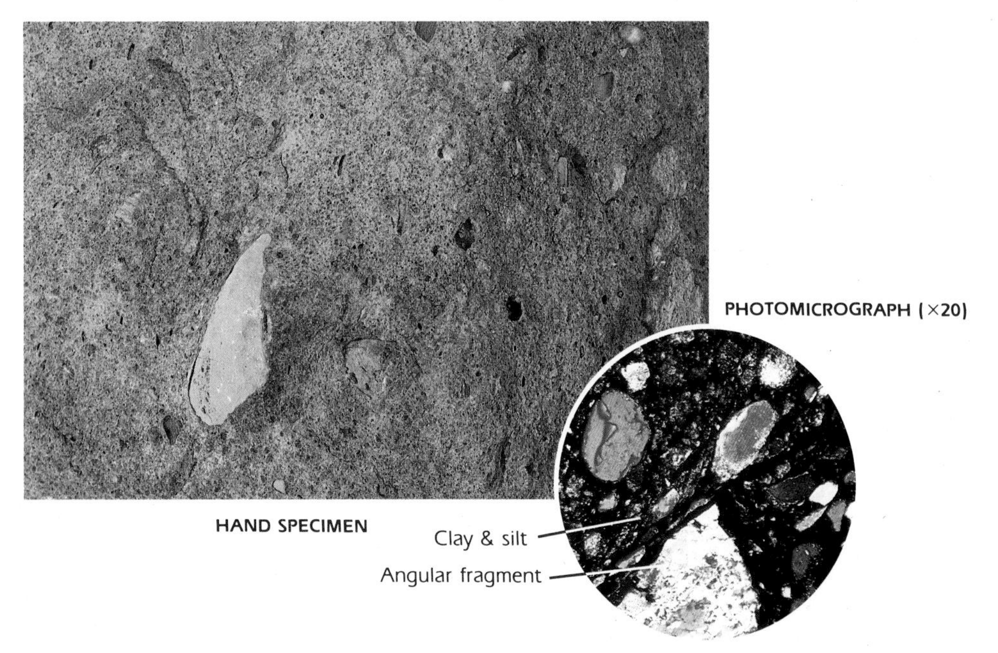

FIGURE 4.6

TILLITE is composed of poorly sorted, unstratified sediment deposited by glaciers. The particles are typically angular. In many deposits, fine-grained sand, silt, and clay-size materials dominate.

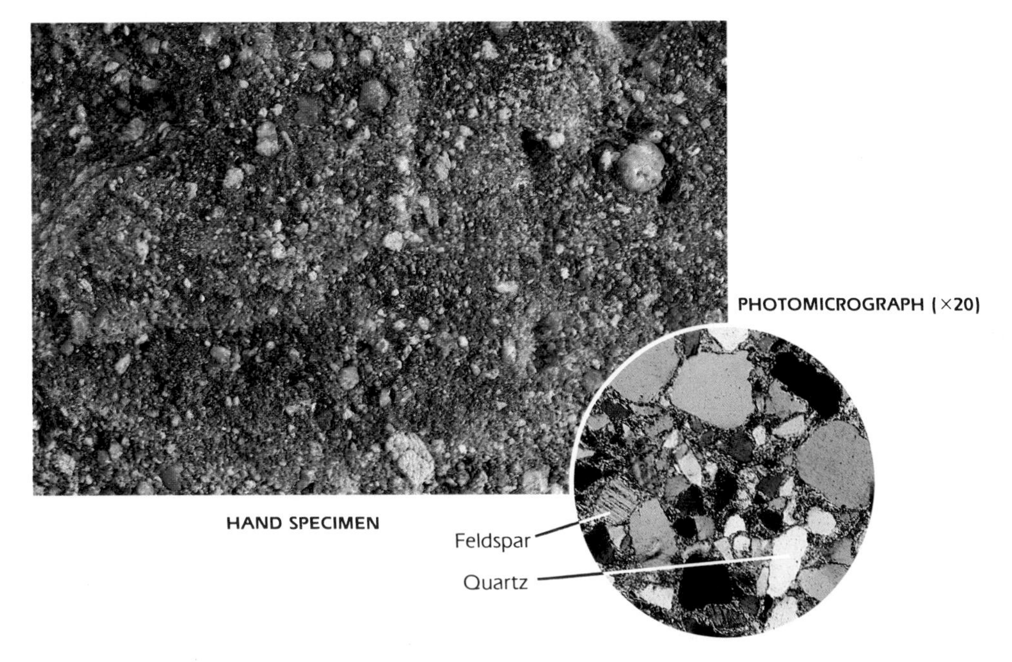

FIGURE 4.7

ARKOSE is a clastic rock composed of at least 25% feldspar. Quartz is the other major constituent. In most arkoses, the grains are in the sand-size range, but some specimens are coarse enough to be termed conglomerates. In this specimen, feldspar is pink, and close inspection reveals its abundance. In thin section, feldspar grains commonly have a plaid pattern and are nearly as abundant as quartz, which appears gray. Calcite cement appears brown in this thin section.

Arkoses form by rapid erosion of a granitic terrain and are commonly deposited in alluvial fans.

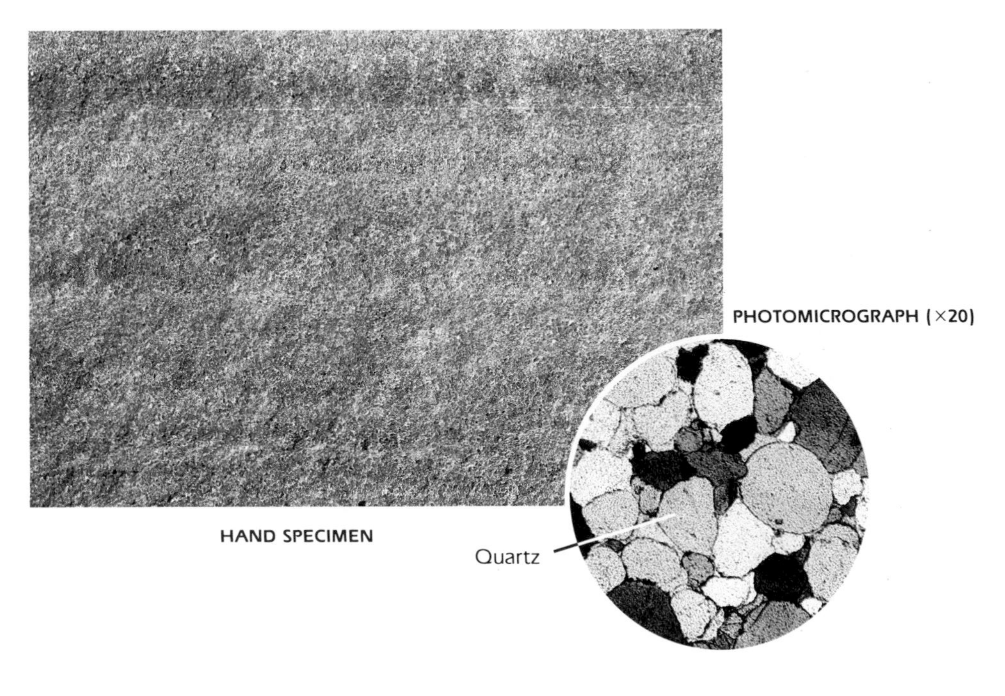

FIGURE 4.8

SANDSTONE is a clastic sedimentary rock consisting mostly of grains ranging from 1/16 to 2 mm in diameter. Grains are generally rounded and show the effects of abrasion, as seen in this thin section. Quartz is usually the dominant mineral, although feldspar, garnet, mica, and other minerals may be present in varying amounts. Calcite, quartz, and iron oxide are the main cementing materials. Sandstones are generally stratified, and are colored buff, red, brown, yellow, or green, depending on impurities and cementing agents.

Sandstones may accumulate in a wide variety of environments, such as beaches, deserts, flood plains, and deltas.

FIGURE 4.9

SHALE is a fine-grained clastic rock consisting of particles less than 1/256 mm in diameter. Most shales are characteristically laminated or thin bedded, and the structure is usually well expressed in both hand specimen and thin section. Quartz, mica, and clay minerals are dominant constituents. Particles are too small to be seen without high magnification. Calcite may be present, usually as cement, in amounts up to 50%.

Eighty percent of exposed sedimentary rocks are shale. The particles accumulate in quiet-water environments, such as shallow-marine waters, lagoons, flood plains, and lakes.

FIGURE 4.10

CRYSTALLINE LIMESTONE (SPARITE) is composed of interlocking calcite crystals large enough to be seen with the unaided eye or with low-power magnification. Well-preserved fossils may be common, and bedding planes are generally distinct. Good rhombohedral cleavage in the calcite crystals is commonly well expressed in a microscopic view.

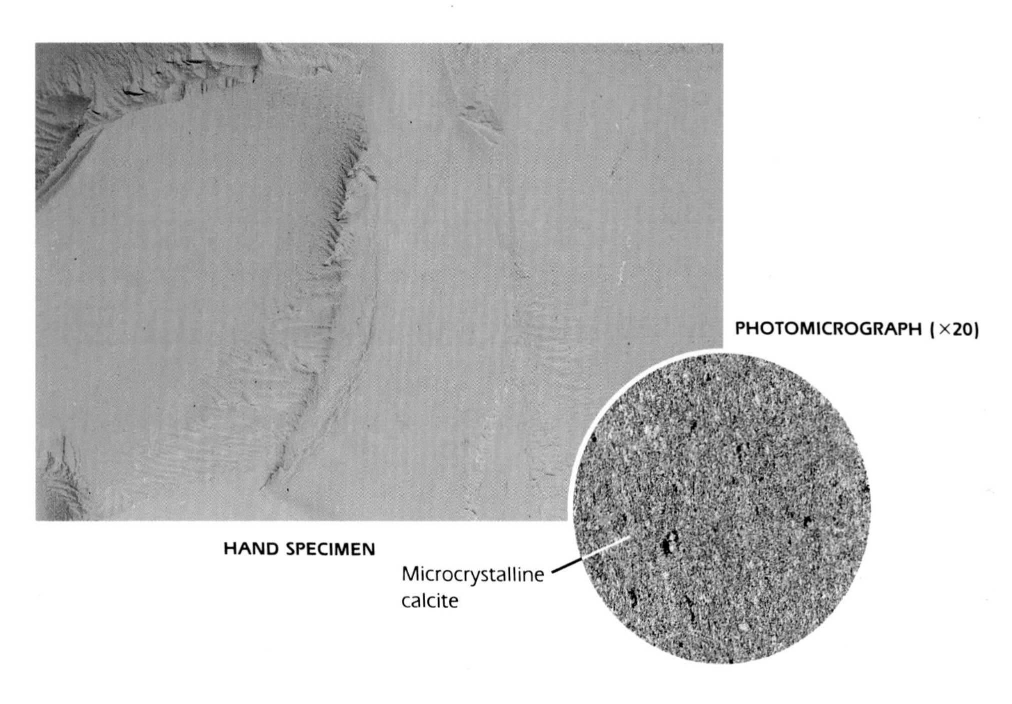

FIGURE 4.11

MICROCRYSTALLINE LIMESTONE (MICRITE) is composed almost entirely of microscopic calcite crystals so small that they are difficult to discern even under the microscope. The rock has uniform texture, is very dense, and characteristically breaks with a conchoidal fracture. It is commonly buff or light yellow, but may range in color to dark gray or black if organic material is abundant. Microcrystalline limestone forms in warm, quiet, shallow seas.

FIGURE 4.12

OOLITIC LIMESTONE is composed of small spheres, which are calcite concretions. The grains range up to 2 mm in diameter and consist of concentric layers formed around a grain nucleus of fine silt, clay, or shell fragment. Calcite commonly cements the spherical particles together to form a rock in which the texture resembles closely a well-rounded quartz sandstone. Stratification, sorting, cross-bedding, and other sedimentary features characteristic of sandstone are also common in oolitic limestone.

Oolites originate in shallow water where currents and waves agitate particles of silt and small shell fragments. These act as nuclei and pick up calcite, which adheres to the grain surface. In this way, multiple layers accumulate on silt particles and shell fragments.

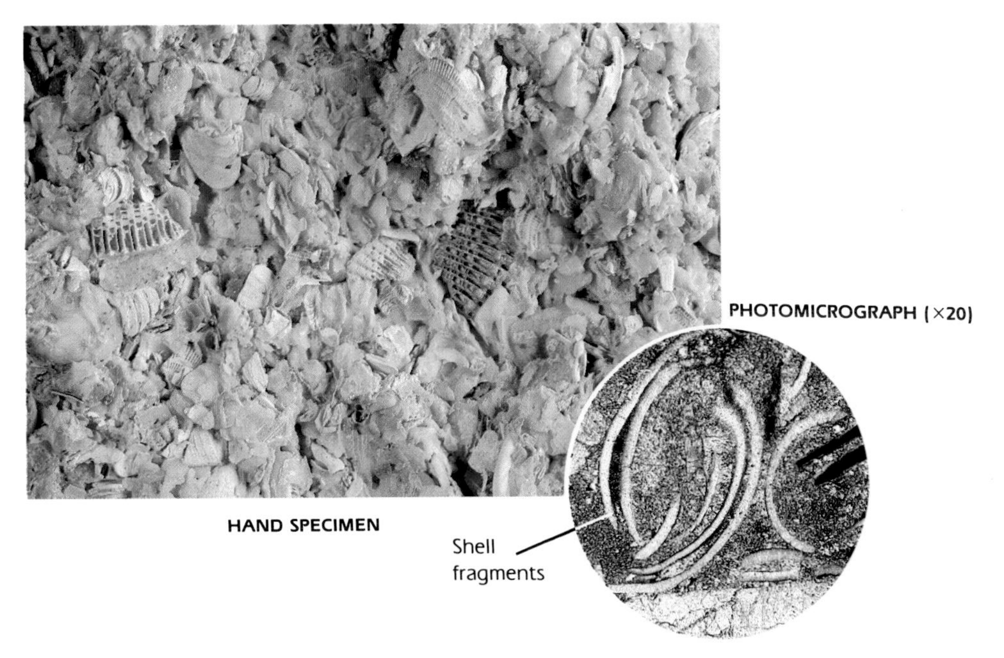

FIGURE 4.13

COQUINA is the name given to a weak, porous, poorly cemented limestone composed almost exclusively of shells and shell fragments. The shell material in a coquina is commonly broken into platy fragments, which are rounded, abraded, and well sorted.

Coquina is thus a particular type of clastic rock composed of fragmental materials that originated in the sedimentary environment, rather than being transported from an eroded landmass. Coquinas are forming today along the southern Atlantic coast and in the Bahama Islands where shell material is washed on shore and accumulates as calcareous sand on a beach or in shallow water.

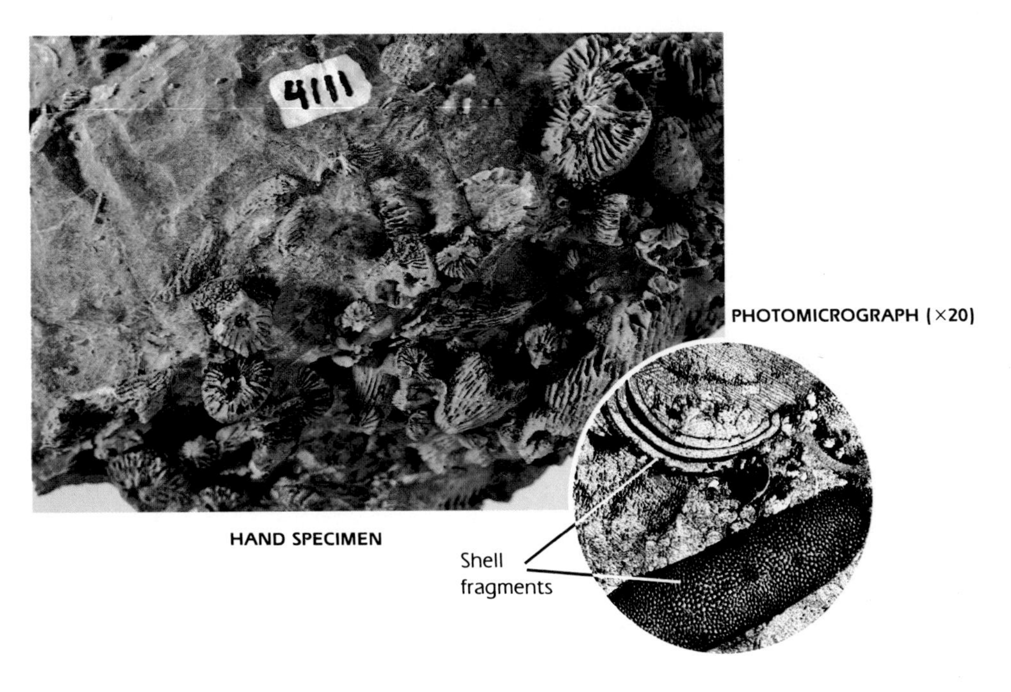

FIGURE 4.14

SKELETAL (FOSSILIFEROUS) LIMESTONE is composed primarily of hard parts of invertebrate organisms such as mollusks, corals, and crinoids. This material is commonly cemented with microcrystalline calcite and may form a dense rock. Fossils often weather out in relief, and show the rock texture and structure. In thin section, details of fossil shell structure are usually well expressed. Fossiliferous limestone forms in warm, shallow seas where marine life is abundant.

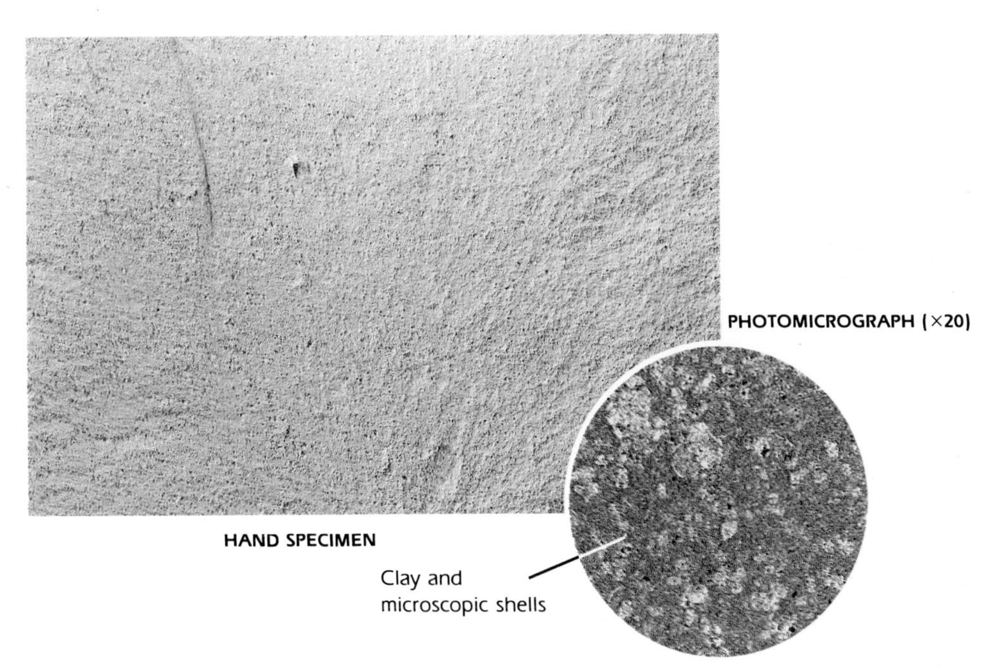

FIGURE 4.15

CHALK is a soft, porous, fine-textured limestone composed of shells of microscopic organisms, mostly foraminifera. It may also contain varying amounts of mud. Coloration is commonly white or buff.

The best known chalks are of Cretaceous age, and include the famous white chalk cliffs on both sides of the English Channel, the Selma chalk of the Gulf Coast, and the Niobrara chalk of Kansas. Chalk is considered a shallow-water deposit, formed by accumulation of the shells of floating organisms.

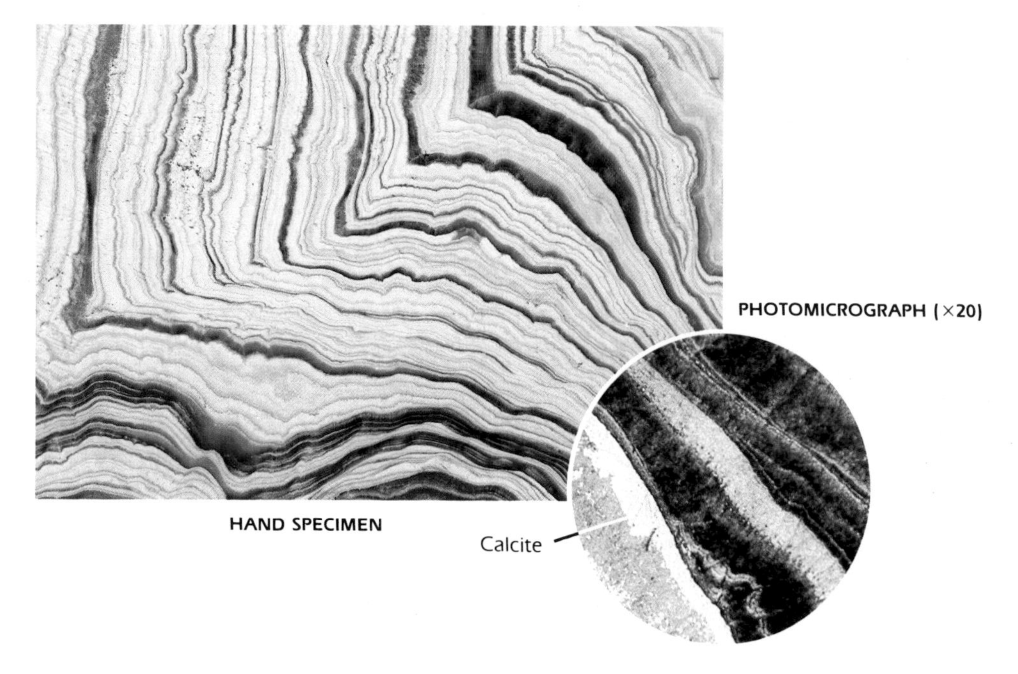

FIGURE 4.16

TRAVERTINE is a calcium carbonate deposit formed in caves and springs. It is characteristically banded with alternating light and dark layers, which result from minor amounts of iron oxide that accumulate during successive periods of deposition. The well-known flowstone, dripstone, stalagmite, and stalactite deposits in caves are all varieties of travertine. Most travertine consists of relatively small deposits of recent age. It is of no great geologic importance, but its beauty and variety of form make it memorable.

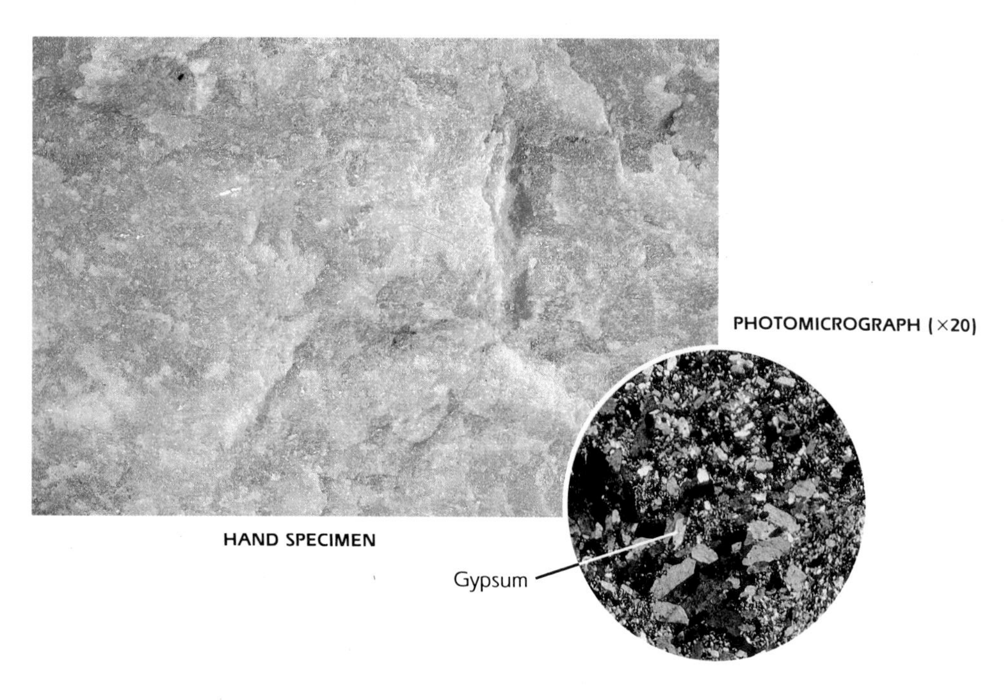

FIGURE 4.17

ROCK GYPSUM is a chemical precipitate composed almost exclusively of aggregates of the mineral gypsum. It is commonly white or delicately colored in various shades of orange or light red due to small amounts of iron oxide. One of its distinctive characteristics is that it can be scratched with a fingernail. Gypsum is commonly massive, but in some deposits, thin, delicate laminae form as a result of seasonal influxes of clay. Gypsum originates from evaporation of saline lakes or sea water trapped in restricted bays. The presence of this mineral thus indicates an arid or semiarid climate at the time of formation.

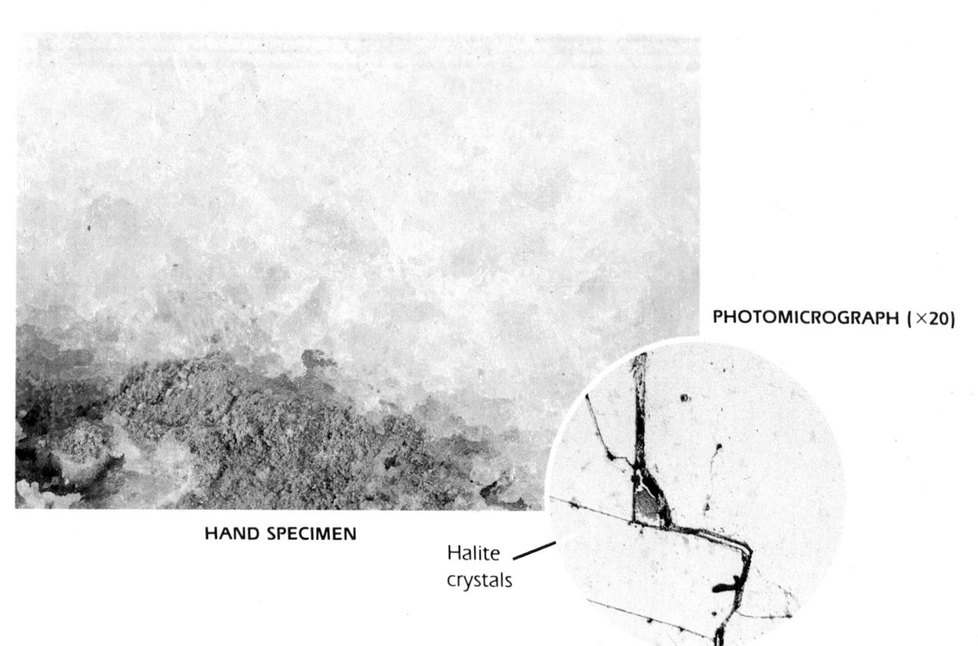

FIGURE 4.18

ROCK SALT is composed of the mineral halite. Crystals may be fine, medium, or coarse, and are generally colorless or stained various shades of red by iron oxide or red clay.

Rock salt is an evaporite deposit and originates in a manner similar to gypsum. Thick salt deposits are found in Michigan, New York, and Kansas, and in relatively recent deposits from saline lakes in the arid western states. An interesting and important characteristic of salt is that it flows at relatively low temperatures and pressures. Salt from deeply buried strata may rise as piercement plugs or salt domes.

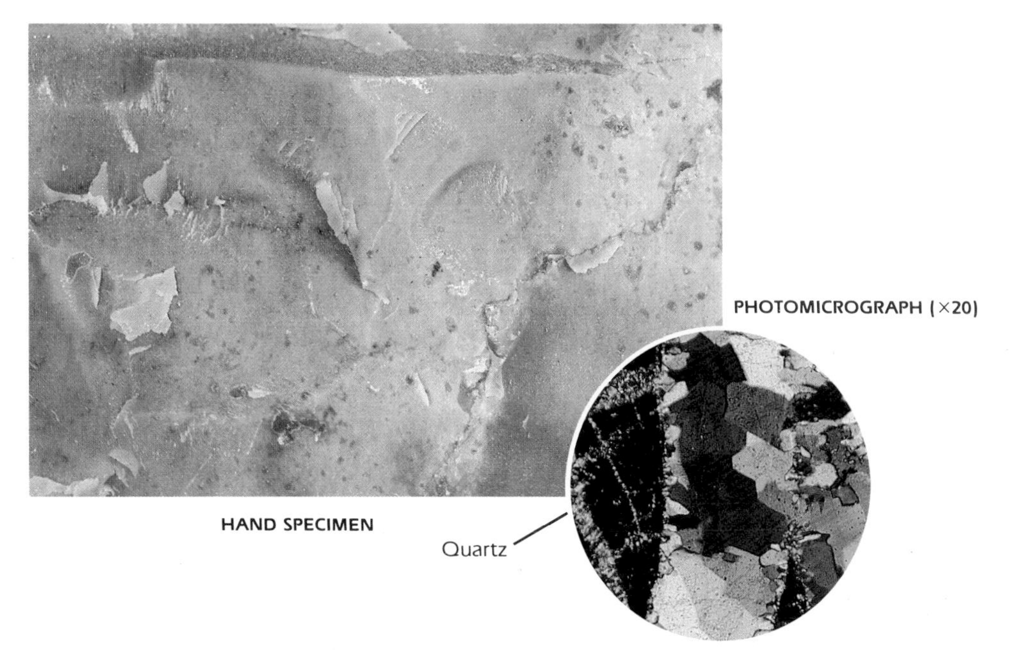

FIGURE 4.19

CHERT is a common chemical rock composed of microcrystalline quartz. Coloration may be white or various shades of gray, green, blue, red, yellow, or black. Under high magnification, a fibrous or granular texture can be seen. Chert commonly occurs as nodules in limestone. Like other varieties of quartz, such as flint and opal, chert is characterized by hardness of 7 and conchoidal fracture. Migrating ground water may deposit large amounts of chert in the pore spaces of limestone.

FIGURE 4.20

PEAT is a dark brown or black organic deposit produced by partial decomposition and disintegration of plants. It forms in bogs and swamps, and represents the initial stage in the formation of coal. Organic matter constitutes from 70% to 90% of the total accumulation. Mineral compounds are an insignificant component. Peat formation requires rapid plant growth and reproduction and a minimum of microorganism activity. In some swamps, peat deposits may be tens of feet thick and cover many square miles.

FIGURE 4.21

COAL is composed of highly altered plant remains and various amounts of clay. It is opaque and noncrystalline, with coloration ranging from light brown to black. Coalification results from burial and compaction of peat, and coal is classified according to the degree of change. Lignite (brown coal) forms first. Deeper burial and greater compaction produce bitùminous (soft) coal. Prolonged burial and pressure produce anthracite (hard) coal.

SEDIMENTARY ENVIRONMENTS

Sedimentary rocks are derived from the weathering and erosional debris of other rocks. This material is deposited at the earth's surface, under conditions of normal temperature and pressure. The major processes involved in the formation of sedimentary rocks are (1) physical and chemical weathering of the parent rock material, (2) transportation of these weathered products by running water, wind, gravity, or ice, (3) deposition of the material in a sedimentary basin, and (4) compaction and cementation of sediment into solid rock. These processes operate at or near the earth's surface and are therefore observable.

A significant result of the transportation and deposition of sediment is the sorting and segregation of the parent rock material into deposits of similar grain size and composition. Materials similar in size and weight are washed and winnowed by wind and water, and are deposited in a sedimentary environment according to the level of mechanical energy operating at the time of deposition. For example, coarse gravels may be deposited in an alluvial fan near a mountain front or on a river bar, while sand is transported downstream to the sea to be concentrated on a beach. Currents and wave action wash away finer material, which is deposited ultimately in an environment of low mechanical energy, such as a marsh or lagoon. Chemical sorting occurs contemporaneously; the more soluble materials dissolve and are removed in solution.

The term *sedimentary environment* refers to the place where the sediment is deposited, and to the physical, chemical, and biological conditions that exist in that place. Figure 4.22 shows in a general way the regional settings of some major sedimentary environments. The following brief summary describes some of the important characteristics of the sedimentary rocks formed in each environment.

ALLUVIAL FANS. Alluvial fans are stream deposits that accumulate near a mountain front in a dry basin. They typically contain poorly sorted coarse gravel and boulders. Fine-grained sand and silt may be deposited near the margins of the fan.

FLOOD PLAINS. The great rivers of the world typically meander across a flat flood plain before reaching the sea; a considerable amount of sediment is deposited on these plains. Rocks formed in a flood plain environment are commonly channels of sandstone deposited on the point bar of a meander and enclosed in a shale deposited on the flood plain.

EOLIAN ENVIRONMENTS. Wind is an effective sorting agent and will selectively transport sand. It leaves behind gravel but lifts dust-sized particles high in the atmosphere and transports them thousands of miles before they come to earth to accumulate as a thin blanket of loess. Windblown sand commonly accumulates in dunes that are characterized by well-sorted, fine grains. The dominant sedimentary structure in eolian environments is large-scale cross-bedding.

GLACIAL ENVIRONMENTS. Glaciers transport but do not effectively sort material. The resulting deposit is an unstratified accumulation of boulders, gravel, sand, and fine silt.

DELTAS. A delta is a large accumulation of sediment that is deposited at the mouth of a river. Deltas are one of the most significant sedimentary environments and include a number of subenvironments such as stream channels, flood plains, beaches, bars, and tidal flats. The deltaic deposit as a whole consists of a thick accumulation of silt, mud, and sand.

SHORELINE ENVIRONMENTS. Beaches, bars, and spits commonly develop along low coasts and partly enclose quiet-water lagoons. The sediment in these environments is well washed by wave action and is typically clean, well-sorted quartz sand. Behind the bars and adjacent to the beaches, fine silt and mud are often deposited as tidal flats.

ORGANIC REEFS. An organic reef is a solid structure composed of the shells and secretions of marine organisms. The reef framework is typically built by corals and algae, but many other types of organisms contribute to the reef community. Together, these organisms produce a highly fossiliferous limestone.

SHALLOW-MARINE ENVIRONMENTS. Shallow seas are widespread along the continental margins, and in the past were even more extensive than today. Sediments deposited in shallow-marine waters form extensive layers of well-sorted sand, shale, and limestone, which typically occur in a cyclical sequence as a result of shifting environments from changes in sea level.

DEEP-OCEAN ENVIRONMENTS. The deep ocean adjacent to the continents receives a considerable amount of sediment transported from the continental margins by turbidity currents. As a current moves across the deep-ocean floor, its velocity decreases gradually, and the sediment carried in suspension settles. The resulting deposit is a widespread sediment layer in which the grain size grades from coarse at the base of the layer to fine at the top. Deep-sea deposits are thus characterized by a sequence of graded beds.

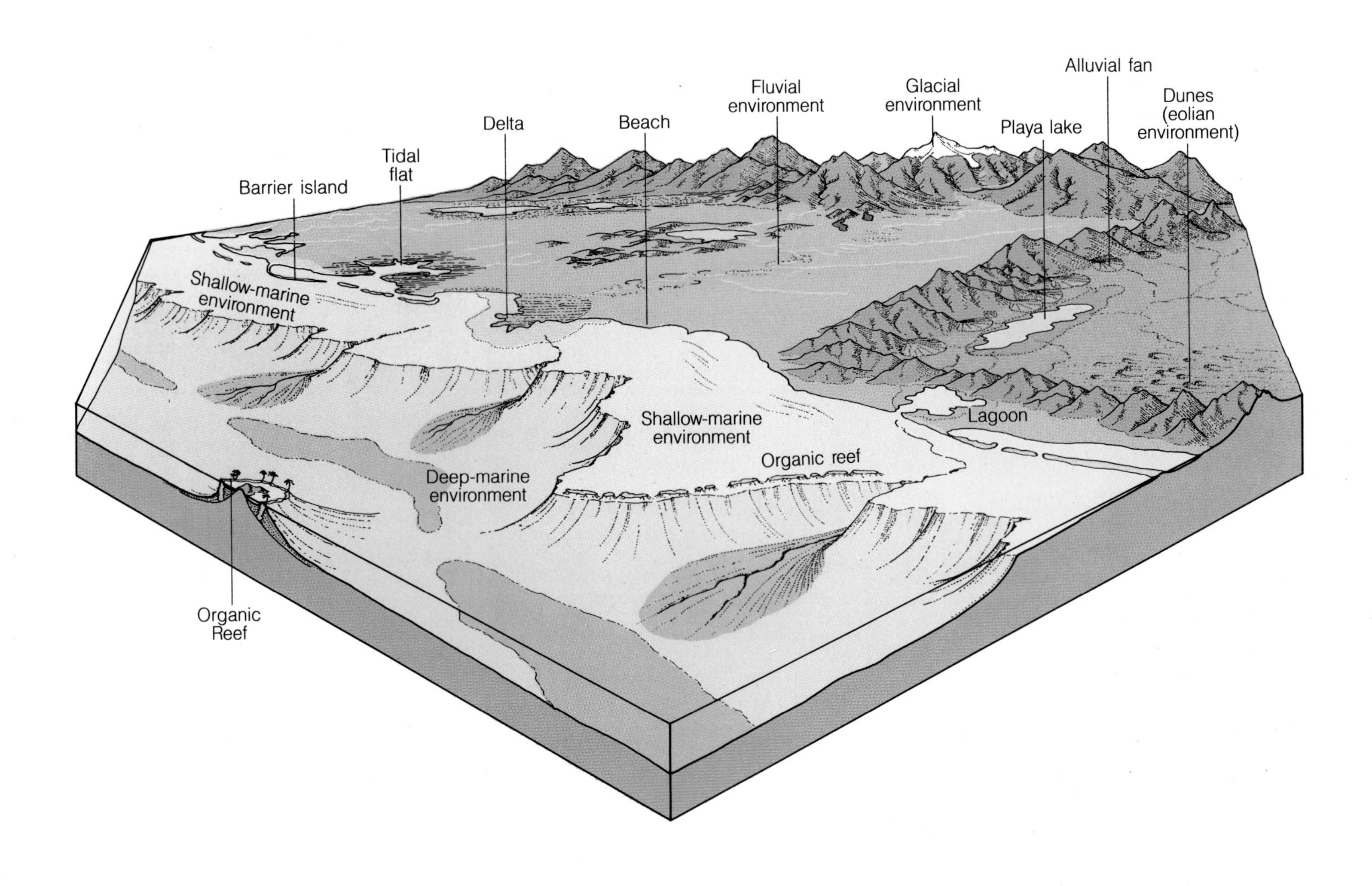

FIGURE 4.22

ENVIRONMENTS OF SEDIMENTATION.

PROCEDURES

To identify the specimens provided by your instructor, proceed as follows:

1 Determine whether the rock is composed of calcium carbonate. (Rocks composed mostly of $CaCO_3$ will effervesce in dilute HCl; other rocks will not.)

2 Determine whether the rock has a clastic (fragmental) or a nonclastic texture. (Use a hand lens or a microscope if one is available.)

3 Determine the grain size and matrix.

- **a** Clastic—Decide whether the grains are predominantly gravel, sand, silt, or clay size. Note the abundance of matrix material.
- **b** Crystalline carbonates—Decide whether the texture is coarse, medium, or fine crystalline.
- **c** "Clastic" carbonates—Decide whether the texture is predominantly oolitic or fossiliferous.

4 Determine the composition of the matrix.

5 With the above information, refer to Figure 4.3 to find the rock name.

PROBLEMS

1 Identify the specimens provided by your instructor.

2 Compare each specimen with the illustrations and descriptions on pages 42–47. How does your specimen differ from the illustration of the same rock type?

3 Are the differences of fundamental importance (i.e., differences of composition or texture), or are they minor (i.e., a difference in color or overall shape)?

4 List the environments in which shale accumulates. What does each of these have in common with the others?

5 What evidence indicates that sandstone forms at the earth's surface?

6 How is $CaCO_3$ extracted from sea water?

7 What is the origin of clay minerals?

8 What rock types commonly occur as fragments in sandstone?

9 Why is olivine a rare mineral in sandstone—even on the beaches of volcanic islands?

10 What properties are common to all limestones regardless of texture?

11 Many limestones are dense, fine grained, and black. How can you distinguish between a limestone and a basalt?

12 How do oolites originate? Where are they deposited?

13 Where do coquinas form?

14 How are rock salt and rock gypsum formed?

15 What is an arkose? How does it differ from a quartz sandstone?

16 What is the major source rock of an arkose?

17 Study the environments shown in Figure 4.22, and list the rock types likely to be formed in each major environment.

18 Why does sandstone form in so many different environments?

19 What geologic events are indicated by a clean, well-sorted quartz sandstone?

20 How would these events differ in the formation of an arkose?

21 Many limestones have a crystalline texture. What features of a limestone indicate that it is not an igneous rock?

22 What are the major processes involved in the genesis of sedimentary rocks?

23 What is the difference between a tuff and a quartz sandstone?

Exercise 5

METAMORPHIC ROCKS

OBJECTIVE

To recognize the major metamorphic rock types and to understand the significance of their texture and composition.

MAIN CONCEPT

Metamorphic rocks are classified on the basis of texture and composition. Two main groups are recognized: (1) foliated and (2) nonfoliated. Regional metamorphism occurs in the roots of mountain belts formed at convergent plate margins. Contact metamorphism develops at the margins of igneous intrusions.

SUPPORTING IDEAS

1 Foliation results from recrystallization and the growth of new minerals.

2 The three main types of foliation are: (a) slaty cleavage, (b) schistosity, and (c) gneissic layering.

3 Nonfoliated texture develops by recrystallization of rocks composed predominantly of one mineral, such as sandstone or limestone.

DISCUSSION

Metamorphic rocks are rocks that have undergone a fundamental change as a result of heat, pressure, and the chemical action of pore fluids and gases. The parent rock material, before metamorphism, might include marine sediments that formed at surface temperatures and pressures, interbedded lava that crystallized at high temperatures, and granite that subsequently intruded the entire sequence. These materials would be unstable at the prevailing temperatures and pressures in the crustal interior, especially in an orogenic belt where lithospheric plates collide. Under conditions of increased temperature or pressure or both, the parent rock reacts in such a way that a new assemblage of minerals is produced—an assemblage that is stable under conditions of higher temperature and pressure. Metamorphic changes always tend to establish equilibrium between the rock material and the prevailing temperature and pressure. The effects of metamorphism include: (1) chemical recombination and growth of new minerals, with or without the addition of new elements from circulating fluids and gases, (2) deformation and rotation of the constituent mineral grains, and (3) recrystallization of minerals to form larger grains. The net result is rock of greater crystallinity, increased hardness, and new structural features that commonly exhibit the effects of flow or other expressions of deformation. All three major rock types can be metamorphosed, but intrusive igneous rocks and previously metamorphosed rocks are less affected by metamorphic processes than are sedimentary rocks that developed at the earth's surface.

METAMORPHIC TEXTURE

FOLIATION

Foliation is a planar element in metamorphic rocks. It may be expressed (1) by closely spaced fractures (slaty cleavage), (2) by the parallel arrangement of platy minerals (schistosity), or (3) by alternating layers of different mineral composition (gneissic layering). Foliation is usually developed during metamorphism by directed stresses that cause differential movement or recrystallization. It is a fundamental characteristic of metamorphic rocks and is a basic criterion of the classification system. The various types of foliation are shown in Figure 5.1.

A. SLATY CLEAVAGE is a type of foliation expressed by the tendency of the rock to split into thin layers. It results from the parallel orientation of microscopic grains of mica, chlorite, or other platy minerals.

B. SCHISTOSITY is a type of foliation resulting from the parallel to subparallel orientation of large platy minerals such as mica, chlorite, and talc.

C. GNEISSIC LAYERING is a coarse-grained, foliated texture in which the minerals are segregated into layers that can be as thick as several centimeters. Light-colored layers commonly contain quartz and feldspar, and dark-colored layers are commonly composed of biotite and amphibole.

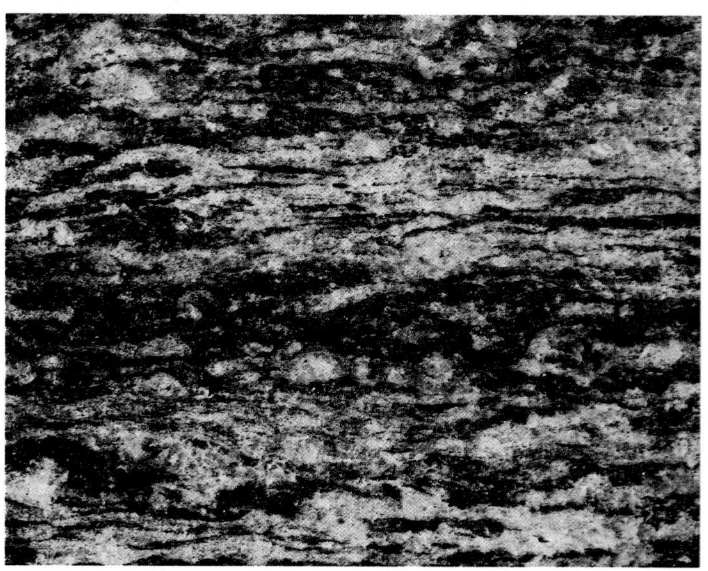

D. GNEISSIC LAYERING in some rocks may be thin discontinuous lenses that impart a distinctive fabric to the rock.

FIGURE 5.1 **EXAMPLES OF FOLIATED TEXTURE** (actual size).

A. STRETCHED PEBBLES result from the metamorphism of conglomerate and impart a distinctive linear fabric to the rock. Strong foliation does not develop.

B. RECRYSTALLIZATION results from metamorphism and may develop a massive, nonfoliated texture as shown here. Calcite crystals have grown much larger than the original grains but are essentially equidimensional. Texture is therefore granular and coarsely crystalline, but nonfoliated.

FIGURE 5.2 **EXAMPLES OF NONFOLIATED TEXTURE** (actual size).

SLATY CLEAVAGE. Slaty cleavage is a type of foliation expressed by the tendency of a rock to split along parallel planes. Do not confuse this planar feature with bedding planes, which are a sedimentary structure. Slaty cleavage results from the parallel orientation of microscopic platy minerals such as mica, talc, or chlorite. In a metamorphic environment, these minerals grow with flat surfaces perpendicular to the applied forces. The perfect cleavage within each tiny mineral grain is thus oriented in the same direction. This creates definite planes of weakness throughout the entire rock and causes the rock to break along nearly parallel planes. Such a surface is shown in Figure 5.1A. Note that slaty cleavage commonly cuts across bedding planes. It is the product of a relatively low intensity of metamorphism.

SCHISTOSITY. Schistosity develops from more intense metamorphism. Mica, chlorite, and talc form visible crystals as a result, and the rock develops a distinctly planar element (Figure 5.1B). Schistosity is thus similar to slaty cleavage, but the platy mineral crystals are much larger, and the entire rock appears coarse grained. The increase in crystal size represents a higher grade of metamorphism in which garnet, amphibole, and other nonplaty minerals also develop. Schistosity typically occurs in mica-rich rocks.

GNEISSIC LAYERING. Gneissic layering is a type of foliation in which the planar element is produced by alternating layers of different mineral composition. Rocks with gneissic layering are characteristically coarse grained and represent a higher grade of metamorphism in which the minerals are recrystallized, stretched, crushed, and rearranged completely. Feldspar and quartz commonly form light-colored layers that alternate with dark layers of ferromagnesian minerals (Figure 5.1C).

NONFOLIATED TEXTURES

Some metamorphic rocks do not possess foliation but appear massive and structureless except for elongated grains or other linear features resulting from directional stress. Examples of nonfoliated texture are shown in Figure 5.2. Nonfoliated rocks are commonly formed from a parent rock composed largely of a single mineral, such as sandstone or limestone, in environments of contact metamorphism where there is little directed stress.

Lineation, such as the elongated pebbles in Figure 5.2A, may develop in nonfoliated rocks. The pebbles were originally round, but have been stretched by the directed stress within the rock. Sand grains in a metamorphosed sandstone often show similar expressions of deformation, or the grains may be fused in a dense, compact mass of interlocking particles. Deformation of limestone will produce streaks of organic debris.

CLASSIFICATION OF METAMORPHIC ROCKS

The formation of metamorphic rocks is so complex that formulating a satisfactory classification system is difficult. The most convenient scheme is to group metamorphic rocks by structural feature, with further subdivision based on composition. Using this classification, two major groups of metamorphic rocks are recognized: (1) those that are foliated (possess a definite planar structure), and (2) those that are nonfoliated, that is, massive and structureless. The foliated rocks can then be subdivided further according to type of foliation. Finally, a large variety of rock types can be recognized in each group according to the dominant minerals. The basic framework for this classification system is shown in Figure 5.3, and the major types of metamorphic rocks are shown in Figures 5.4–5.12.

FIGURE 5.3

CLASSIFICATION OF METAMORPHIC ROCKS.

Texture			Composition	Rock Name
FOLIATED	NONLAYERED	Very fine grained	CHLORITE, MICA, QUARTZ	SLATE
FOLIATED	NONLAYERED	Fine grained	CHLORITE, MICA, QUARTZ, FELDSPAR, AMPHIBOLE	PHYLLITE
FOLIATED	NONLAYERED	Coarse grained	CHLORITE, MICA, QUARTZ, FELDSPAR, AMPHIBOLE, PYROXENE	SCHIST
FOLIATED	LAYERED	Coarse grained	MICA, QUARTZ, FELDSPAR, AMPHIBOLE, PYROXENE	GNEISS
NONFOLIATED	Coarse grained		Deformed grains of any rock type	METACONGLOMERATE
NONFOLIATED	Fine to coarse grained		Quartz	QUARTZITE
NONFOLIATED	Fine to coarse grained		Calcite or dolomite	MARBLE

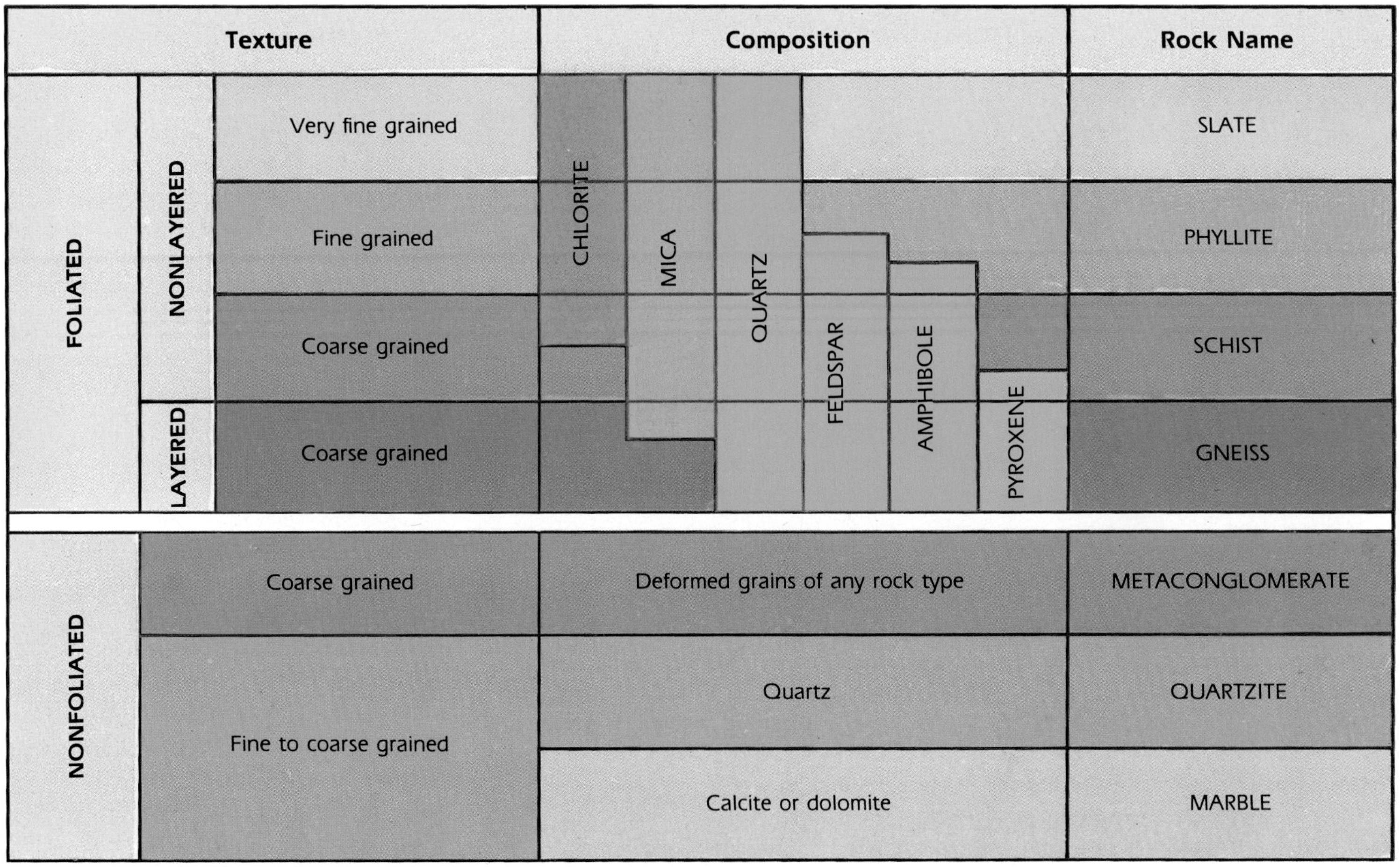

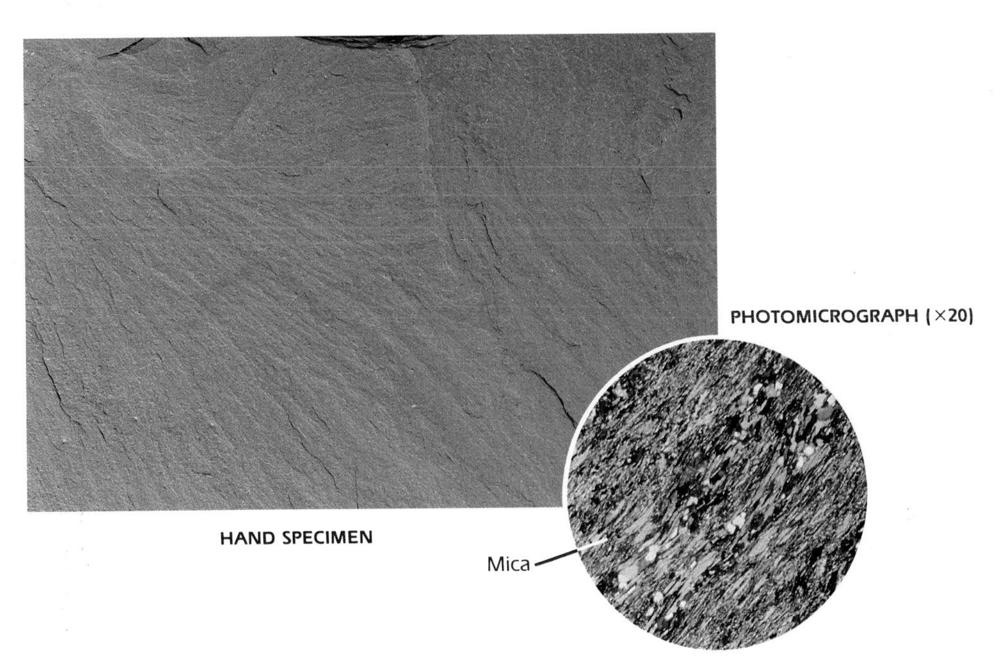

FIGURE 5.4

SLATE is a fine-grained metamorphic rock possessing a type of foliation known as slaty cleavage. In many slates, traces of the original bedding planes in the shale are expressed by subtle changes in color or grain size, but slaty cleavage is still the most obvious structure. The constituent minerals are commonly so small that they can be seen under high magnification only. Slates are characteristically dense and brittle. Coloration may be gray, black, red, or green. They are low-grade metamorphic rocks derived principally from the metamorphism of shale.

PHYLLITE (see photomicrograph) is similar to slate but is distinguished by a satin luster, or sheen, developed on the planes of foliation. The luster results from the growth of larger mica crystals than are commonly found in slate.

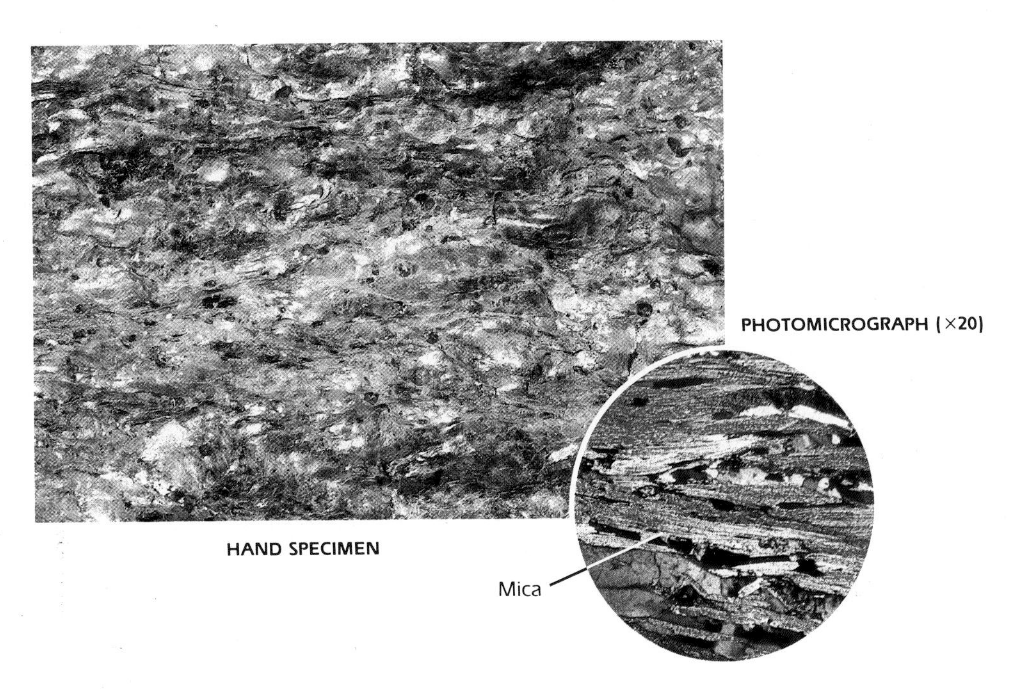

FIGURE 5.5

SCHIST is a metamorphic rock in which foliation is due to parallel arrangement of relatively large crystals of platy minerals. Muscovite, biotite, chlorite, and talc are the important platy constituents. Feldspars are rare, but quartz, garnet, and hornblende are common accessory minerals.

Schists may be further classified on the basis of the more important minerals present. The most common types are chlorite schists, muscovite schists, hornblende-mica schists, and garnetiferous mica schists. The unifying characteristic, regardless of composition, is that foliation results from parallel arrangement of relatively large platy minerals. Schists are produced by a metamorphism of higher intensity than that which produces slates. Under such conditions, a variety of parent rock types, including basalt, granite, sandstone, and shale, may be converted to schists.

FIGURE 5.6

BLUESCHIST results from the high-pressure, low-temperature metamorphism of oceanic crust (basalt and oceanic sediment) in the subduction zone below a trench. The rapid and deep subsidence of the cool lithospheric slab at a subduction zone creates the high pressure of relatively deep burial. The low temperature persists because the descending plate heats up slowly while the pressure is increasing rapidly.

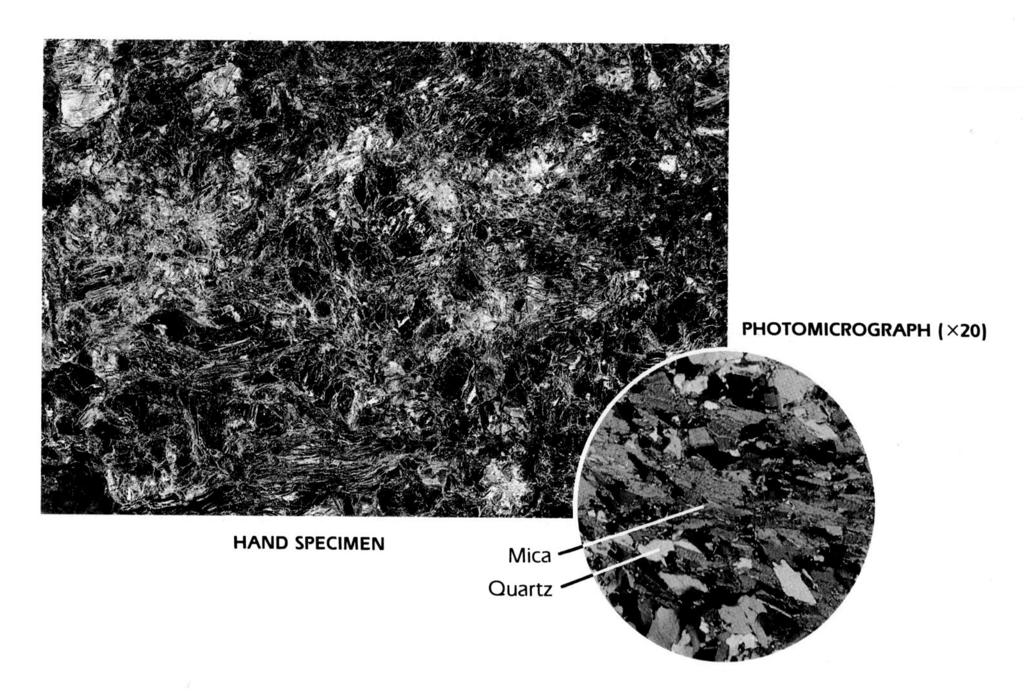

FIGURE 5.7

GREENSCHIST results from the high-pressure, high-temperature metamorphism of a variety of rock types. It forms in the roots of mountain belts where high pressure is created at depths similar to those where blueschist is formed. In the mountain roots, however, high temperature also occurs as a result of ascending magma generated by melting at convergent plate margins. Under some high-temperature conditions, blueschist is metamorphosed into greenschist.

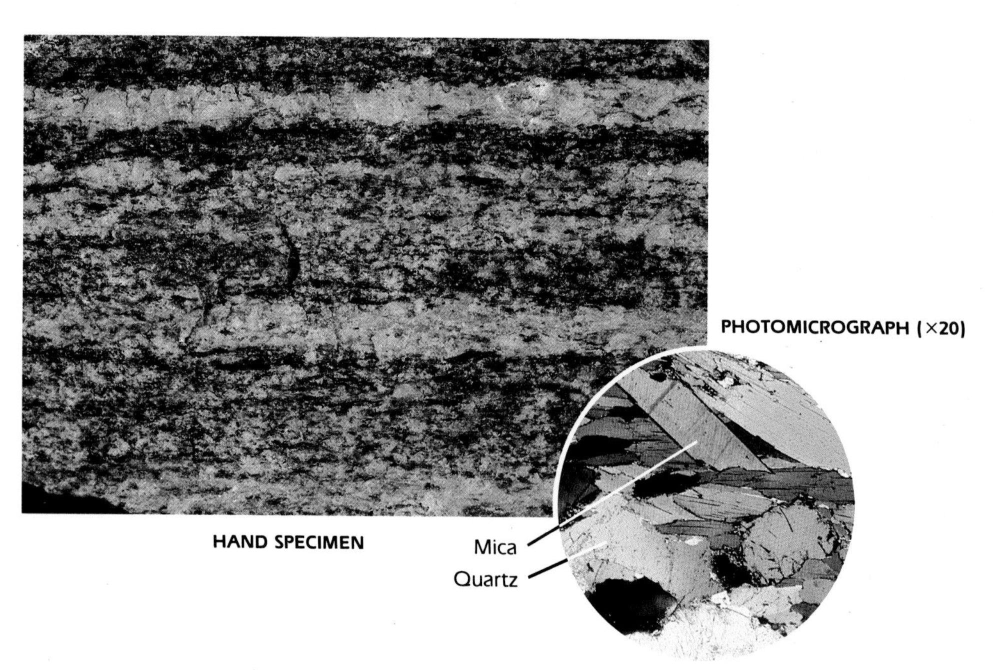

FIGURE 5.8

GNEISS is a metamorphic rock in which foliation results from layers composed of different mineral groups. Feldspar and quartz are the chief minerals in gneiss, with minor amounts of mica, amphibole, and other ferromagnesian minerals. Gneiss has a coarse-grained, granular texture. It resembles granite in composition, but is distinguished from granite by the foliation.

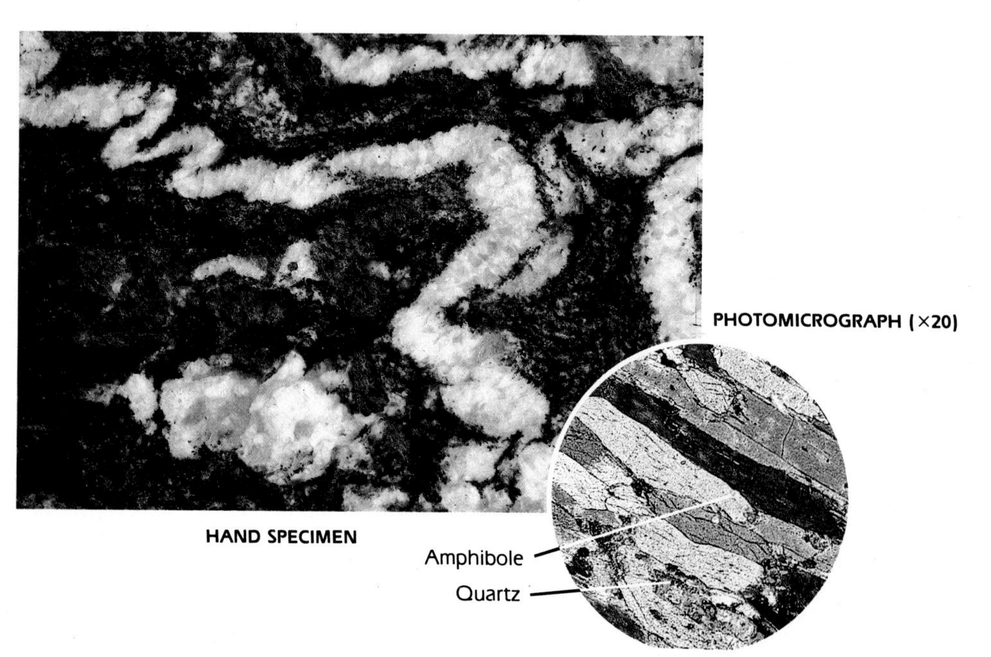

FIGURE 5.9

GNEISS may be foliated with semicontinuous layers of light and dark minerals (Figure 5.8) or with highly contorted, well-defined layers like those shown here. In many rock bodies, layers expressing foliation are several inches thick.

Gneisses are among the most abundant metamorphic rocks. They represent a high grade of metamorphism, and may originate from various granite-rhyolite rocks or from lower-grade metamorphic and sedimentary rocks.

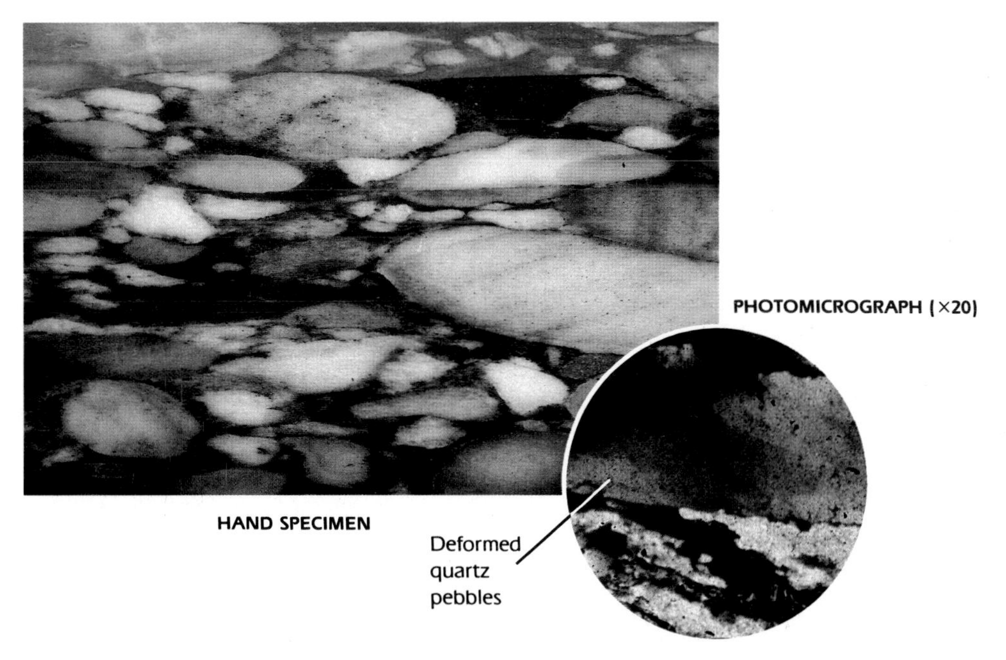

FIGURE 5.10

METACONGLOMERATE is conglomerate that has been altered by heat and pressure to such an extent that individual pebbles are stretched, deformed, and fused. Stretched pebbles commonly show a definite lineation related to the orientation of stresses, but the rock is not strongly foliated. Most metaconglomerate is so indurated that the rock fractures across the pebbles as easily as around them. In this specimen, the matrix of sand grains has been squeezed and fused to the larger pebbles, so the pebble borders appear in the photograph to be blurred, or out of focus. Under a microscope, the sand-silt matrix will show deformation similar to that exhibited by the pebbles.

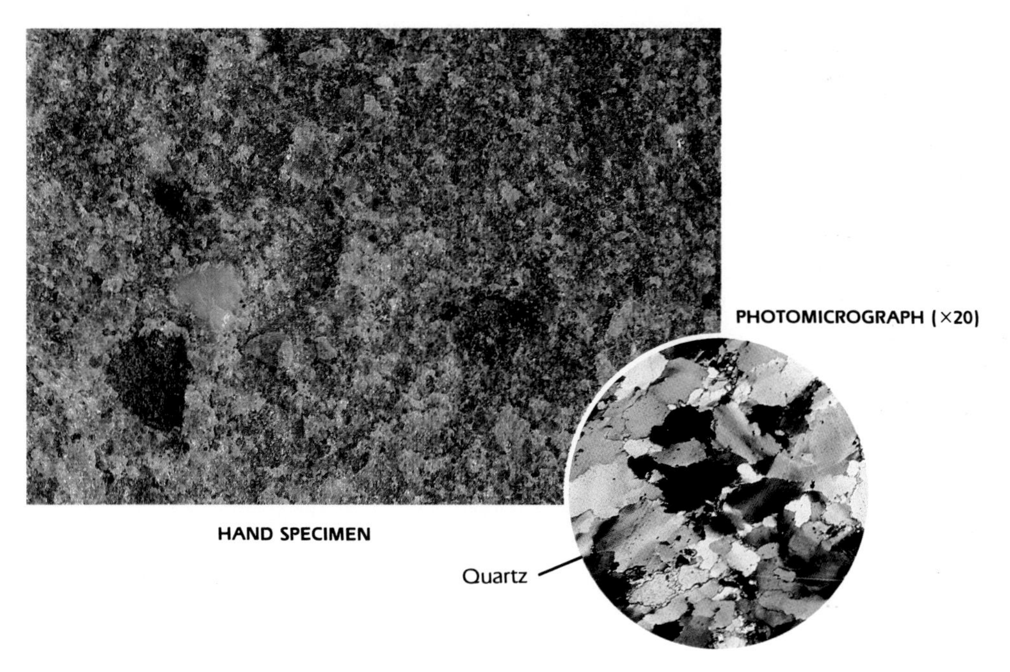

FIGURE 5.11

QUARTZITE is a nonfoliated metamorphic rock composed principally of quartz. In some deposits, quartz is the only significant mineral present. The individual grains of quartzite are stretched, deformed, interlocked, and fused, so the rock breaks across the grains indiscriminately. Pure quartzite is derived from quartz sandstone, but some quartzites may contain as much as 40% other minerals, mica being one of the most abundant. Although quartzite is nonfoliated, some formations that were originally interbedded with shale contain incipient slaty cleavage or relict bedding, which imparts a planar element to the rock.

FIGURE 5.12

MARBLE is a nonfoliated metamorphic rock composed principally of calcite or dolomite. The crystals are commonly large and interlock to form a dense crystalline structure. Bands or streaks of organic impurities resulting from flowage or extreme deformation are common in some deposits. Coloration may be white, pink, blue gray, or brown. Like limestone, marble is characterized by its softness and its effervescence with hydrochloric acid.

THE ORIGIN OF METAMORPHIC ROCKS

REGIONAL METAMORPHISM. Regional metamorphism results from the combined action of high temperatures and directed stresses. It is common where lithospheric plates collide and the rocks are folded, crushed, sheared, and stretched. Rocks in the roots of orogenic belts are generally altered by recrystallization, growth of new minerals, and reorientation of existing minerals.

Careful field and laboratory studies indicate that certain mineral assemblages are indicative of varying degrees of metamorphism. Chlorite, talc, and mica, for example, form under less extreme temperatures and pressures, whereas garnet and amphibole develop under more intense conditions.

CONTACT METAMORPHISM. Contact metamorphism results from high temperatures and vigorous solution activity concentrated near the contacts of a cooling magma. This type of metamorphism is localized and diminishes rapidly away from the intrusive body. Heat and the chemical activity of fluids associated with the magma are the major metamorphic agents. The affected rocks generally recrystallize to form hard, massive bodies.

METAMORPHISM AND TECTONICS

Most metamorphic rocks were probably formed in orogenic belts at convergent plate margins. These rocks represent an important phase in the growth of continents (Figure 5.13). The major steps in the process of continental accretion can be summarized as follows:

Sediment derived from erosion of a continent (or island arc) is transported to the trailing margin of the continent and deposited in a geocline. Convecting currents in the asthenosphere may change the direction of plate movement, and ultimately the trailing margin with the geoclinal sediments collides with another plate. The geocline is thus deformed into a mountain range. Rocks in the roots of the mountain are subjected to high pressure because of deep burial and to high temperatures because of the generation and migration of magma that is intruded into the mountain belt. The tremendous horizontal pressures of the converging plates cause the rocks in the roots of the mountain belt to partly melt and recrystallize. New minerals grow under the horizontally directed stresses and typically develop a foliation that is perpendicular to the stress (i.e., essentially vertical). High-grade metamorphism of geoclinal sediments forms gneisses and schists deep in the mountain roots. Slates develop at shallow depths. Blueschists result from the low-temperature, high-pressure metamorphism of deep-marine shale deposited on the continental margins. Greenschists form in the zone of deep burial as a result of high pressure and high temperature. Some completely melted rock material may rise to the surface and cool as dikes injected between the planes of foliation. Erosion of a mountain range causes the orogenic belt to rise due to isostatic adjustment. Ultimately, the metamorphic rock formed deep in the mountain root is exposed at the earth's surface and forms a new segment of the continental shield.

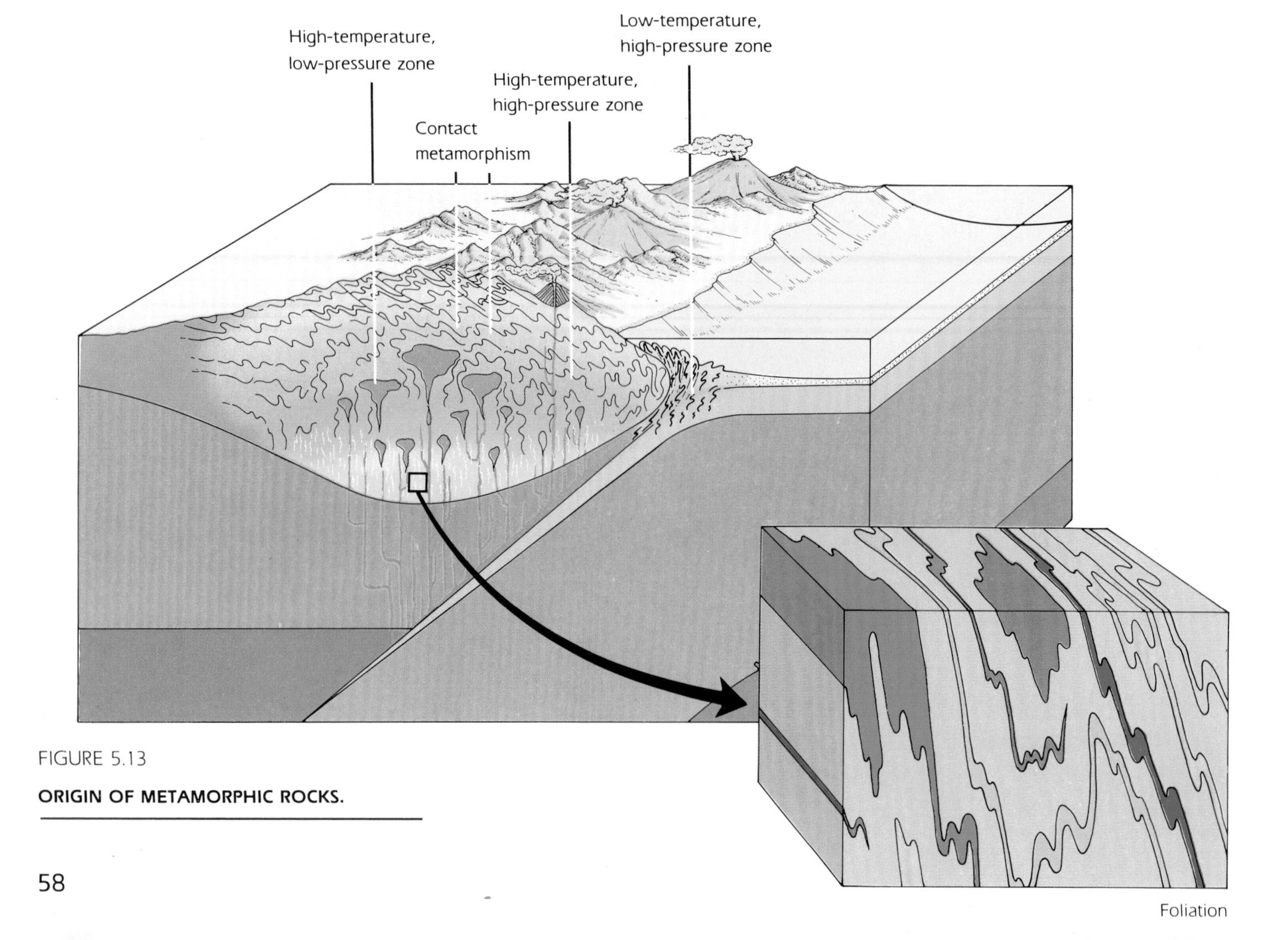

FIGURE 5.13

ORIGIN OF METAMORPHIC ROCKS.

PROCEDURES

To identify metamorphic rock specimens, use the following procedures:

1 Determine whether the rock is foliated or nonfoliated.

2 If the rock is foliated, determine the type of foliation.

3 Identify the minerals present.

4 With this information, refer to Figure 5.3 and identify the rock by name.

5 On the basis of texture and composition, determine, if possible, the original rock type.

PROBLEMS

1 Identify the rock specimens provided by your instructor.

2 What are the major differences between regional and contact metamorphism?

3 How is the orientation of foliation related to stress?

4 What metamorphic rocks are likely to contain garnet?

5 Why do clay minerals change to mica in the metamorphism of shale?

6 Is contact metamorphism likely to produce foliation? Explain.

7 How can you distinguish between the following rock types?

a Marble and quartzite

b Slate and shale

c Conglomerate and metaconglomerate

d Granite and gneiss

Exercise 6

MAPS AND AERIAL PHOTOS

OBJECTIVE

To become familiar with the major types of aerial photographs, remote sensing imagery, and topographic maps used in the study of the earth's surface features, and to learn how to analyze and interpret them.

MAIN CONCEPT

Topographic maps, aerial photographs, and remote sensing imagery are models of portions of the earth's surface and are fundamental tools of geologic research.

SUPPORTING IDEAS

1 Aerial photographs are important in geologic work because of the vast amount of surface detail recorded on them.

2 Landsat imagery provides important regional information available in no other way.

3 Various remote sensing devices, such as radar imagery, provide specialized geologic information.

4 Topographic maps show the configuration of the earth's surface by means of contour lines. These are maps of fundamental importance in geology, because accurate measurements of both horizontal and vertical distances can be made from them.

DISCUSSION

Study of the earth's surface features and the processes that form those features presents a major problem of scale. Mountain ranges, plateaus, and drainage basins are all too large to be seen from any single viewpoint, or even from a thousand different viewpoints. To study such large features, geologists use various types of aerial photography, remote sensing imagery, and topographic maps. Each is, in a real sense, a type of scale model showing various aspects of the size, shape, and spatial relationships of the earth's surface features. The great value of these models is that they provide a regional perspective from a vertical view. They represent the reduction of vast amounts of data to a model the size of a piece of paper that can be analyzed and managed easily. Topographic maps, aerial photographs, and remote sensing are therefore basic tools of the geologist and are indispensible to the study of landforms and their origin. Just as the microscope gives the biologist a new perspective, or the telescope aids the astronomer, so these models help us to study geologic features that cannot be seen from viewpoints on the earth's surface.

AERIAL PHOTOGRAPHY AND REMOTE SENSING IMAGERY

AERIAL PHOTOGRAPHS

Aerial photographs have long been a fundamental tool in geologic studies because they show the surface features of the earth in remarkable detail from a vertical perspective. To beginning students, however, a vertical view of the earth's surface is new and unnatural, and many features may appear at first glance to be unfamiliar or exotic patterns, quite unlike the features that we are used to seeing from the ground. Roads, buildings, farmlands, and lakes may be recognizable, but some experience is needed to identify various types of terrain, rock bodies, and other geologic features.

STEREOSCOPIC VIEWING. One of the great advantages of aerial photographs is that they can be viewed stereoscopically so an image of the landscape appears in three dimensions. This is possible because aerial photographs are taken in sequence along a flight line with approximately 60% overlap of each photo in the flight direction. When two adjacent photos along a flight line are viewed through a stereoscope, the brain combines the two images to produce one three-dimensional image of the surface. Hills and valleys appear to stand out in bold relief. Stereoscopic viewing of

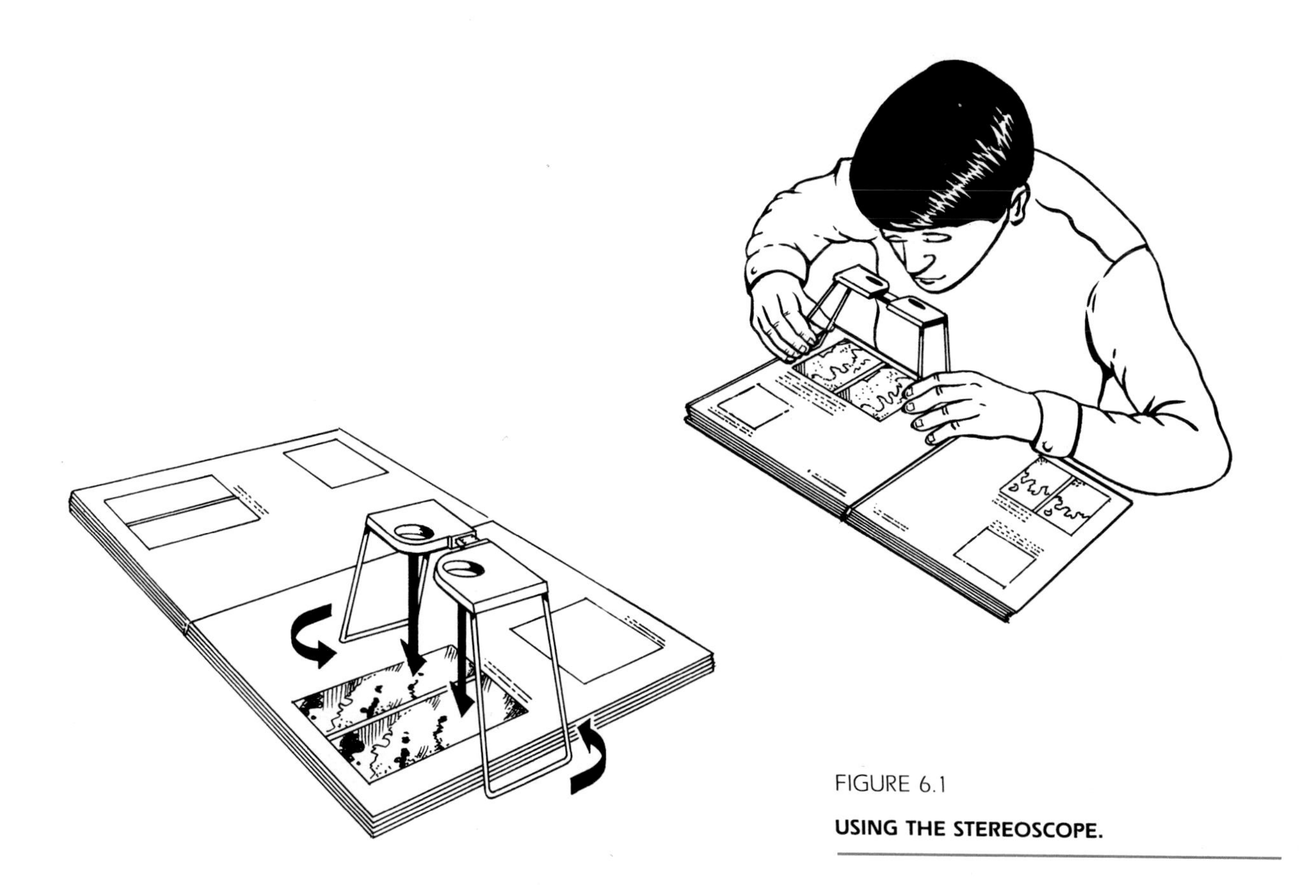

FIGURE 6.1

USING THE STEREOSCOPE.

the landscape is like having a personal view from directly above, and even has an added advantage because photographs can be used to make notations and for mapping.

Many of the vertical aerial photographs in the manual are printed in pairs to form stereograms (see Figure 17.12). To view a stereogram, simply place a lens stereoscope over the stereogram so your nose is directly above the line separating the two adjacent photos (Figure 6.1). The axis of the stereoscope (an imaginary line connecting the center of each lens) should be perpendicular to the line separating the photos. If stereovision is not attained immediately, rotate the stereoscope slightly, as shown by the arrows in Figure 6.1, or adjust the lens separation on the stereoscope to better fit your eyes. With a little practice, you can become efficient at stereoscopic viewing, which will provide an important new method of studying the surface features of the earth. Figures 6.2A–H show some different types of aerial photographs and the common geologic features seen on them.

PROBLEMS

1 List the advantages and the limitations of the following types of aerial photographs in geologic studies.

- **a** Low altitude
- **b** High altitude
- **c** Color infrared

2 Using a lens stereoscope, inspect the stereogram in Figure 6.2H. What are the advantages and limitations of stereoaerial photographs in geologic studies?

LANDSAT IMAGERY

On 23 July 1972, the United States launched the first Earth's Resources Observation System (EROS) and began collecting, on a repetitive basis, satellite images of the earth's surface. Three similar Land Satellites (Landsats 2, 3, and 4) have since been launched in near-polar orbits 570 mi (918 km) above the earth. These satellites are capable of obtaining a variety of images of the earth's surface. Landsat images can be manipulated in false color and further enhanced by computer to make available a tremendous amount of data about the earth's surface. Other remote sensing images are produced by radar, which is capable of "looking through" cloud cover to record remarkably clear images of surface features. Examples of various types of Landsat and remote sensing images are shown in Figures 6.3A–6.3I.

PROBLEMS

Study the examples in Figure 6.3, and list the advantages and limitations of the following types of remote sensing imagery:

1 Black and white Landsat imagery

2 Color computer-enhanced Landsat imagery

3 Radar imagery

0 0.2 miles

A. LOW ALTITUDE: CITY.

0 1 mile

B. INTERMEDIATE ALTITUDE: CITY AND PORT FACILITIES.

0 1 mile

C. INTERMEDIATE ALTITUDE: VOLCANO.

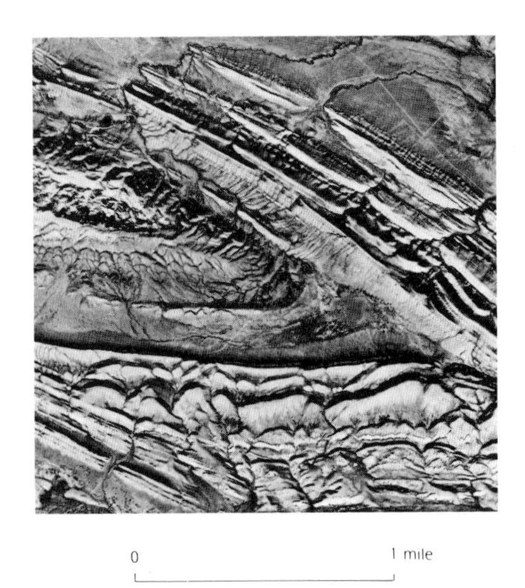

0 1 mile

D. INTERMEDIATE ALTITUDE: FOLDED SEDIMENTARY ROCKS.

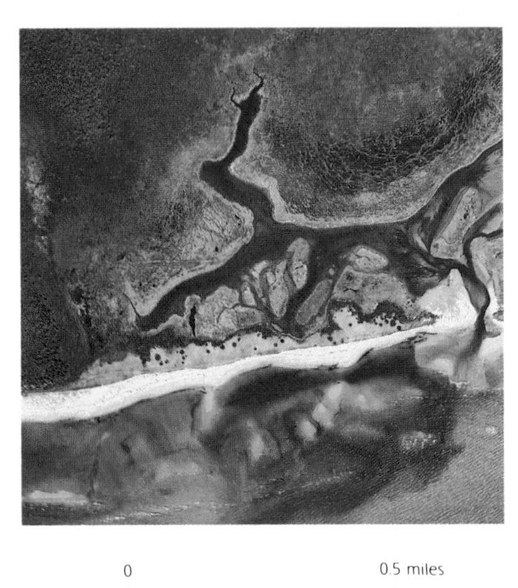

0 0.5 miles

E. LOW ALTITUDE, NATURAL COLOR: SHORELINE.

0 1 mile

F. HIGH ALTITUDE, COLOR INFRARED: GLACIER.

0 1 mile

G. HIGH ALTITUDE: TOWN, RIVER AT FLOOD STAGE.

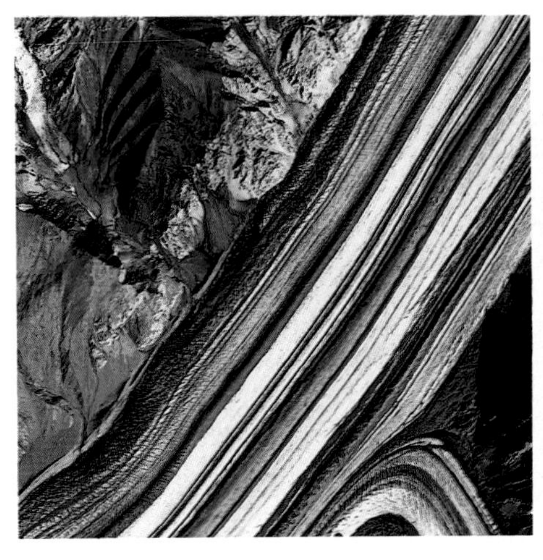

0 1 mile

H. STEREOGRAM, HIGH ALTITUDE, COLOR INFRARED: GLACIER.

FIGURE 6.2 **EXAMPLES OF AERIAL PHOTOGRAPHS.**

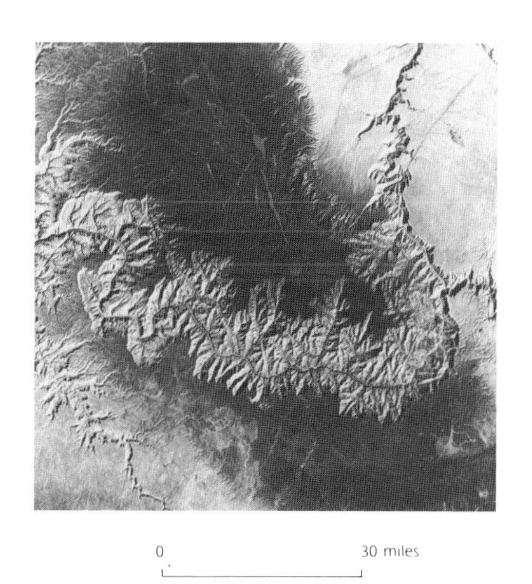

0 30 miles

A. LANDSAT IMAGE: Grand Canyon, Arizona.

0 30 miles

B. SKYLAB PHOTOGRAPH (natural color): Lake Powell, Utah.

0 30 miles

C. LANDSAT IMAGE: Valles Caldera, New Mexico.

0 2 miles

D. LANDSAT IMAGE Death Valley.

0 0.5 miles

E. LANDSAT IMAGE (enlarged and computer-enhanced): Death Valley.

0 1 mile

F. LANDSAT IMAGE (enlarged and computer-enhanced): Wyoming.

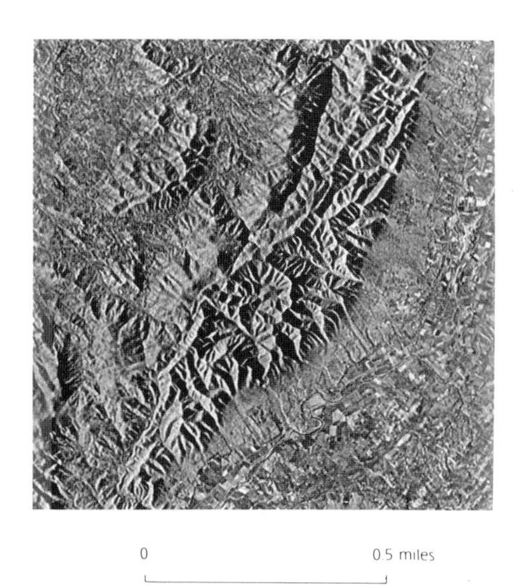

0 0.5 miles

G. RADAR IMAGE: Appalachian Mountains, Pennsylvania.

0 5 miles

H. RADAR IMAGE: Volcano.

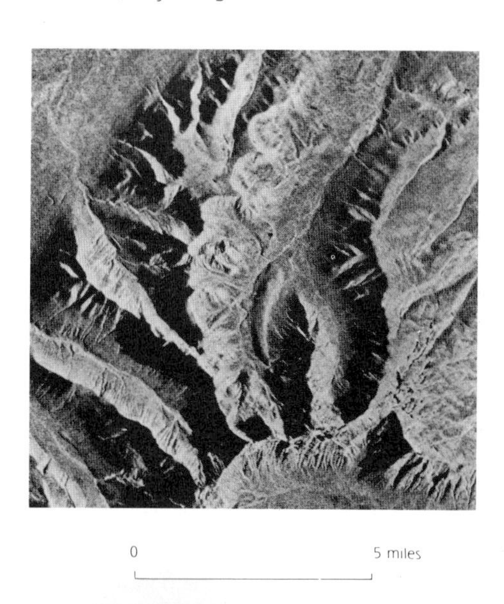

0 5 miles

I. RADAR IMAGE: Glaciated terrain.

FIGURE 6.3 **TYPES OF REMOTE SENSING IMAGERY.**

TOPOGRAPHIC MAPS

Topographic maps are maps that show the configuration of the earth's surface by means of contour lines. They are constructed with great accuracy to scale, so areas, elevations, slope angles, and volumes can be measured easily. These maps are therefore an indispensible tool for geologic studies and are critical to other endeavors as well (e.g., construction planning, making transportation routes, and military operations). Indeed, they are of fundamental importance to the economic development and environmental protection of the nation. Topographic maps are accurate terrain models, and as such show infinitely more than location of features, distance, and direction. As scale models, they contain a vast store of information that may not be readily apparent to the casual observer. The extraction and use of this information is available to anyone with the skill to use this type of map. Users are limited only by their own skills and knowledge.

LOCATING FEATURES ON A MAP

An important function of a map is to provide the means of defining the location of a point or area, so a specific position can be found again at a later date, communicated to others, or recorded for scientific, legal, or political purposes. The most effective way to do this is by a system of grid lines (Figure 6.4). By international agreement, the earth's surface is divided into a series of north-south and east-west grid lines. The north-south lines are called lines of longitude, or meridians. They represent segments of arc on the earth's equator measured in degrees, minutes, and seconds. The line of zero longitude, the *prime meridian,* passes through Greenwich, England. All other longitude lines are measured as east or west of the prime meridian to the 180-degree line of longitude known as the *international date line.*

The other dimension of the grid is latitude. Lines of latitude circle the globe parallel to the equator, which is zero latitude. Degrees of latitude are measured north or south of the equator up to 90-degrees latitude at either pole. The system of latitude lines also divides the earth into two hemispheres, the Northern Hemisphere and the Southern Hemisphere.

PROBLEM

Study Figure 6.4 and determine the latitude and longitude of points *K*, *L*, *M*, and *N*.

Latitude and longitude lines are excellent for large regional references but are cumbersome when used to subdivide the land surface on a local scale. Most of the United States west of the Ohio and Mississippi rivers has been surveyed and subdivided in accordance with the U.S. Federal Rectangular Survey, or the Land Office Grid System (Figure 6.5).

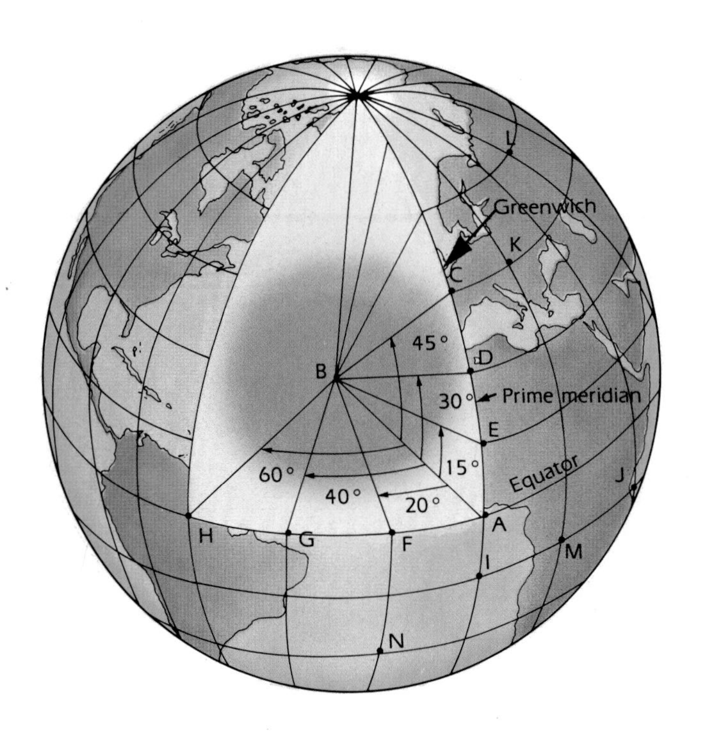

FIGURE 6.4

MEASUREMENTS OF LATITUDE AND LONGITUDE.

The equator forms a great circle on which angles of longitude are measured. The prime meridian is a north-south line through Greenwich, England. Point **F,** in the Atlantic Ocean, if projected to the center of the earth and back to the prime meridian, would describe angle **FBA,** which is 20 degrees. All points lying on the north-south line through point **F** would thus be 20 degrees west of the prime meridian. Point **G,** near the east coast of South America, if projected to the center of the earth and back to the prime meridian at point **A,** would describe angle **GBA,** which is 40 degrees. All points lying on a north-south line through point **H** would be longitude 60 degrees west. Point **J** is 40 degrees east longitude.

Angles of latitude are measured on a great circle passing through the poles, with the equator as a reference line. Point **C** in Europe, if projected to the center of the earth and back to the equator, would describe angle **CBA,** which is 45 degrees. All points lying on a line that passes through point **C** parallel to the equator would be 45 degrees north of the equator. This includes the line passing through the northern part of the United States. Similarly, point **D** in northern Africa, if projected to the earth's center and back to the equator, would describe angle **DBA,** which is 30 degrees. All points on a line parallel to the equator passing through point **D** would be 30 degrees north latitude. Point **E** is 15 degrees north latitude, and point **I** is 15 degrees south latitude.

FIGURE 6.5

LAND OFFICE GRID SYSTEM.

Subdivisions based on the Land Office Grid System involve the location of an initial point from which all subdivision of a state or region is referenced. The latitude line for that point is called the base line, and the longitude line is the principal meridian. From the reference point so designated, grid lines are surveyed at 6-mi intervals. The north-south lines define strips of land called ranges, and are numbered (i.e., 1, 2, 3, and so forth) east and west of the principal meridian (Figure 6.5A). East-west lines establish strips of land called townships, which are numbered north and south of the base line. Each square of the grid contains 36 mi² and is referred to as a township. Its location is designated by the numbers of the township and the range tiers it contains. An example would be Township 2 North (T2N), Range 3 West (R3W).

A township is further subdivided into 36 sections. Each section is 1 mi² and is numbered as shown in Figure 6.5B. The individual sections in turn are divided into quarters and eighths (Figure 6.5C).

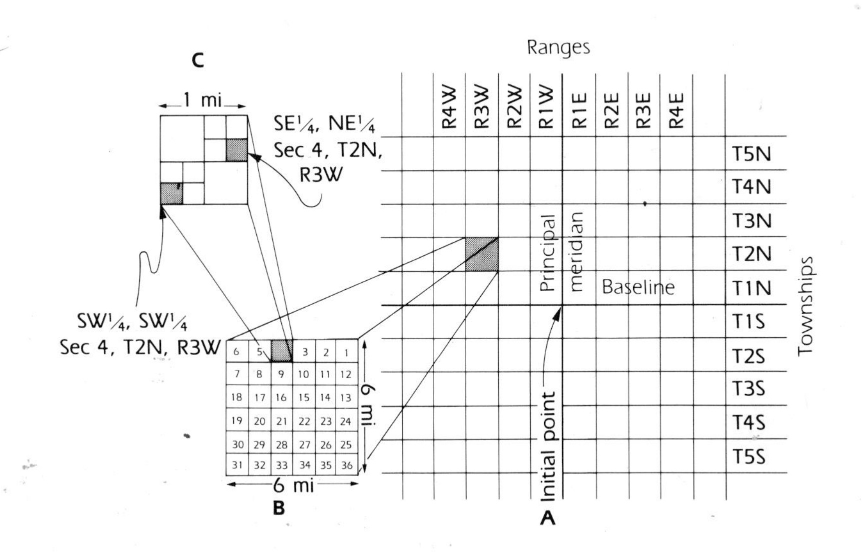

CONTOUR LINES

The great value of topographic maps is that they show the shape and elevation of the land surface. This is done by means of contour lines. A contour line is an imaginary line on the earth's surface connecting points of equal elevation. A contour line may also be described as a line traced by the intersection of a level surface with the ground. Natural expressions of contour lines are elevated shorelines, cultivated terraces, and patterns produced by contour plowing.

To clarify the idea of contours, consider an island in a lake and the patterns made on it when the water level recedes. The shoreline represents the same elevation all around the island and is thus a contour line (Figure 6.6A). Suppose that the water level of the lake drops 10 ft and that the position of the former shoreline is marked by a gravel beach (Figure 6.6B). Now there are two contour lines, the new lake level and the old stranded beach, each depicting accurately the shape of the island at these two elevations. If the water level should continue to drop in increments of 10 ft, with each shoreline being marked by a beach, additional contour lines would be formed (Figures 6.6C and 6.6D). A map of the raised beaches is in essence a contour map (Figure 6.6E), which represents graphically the configuration of the island.

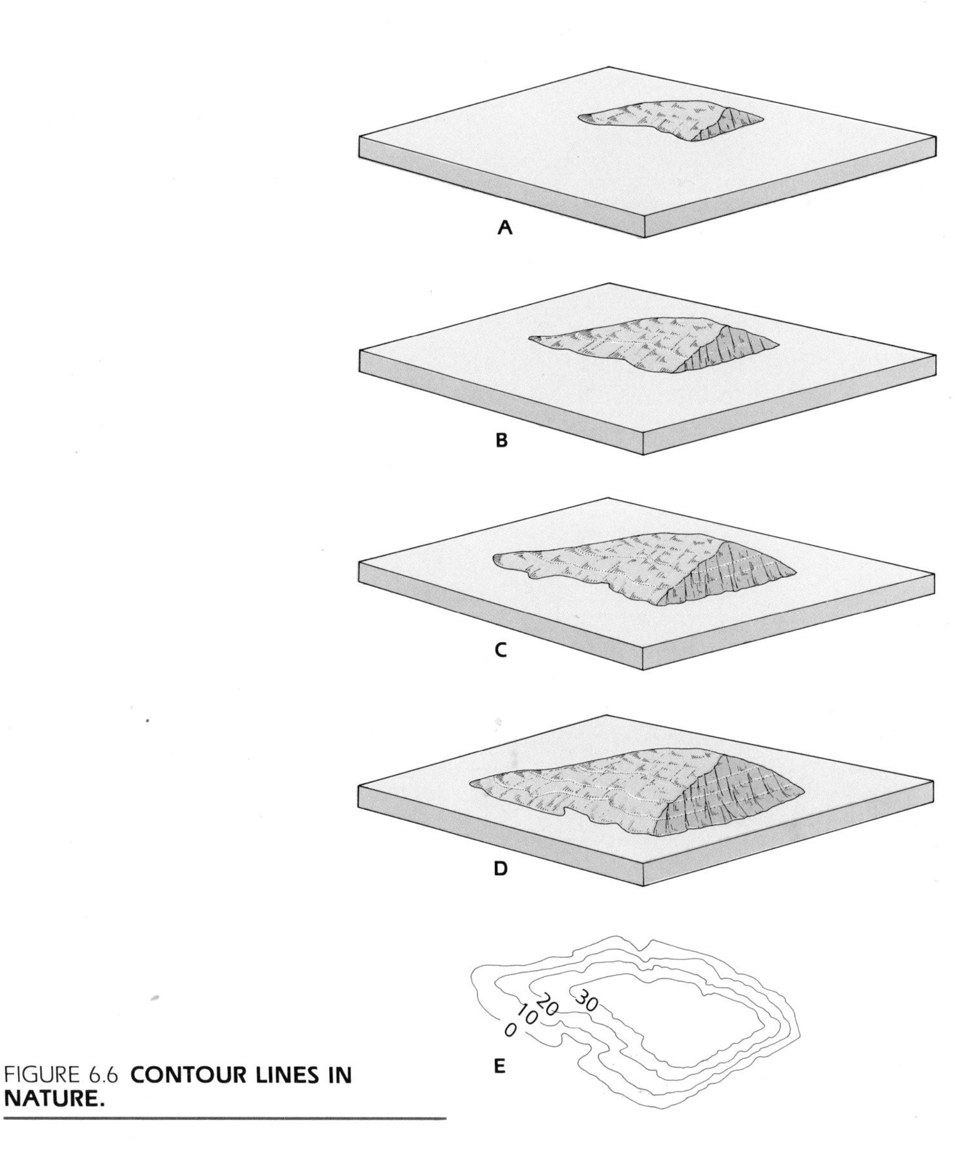

FIGURE 6.6 **CONTOUR LINES IN NATURE.**

CHARACTERISTICS OF CONTOURS

Learning to read contour maps effectively is not an easy task. Nature has few lines reminiscent of contour lines, and the visualization of a three-dimensional surface represented by contours only requires careful study and practice. After some training, however, you should be able to visualize the landscape represented by a contour map, and you will obtain more reliable information from a contour map than from any other type of map or imagery.

Perhaps the best way to learn to read a contour map is to study contour lines in stereoscopic vision (see Figures 6.7–6.10). This should help you to visualize more easily those surface features such as hills, valleys, cliffs, gentle slopes, and depressions when they are portrayed on conventional nonstereo maps. Stereographic contours give you the opportunity to visualize clearly the three-dimensional model that is the focus of regular topographic maps.

PROCEDURES

An effective method is to view the maps with a stereoscope, then to close one eye so you see two dimensions only. Study a specific feature such as a hill, cliff, or gentle slope. With sufficient practice, your experience with three-dimensional contour maps should permit you to visualize easily the three-dimensional aspect of a traditional contour map.

The following basic rules will help you get started:

1 The difference in elevation between adjacent contours is constant on any given map and is referred to as the *contour interval*, or CI. The most frequently used contour interval on 7½-minute and 15-minute quadrangles is 20 ft, although intervals of 5, 10, 20, 40, 50, 100, or 200 ft may be used if necessary to express the topography of the area being mapped.

2 Contour lines trend up a valley, cross the stream, and extend down the valley on the opposite side. The lines thus form a "V" pattern, with the apex of the V pointing in an upstream direction (Figure 6.7).

3 Contours never cross or divide. They may appear to merge to express a vertical cliff, but in reality they are stacked one on top of another and only appear to touch (Figure 6.8).

4 The spacing of contour lines reflects the gradient or slope. Closely spaced contours represent steep slopes. Contours spaced far apart represent gentle slopes (Figure 6.8).

5 Hills and knobs are shown as closed contours (Figure 6.9).

6 Closed depressions (basins with no outlets) are shown by closed contours (Figure 6.10), with hachures (short lines pointing downslope, see Figure 6.9). (These are not shown in the stereogram.)

DETERMINING ELEVATIONS

Elevation refers to height (in feet or meters) above sea level and is essentially synonymous with altitude. Specific elevations established by highly accurate surveys are referred to as "benchmarks" and are shown on topographic maps in various ways. Typical benchmark locations are at the center of a town, on a hilltop, and at the bottom of a depression. The benchmarks are printed on the map in black. Spot elevations, which are less accurate, are printed in brown. In addition, the approximate elevation of any point on the map can be interpolated from the contour lines. For example, a point midway between the contours of 1240 ft and 1260 ft would probably have an altitude of 1250 ft, and a point located just below the 1260-ft contour line would probably be at an elevation close to 1258 or 1259 ft. Such approximations are based on the assumptions (1) that the slopes have a constant gradient and (2) that the elevation is proportional to the horizontal distance. These assumptions are not always true, but a careful study of slope trends usually permits one to estimate quite accurately the elevations between contours.

Relief is the difference in elevation between high and low points. You can determine easily the local relief of an area by subtracting the lowest elevation from the highest elevation.

Height and *depth* are measurements made relative to some local feature. For example, a monument might be 555 ft high relative to the ground, but might have an overall elevation (at its top) of 1555 ft if the ground is 1000 ft above sea level.

COLORS AND SYMBOLS

If maps showed all of the natural features as well as the marks of civilization within any given area, they would be a collection of useless clutter. For the sake of clarity, symbols are used to indicate a variety of cultural features (e.g., buildings, political boundaries, roads, and railroads). Contour lines are printed in brown. Streams, rivers, lakes, and other bodies of water are printed in blue. On some maps, features of special importance are shown in red, and vegetation is in green. Map symbols used by the U.S. Geological Survey appear on the inside cover of this manual.

PROBLEMS

1 Using the appropriate letters, label the following features on the map in Figure 6.7.

- **a** Undissected plateau
- **b** Head of the canyon
- **c** Tributary canyon
- **d** Steepest canyon wall
- **e** Minor valleys cut into plateau

2 Using the appropriate letters, label the following features on the map in Figure 6.8.

- **a** Top of mesa
- **b** Steep cliff
- **c** Lowest point on map
- **d** Highest surface on map
- **e** Valleys cut in surface of mesa and butte

3 Using the appropriate letters, label the following features on the map in Figure 6.9.

- **a** Hills
- **b** V-shaped contour pointing upstream
- **c** Closed depression
- **d** Highest point on map
- **e** Steepest slope

4 Show a closed depression with hachure lines.

5 Using the appropriate letters, label the following features on the map in Figure 6.10.

- **a** Lowest depression
- **b** Highest point
- **c** Hills

6 Make a simple sketch map showing contour lines expressing

- **a** Steep cliff
- **b** Rounded hill
- **c** Closed depression
- **d** Stream and stream valley

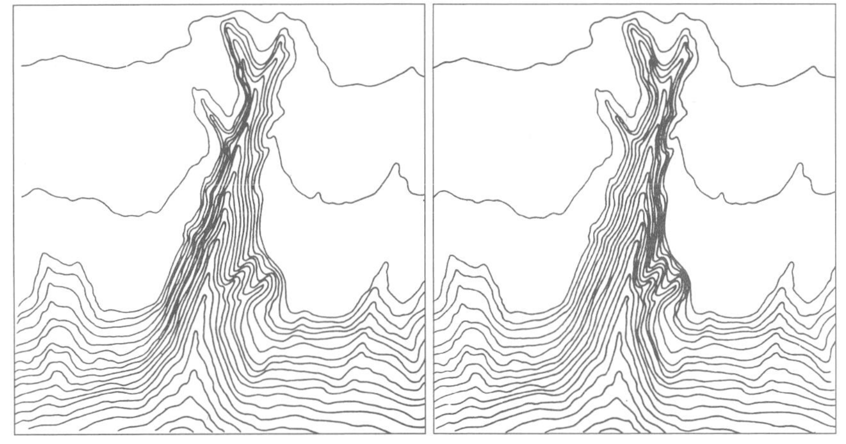

FIGURE 6.7 **CONTOUR MAP OF A CANYON.**

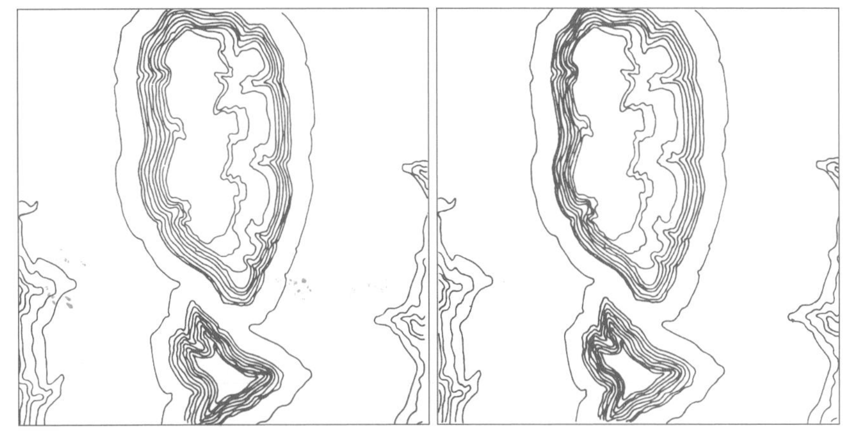

FIGURE 6.8 **CONTOUR MAP OF MESAS AND BUTTES.**

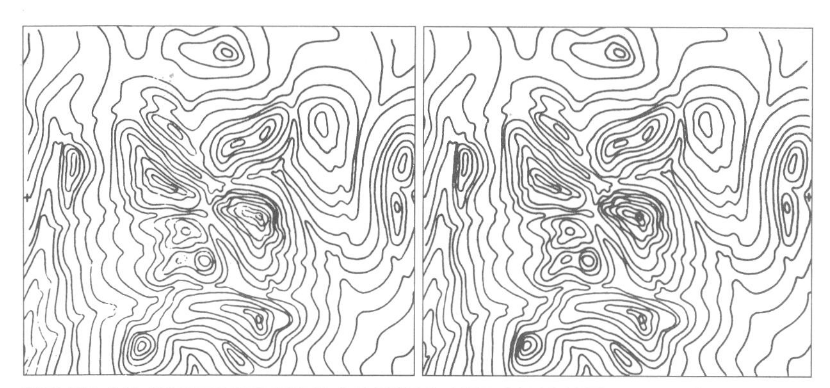

FIGURE 6.9 **CONTOUR MAP OF HILLS AND VALLEYS.**

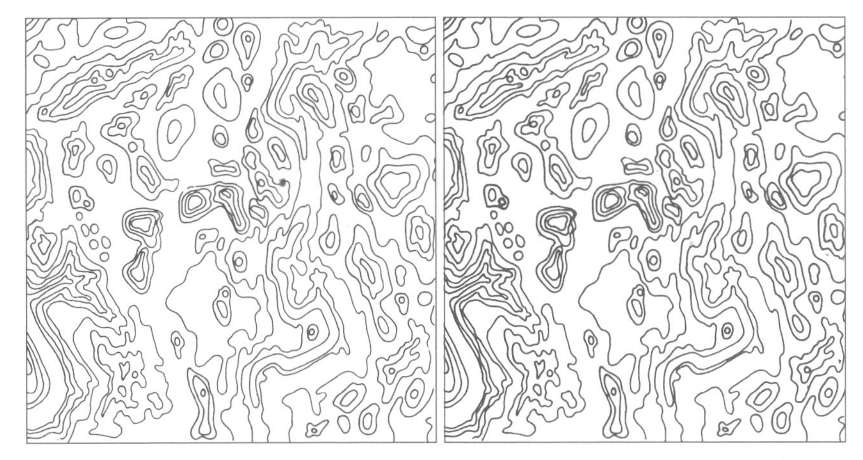

FIGURE 6.10 **CONTOUR MAP OF CLOSED DEPRESSIONS.**

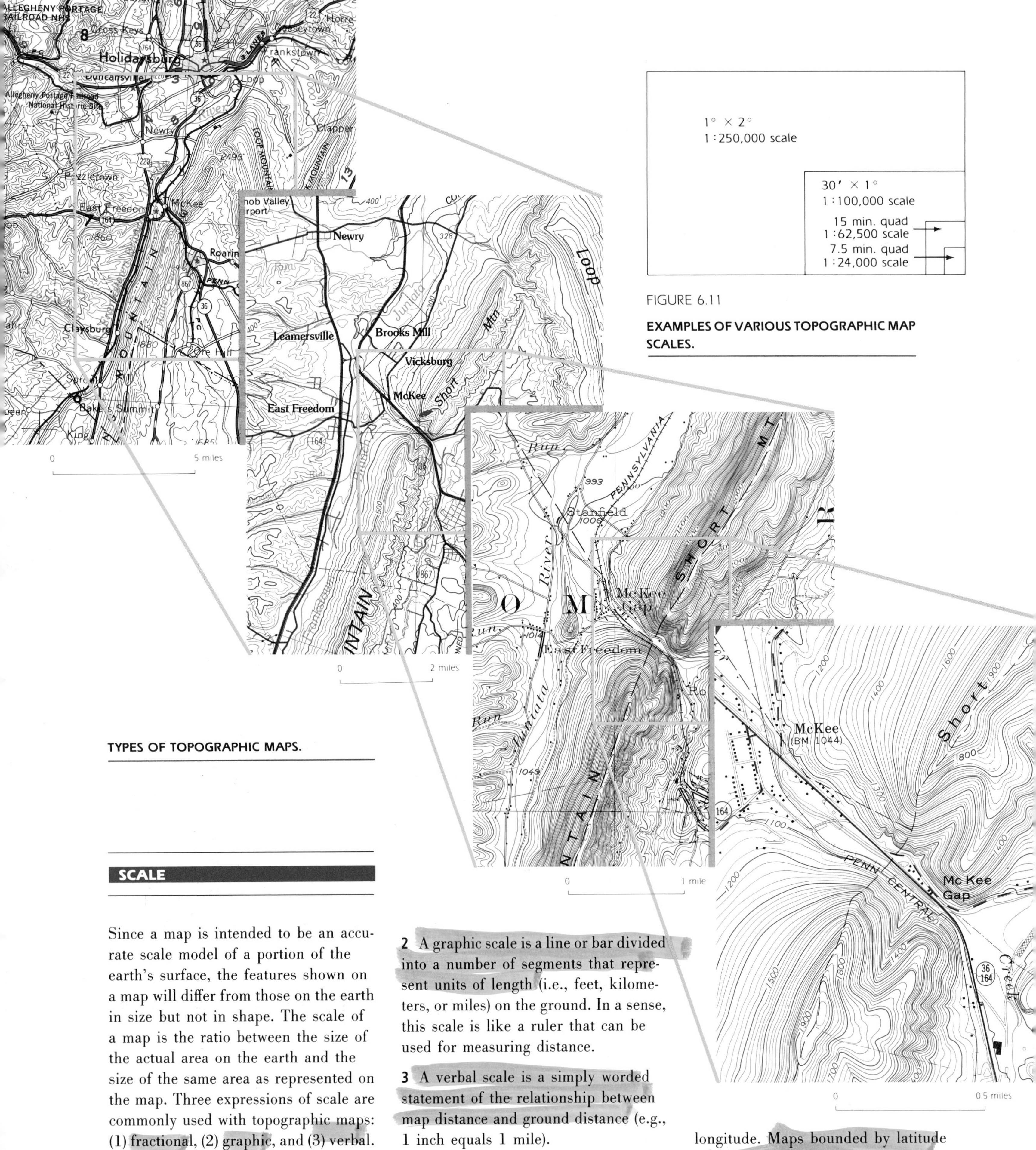

FIGURE 6.11

EXAMPLES OF VARIOUS TOPOGRAPHIC MAP SCALES.

TYPES OF TOPOGRAPHIC MAPS.

SCALE

Since a map is intended to be an accurate scale model of a portion of the earth's surface, the features shown on a map will differ from those on the earth in size but not in shape. The scale of a map is the ratio between the size of the actual area on the earth and the size of the same area as represented on the map. Three expressions of scale are commonly used with topographic maps: (1) fractional, (2) graphic, and (3) verbal.

1 The fractional scale is expressed as a simple fraction (1/62,500) or as a proportion (1 : 62,500). These expressions mean that one unit of distance on the map is equal to 62,500 of the same unit on the ground.

2 A graphic scale is a line or bar divided into a number of segments that represent units of length (i.e., feet, kilometers, or miles) on the ground. In a sense, this scale is like a ruler that can be used for measuring distance.

3 A verbal scale is a simply worded statement of the relationship between map distance and ground distance (e.g., 1 inch equals 1 mile).

The U.S. Geological Survey publishes maps on a variety of scales to meet the needs of supplying both detail and broad coverage. The survey system of subdividing areas for map coverage uses the universal coordinate lines of latitude and longitude. Maps bounded by latitude and longitude are called *quadrangles.* The system provides quadrangles of different size and various scales like those shown in Figure 6.11. Examples of the various scales of topographic maps and the types of landforms expressed by contours are shown in Figure 6.12.

1/ air photo index sheet
air-photo mosaic

COMPARISON OF TOPOGRAPHIC MAPS WITH AERIAL PHOTOGRAPHS

The Menan Buttes area in southern Idaho (Figure 6.13A) offers an excellent opportunity to study the way in which simple topographic forms are expressed by contour lines. By carefully examining aerial photographs of the buttes (Figure 6.13B) with a stereoscope, and by comparing the stereoscopic image with the topographic contours, you can further develop your ability to visualize the land forms expressed on a topographic map. The following problems will help you to compare the map with the photograph.

PROBLEMS

1 Study the volcano of Menan Buttes. The slopes of the crater differ significantly from one side to the other. How are steep slopes expressed by contour lines?

2 How are the more gentle slopes beyond the base of the volcano expressed by contour lines?

3 How are the rugged slopes in the northwest part of the crater expressed by contour lines?

4 How is the closed depression of the crater expressed by contour lines?

5 Study the flood plain area to the south of the butte. How is a relatively flat surface expressed by contour lines?

6 Sketch contour lines of the surface shown in the stereoscopic view. Work directly on the photograph and compare your results with the topographic map.

7 What are the advantages of an aerial photograph compared with a topographic map? What are the advantages of a topographic map?

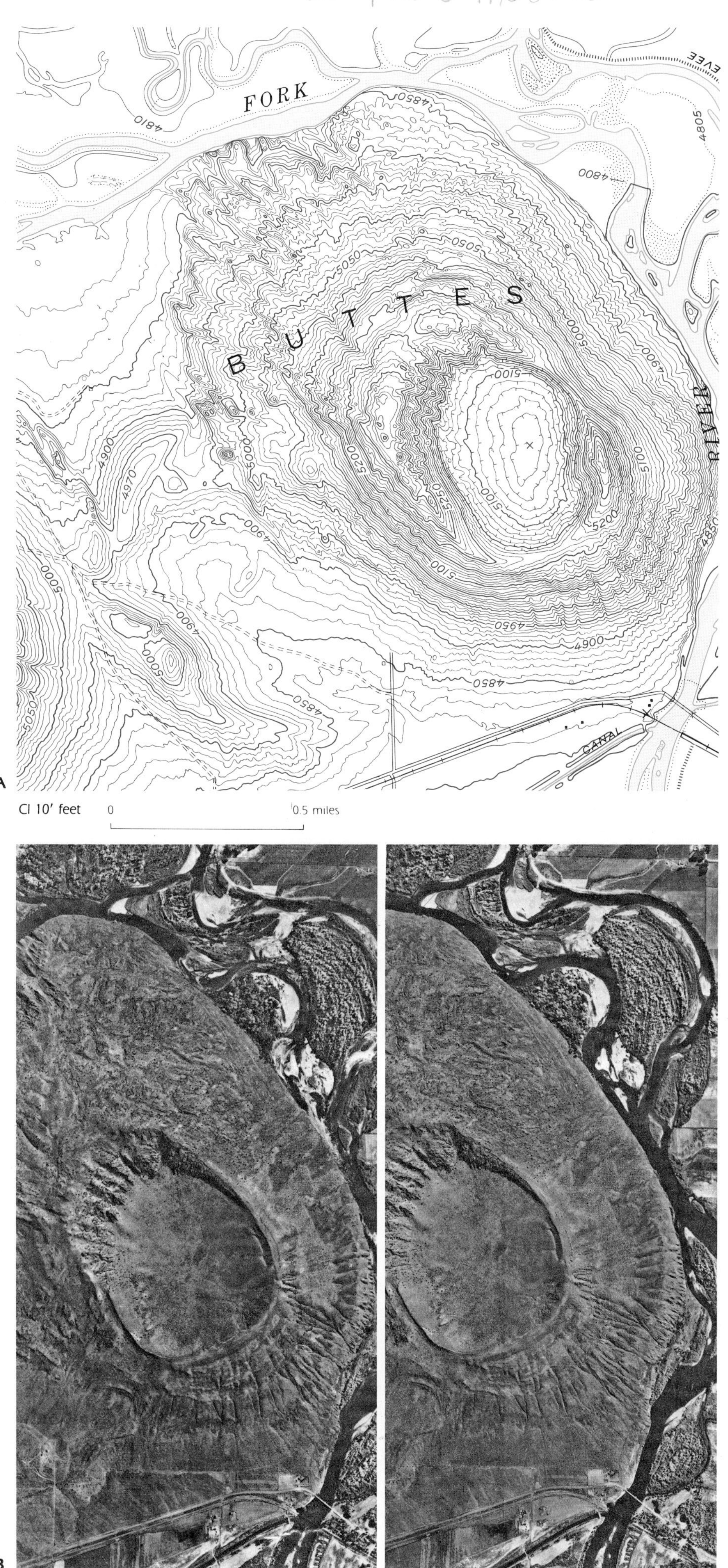

FIGURE 6.13

COMPARING A TOPOGRAPHIC MAP WITH STEREOSCOPIC PHOTOGRAPHS.

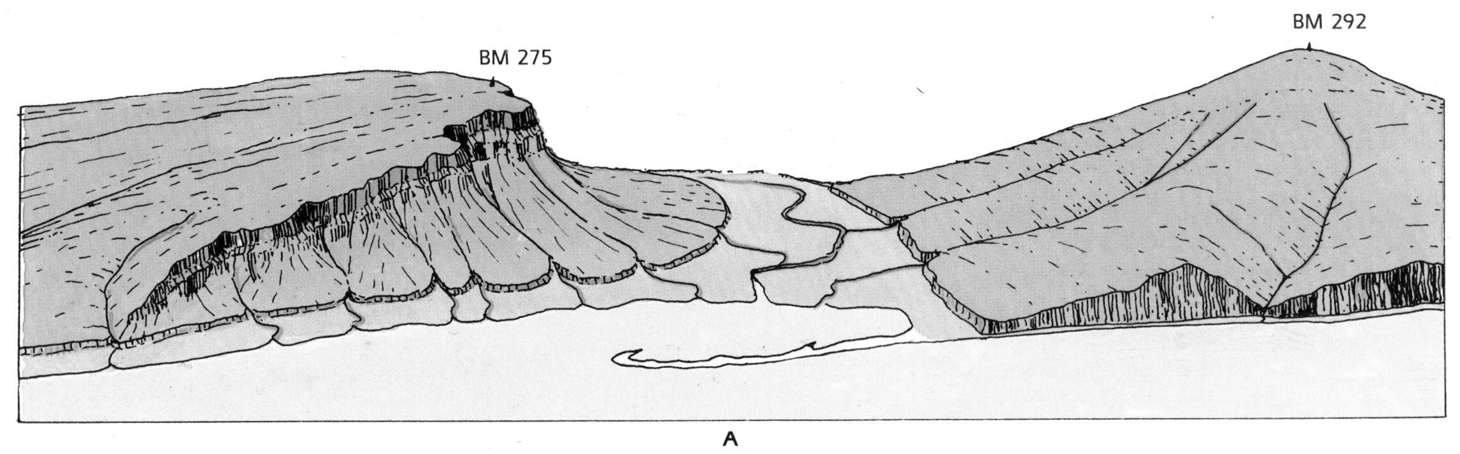

A

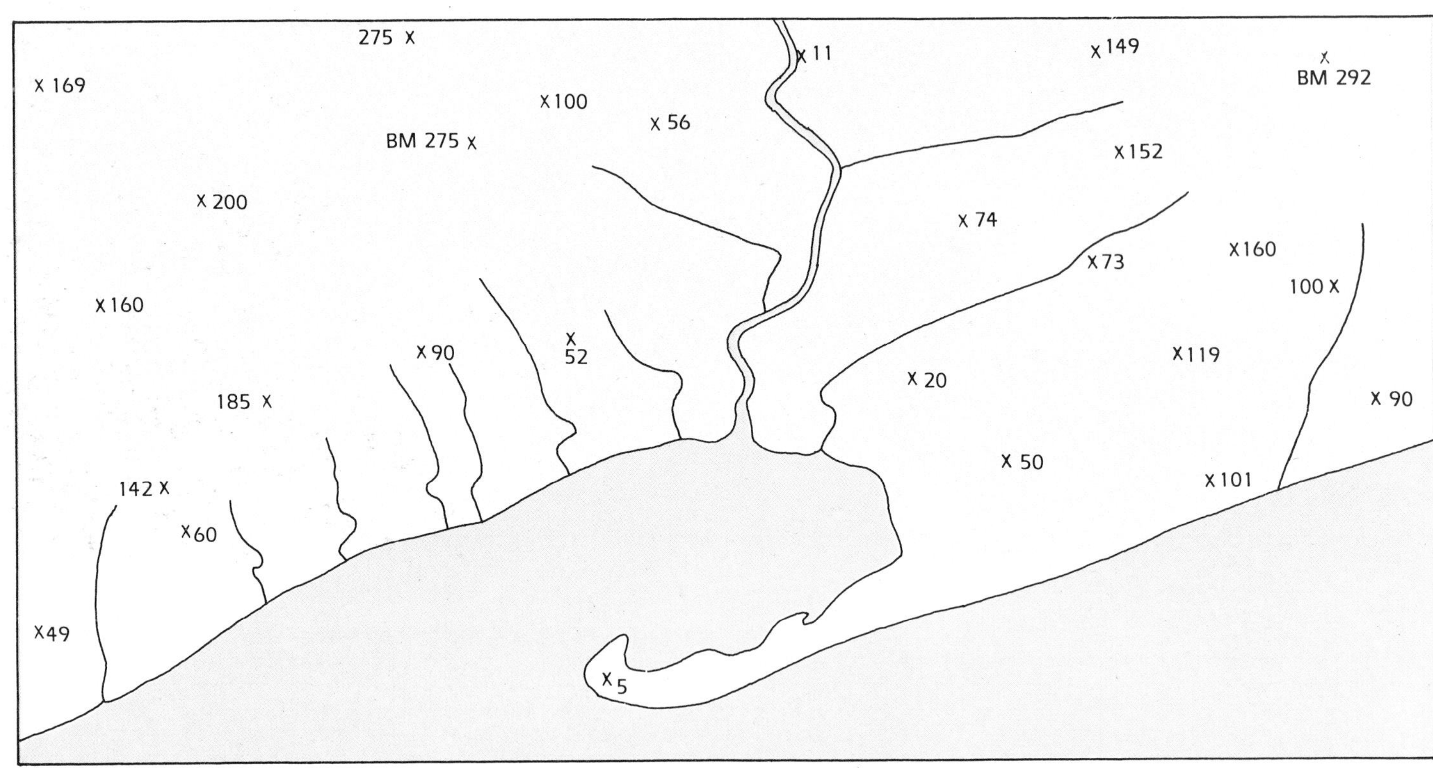

B

FIGURE 6.14: Constructing a Contour Map from Established Elevations

Contour maps were originally constructed by surveying the location and elevation of a number of points in the field. The surveyor then sketched contour lines in the field by extrapolating between the surveyed points. Today, maps are created and updated from stereoscopic aerial photographs, and much information is stored, manipulated, and retrieved by computer. This exercise of constructing your own topographic map from surveyed points is extremely useful; it will help you to observe and understand the surface features depicted on a topographic map.

PROBLEM

1 Construct a contour map of the landforms shown in Figure 6.14. Use a contour interval of 20 ft.

PROCEDURE

First, take time to study the various landforms shown in the sketch (Figure 6.14A). Note that the river flows through a low valley and into a bay, which is partly enclosed by a sand spit only 5 ft above sea level. Both sides of the valley are terraced. The hill on the right has been smoothly eroded to form a gradual slope above a wave-cut cliff. The hill on the left is an inclined tableland and is crossed by a few shallow valleys. A steep cliff formed on the left hill faces the river and the sea. All of these features should be shown by the contours on your map. As you begin, it may be advantageous to sketch lightly a number of contour lines in perspective directly on the diagram (Figure 6.14A). In placing contours on the map, start near the edge at the lowest elevations and work up the major streams. Pay particular attention to the established elevation points, and be sure that all contour lines are in harmony with these. To find the appropriate position of a contour line between two control points, study the picture of the slope and estimate the map position of that slope section.

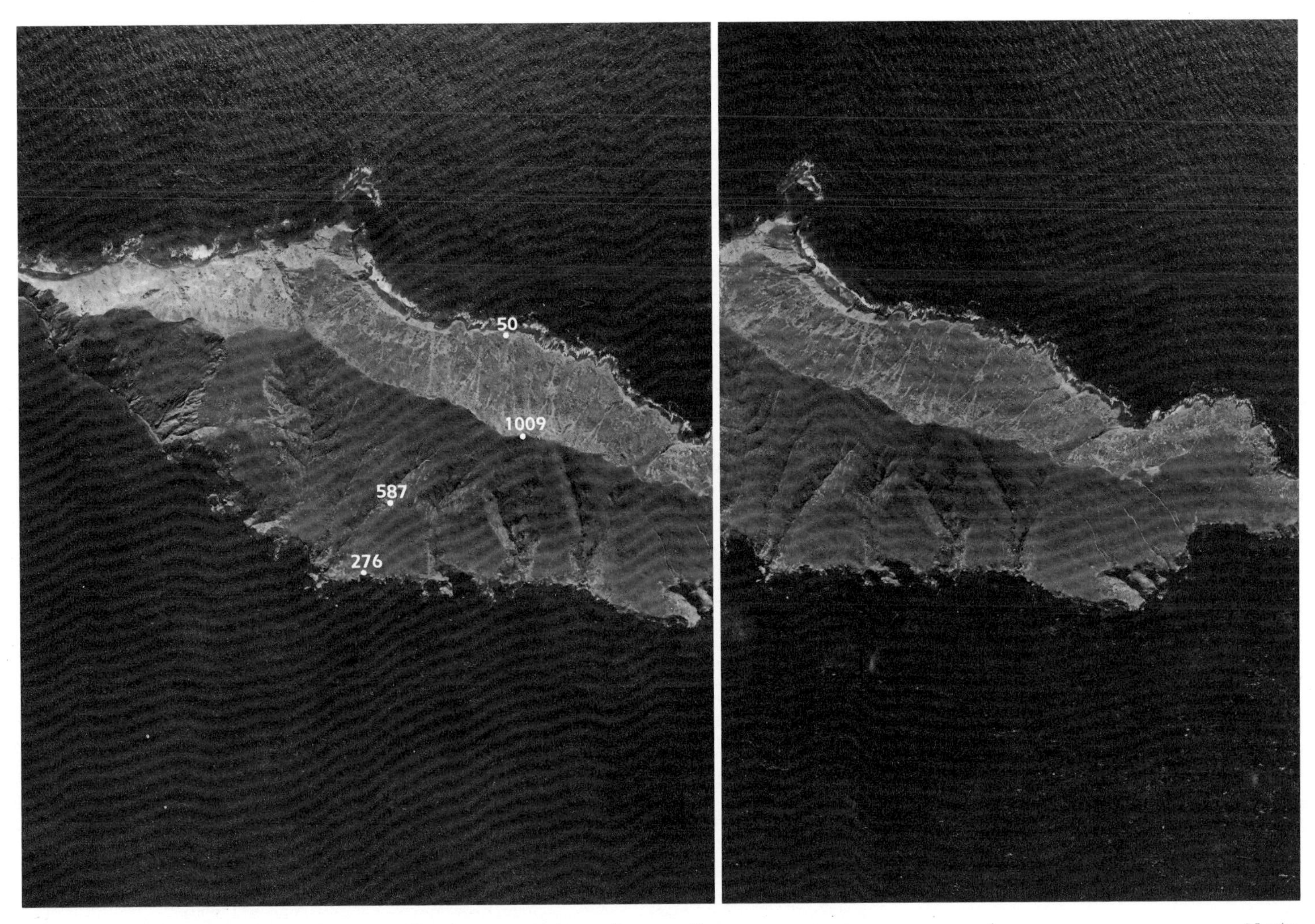

FIGURE 6.15: Constructing a Contour Map from Stereoscopic Aerial Photographs

PROBLEM

1 Construct a contour map of the island area shown in stereoscopic view in Figure 6.15. Use a contour interval of 200 ft.

PROCEDURE

Note the configuration of the shoreline, and remember that the water level is a horizontal surface and that all contours must be parallel to this surface. Note the elevations established by a survey (printed in white). Use a light lead pencil, because you may need to do some erasing before you are satisfied with the result. It is generally best to start with lower elevations and to work up the major streams. Label every fifth contour line.

State briefly how you could increase the amount of topographic detail shown on your map.

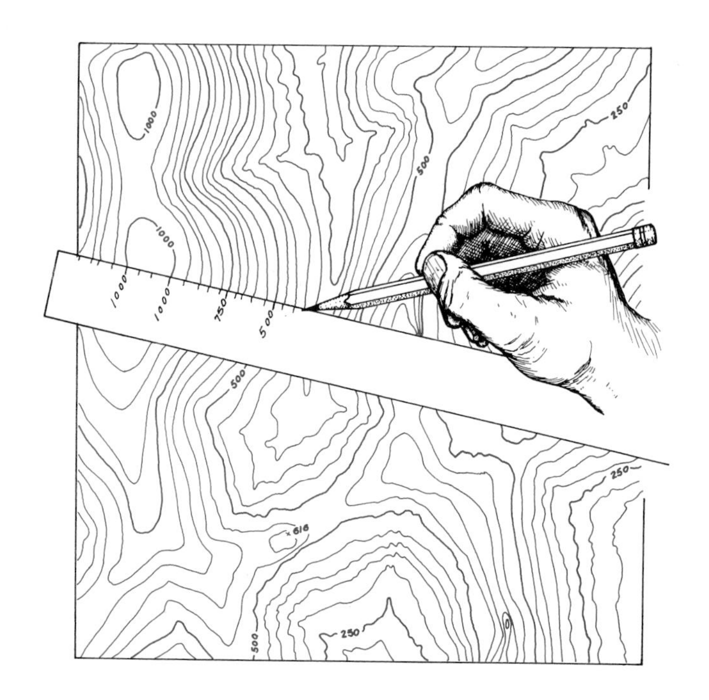

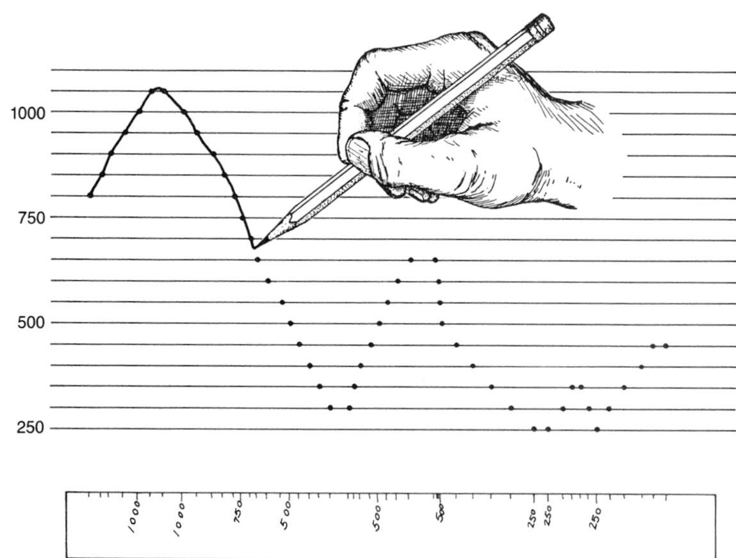

FIGURE 6.16

METHOD OF CONSTRUCTING A TOPOGRAPHIC PROFILE.

THE TOPOGRAPHIC PROFILE

Topographic maps present a view of the landscape as seen from directly above, an excellent perspective from which to examine regional relationships. This view is, however, unnatural to us, for we are accustomed to seeing hills and valleys from a horizontal perspective. In detailed studies of landforms, it may be desirable to construct a profile, or cross section, through certain critical areas so that various features can be analyzed from a more natural viewpoint. Such a profile can be constructed quickly and accurately along any straight line on a contour map, as illustrated in Figure 6.16. The appearance of the profile will vary, depending on the spacing of the horizontal lines on the profile paper. If the vertical scale is the same as the horizontal scale, the profile will be nearly flat. For this reason, the vertical scale is usually exaggerated to show local relief.

PROBLEM

1 Construct a topographic profile along line **A-A′** across the map in Figure 6.17A.

PROCEDURE

1 Lay a strip of paper along the line for which the profile is to be constructed.

2 Mark on the paper the exact place where each contour, stream, and hilltop crosses the profile line.

3 Label each mark with the elevation of the contour it represents. If contour lines are closely spaced, it is sufficient to label the index contours only.

4 Prepare a vertical scale (Figure 6.17B) on profile or graph paper by labeling horizontal lines to correspond to the elevation of each index contour line.

5 Place the paper with the labeled marks at the bottom of the profile paper and project each contour onto the horizontal line of the same elevation.

6 Connect all of the points with a smooth line.

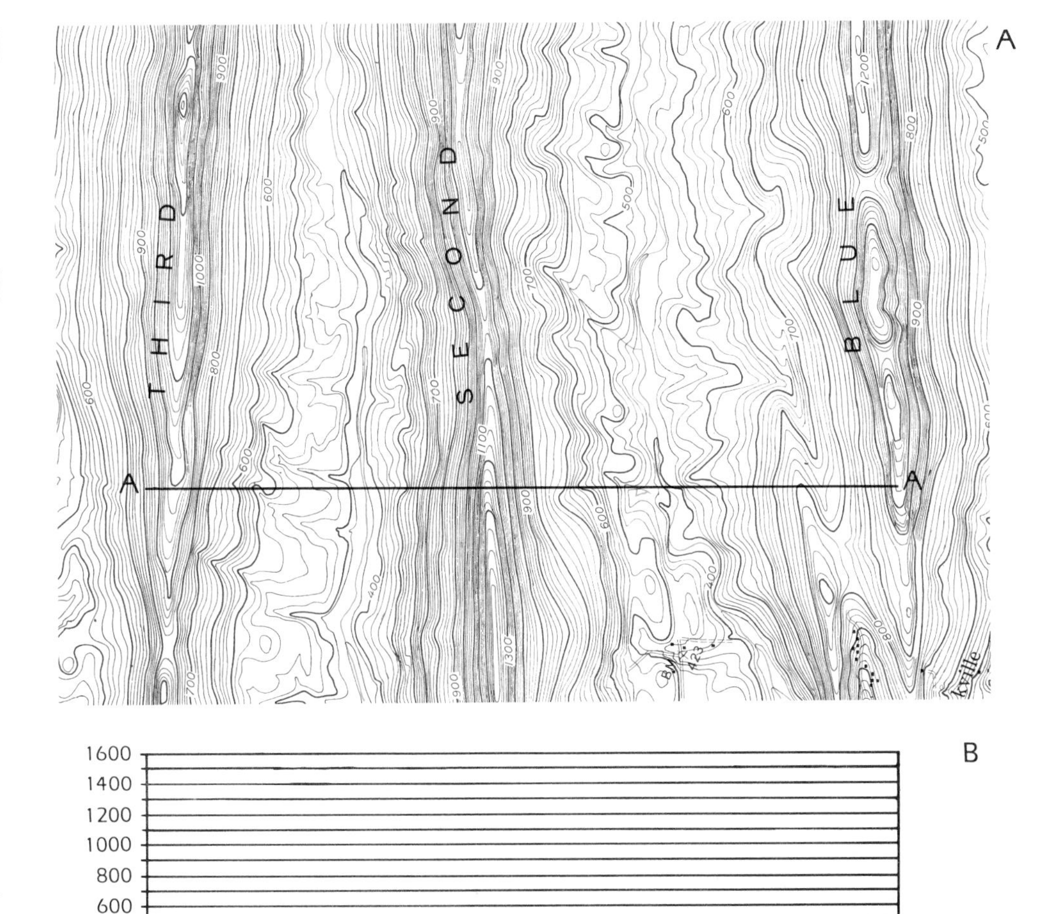

CI 20 ft 0 1 mile

FIGURE 6.17 **CONSTRUCTING A TOPOGRAPHIC PROFILE ALONG A LINE ON A MAP.**

COMPUTER-GENERATED PROFILES

We can now store the vast amount of material contained on a topographic map in a computer and retrieve the data in various forms. One useful form is to have the computer construct a large number of closely spaced cross sections at right angles; this produces a model of a terrain in perspective (Figure 6.18). The vertical scale can be exaggerated and the block diagram rotated, so an observer can view the terrain from any perspective. This permits the study of many subtle surface features that might not be readily apparent from map studies alone.

Because computers can store, update, retrieve, and manipulate vast amounts of data, the U.S. Geological Survey is transferring many maps to computer tape. The tape can then be purchased and used in lieu of the map itself.

FIGURE 6.18

COMPUTER-GENERATED DIAGRAM.

Exercise 7

LANDFORMS OF THE UNITED STATES

OBJECTIVE

To review the relative size and spatial relationships of the major physiographic provinces of the United States in order to understand better the regional setting and geologic significance of local landforms.

MAIN CONCEPT

The major provinces of the United States can be recognized on a landforms map on the basis of elevation, topography, patterns of surface features, and origin.

SUPPORTING IDEAS

Landforms, river systems, and coasts commonly have many distinctive characteristics that reflect certain facts about their origin and history.

DISCUSSION

The major physiographic provinces of the United States can be classified on the basis of landforms, structure, rock types, and geologic history. A summary of the major divisions is as follows:

INTERIOR LOWLANDS. The Interior Lowlands extend from Ohio to Kansas. The surface of this region is generally between 500 to 1000 ft above sea level (light green on the map in Figure 7.1). This province is part of the stable platform in which the rocks are mostly Paleozoic sedimentary strata, warped into broad domes and basins. In several areas, Precambrian crystalline rocks are exposed in the core of eroded domes.

GREAT PLAINS. The Great Plains extend from Kansas and the Dakotas westward to the front of the Rocky Mountains. Most of the area is 1000 to 2000 ft above sea level (light tan on the map in Figure 7.1). This province also is part of the stable platform and is covered with horizontal Mesozoic and Cenozoic sedimentary rocks.

APPALACHIAN MOUNTAINS. The Appalachian Mountains are an old mountain belt extending from central Alabama to eastern Canada. The belt is subdivided into four major provinces: (1) the Appalachian Plateau, (2) the Valley and Ridge, (3) the Blue Ridge, and (4) the Piedmont.

1 *The Appalachian Plateau* is a plateau located west of the Valley and Ridge Province and is dissected mostly by the drainage system of the Ohio River. The area is underlain by gently dipping sedimentary rocks and is more than 1000 ft above sea level (tan on the map in Figure 7.1).

2 *The Valley and Ridge Province* is an elongate belt of folded Paleozoic sedimentary rocks that are deeply eroded. The resistant rock layers have formed extensive zigzag ridges, and the nonresistant formations have been eroded to form intervening valleys.

3 *The Blue Ridge Province* is an elongate mountainous area located east of the Valley and Ridge and extending from northern Georgia to southern Pennsylvania. The rocks are mostly complex metamorphics.

4 *The Piedmont Province* is an eroded, rolling upland carved on highly deformed igneous and metamorphic rocks. It lies between the Coastal Plains on the east and the Blue Ridge on the west.

COASTAL PLAINS PROVINCE. The Coastal Plains are a wide lowland extending from New York to Mexico. The area is generally less than 500 ft above sea level (dark green on the map), and consists of gently dipping sedimentary rocks eroded to form a series of cuestas.

NEW ENGLAND PROVINCE. The New England Province is a mountainous area eroded on metamorphic and igneous rocks. The region has been glaciated.

ROCKY MOUNTAINS (CORDILLERAN). The Rocky Mountains are part of a complex mountain range located west of the Great Plains. The range is subdivided into a number of provinces based on structure, rock type, and topography of the area.

1 *The Northern Rockies* are folded and faulted mountains composed of sedimentary, metamorphic, and igneous rocks. They extend from central Idaho to Canada.

2 *The Southern Rockies* consist of north-south trending, elongate crustal upwarps that extend from central New Mexico to Wyoming. Precambrian crystalline rocks are exposed in the core of the upwarps.

3 *The Central Rockies* are faulted and upwarped mountain ranges in central Utah and western Wyoming. Precambrian rocks are commonly exposed in the core of the upwarps.

BASIN AND RANGE. The Basin and Range province consists of numerous ranges, which are mostly the result of block faulting, separated by basins filled with sedimentary deposits of Cenozoic age. This province extends throughout Nevada, western Utah, and the southern parts of California, Arizona, New Mexico, and west Texas.

COLORADO PLATEAU. The Colorado Plateau consists of nearly horizontal or gently tilted layers of sedimentary rock. It is located in Utah, Arizona, New Mexico, and Colorado, and is dissected into large canyons by the drainage system of the Colorado River. Volcanic rocks are common along the plateau margins.

COLUMBIA PLATEAU. The Columbia Plateau, located in eastern Washington and Oregon and southern Idaho, is composed of extensive basaltic lava flows.

CASCADE MOUNTAINS. The Cascade Mountains, a north-south trending chain of recent composite volcanoes, extends from northern California to the Canadian border.

SIERRA NEVADA. The Sierra Nevada, a tilted fault block of granitic batholiths, are in eastern California.

PACIFIC COASTAL RANGES. The Pacific Coastal Ranges are complexly folded and faulted masses of sedimentary and metamorphic rocks. The ranges are located in the Pacific coastal states.

PROBLEMS

1 Draw the boundaries of the following physiographic provinces on Figure 7.1:

- a Coastal Plains
- b Piedmont
- c Blue Ridge
- d Valley and Ridge
- e Great Plains
- f Basin and Range
- g Colorado Plateau
- h Northern Rockies
- i Sierra Nevada
- j Cascade Mountains
- k Columbia Plateau

2 Briefly compare and contrast the eastern and western coasts of the United States.

3 Judging from the topography of southern Missouri, would you expect this area to be (a) a basin, (b) a domal upwarp, (c) a glaciated terrain, (d) a part of the Coastal Plains, or (e) a part of the Appalachian Plateau?

4 Study the High Plains in the Texas Panhandle. Judging from the drainage patterns, is this area (a) growing larger, (b) growing smaller, or (c) stabilized and unchanging?

5 What is the maximum elevation of the state of Florida?

6 What is the lowest approximate elevation in California?

7 Circle the major volcanoes in the Cascade Mountains.

8 What do the small circular features in the Colorado Plateau in northern Arizona represent?

9 Label the following rivers:

- a Mississippi
- b Missouri
- c Ohio
- d Rio Grande
- e Colorado
- f Columbia

10 Why is there no large, well-integrated drainage system in Florida?

11 Describe the drainage system of the Coastal Plains.

12 Describe the drainage system of the Great Plains.

13 Describe the drainage of the Basin and Range.

14 Study the map and locate the following features:

- a Wisconsin Highlands: A large domal upwarp with crystalline rocks exposed in the core. Elevation ranges between 1000 and 2000 ft.
- b Adirondack Mountains: A large, circular domal upwarp in northern New York in which complex Precambrian igneous and metamorphic rocks are exposed.
- c Black Hills: A large domal upwarp near the Wyoming-South Dakota border in which Precambrian igneous and metamorphic rocks are exposed.
- d Bighorn Mountains: A domal upwarp in north central Wyoming.
- e Ozark Dome: An upwarp located in eastern Missouri.

15 Locate the following:

- a San Andreas Fault, California
- b Grand Canyon, Arizona
- c Snake River Plains, Idaho
- d Death Valley, California
- e Mississippi Delta

FIGURE 7.1 **SHADED RELIEF MAP OF THE UNITED STATES.**

LAKE SUPERIOR
LAKE MICHIGAN
LAKE HURON
LAKE ONTARIO
LAKE ERIE
ATLANTIC OCEAN
ELEVATION TINTS
FEET
METERS
12,000
9,000
5,000
2,000
1,000
500
0
3,658
2,743
1,524
610
305
152
0 (sea level)
SHADED RELIEF
Richard Edes Harrison, 1969
Albers Equal Area Projection
SCALE 1:7,500,000
MILES
KILOMETERS

Exercise 8

STREAM EROSION AND DEPOSITION

OBJECTIVE

To understand the processes of stream erosion and deposition, and to recognize the landforms that these processes produce.

MAIN CONCEPT

Rivers erode the landscape by (1) downcutting, (2) headward erosion, and (3) slope retreat. Deposition of sediment transported by a river system produces (1) flood plains, (2) stream terraces, (3) deltas, and (4) alluvial fans.

SUPPORTING IDEAS

1 Downcutting of a stream channel results from the abrasive action of sand and gravel.

2 Headward erosion occurs at the head of all tributaries and causes the drainage network to migrate upslope.

3 Slope retreat is accomplished by a variety of types of mass movement and by the headward erosion of minor tributaries.

4 A river flowing over a low gradient deposits part of its load to form natural levees, point bars, and backswamps.

5 When a river enters a lake or ocean, it deposits most of its sediment load to form a delta. When a river flows into a dry basin, an alluvial fan results.

DISCUSSION

Running water is by far the most important agent of erosion on our planet, and most of the landscape is sculptured in some way by streams and rivers. As a result, stream valleys are the most widespread landform on the continents. An understanding of the basic processes by which stream valleys form is thus essential to the study of physical geology.

PROCESSES OF STREAM EROSION

The basic process of stream erosion is the abrasive action of sand and gravel as it is moved by running water. By this process of downcutting a stream deepens its channel. In addition, every stream tends to erode headward and to extend its channel upslope until the stream reaches the divide. As the drainage network grows larger by headward erosion and deeper by downcutting, the valley walls become subject to a variety of slope processes such as creep, debris flows, and landslides. These processes, plus the erosive activities of minor tributaries, cause the valley slopes to recede from the river channel.

FIGURE 8.1: **Mesa Verde, Colorado**

PROBLEMS

1 To visualize a cross-sectional view of this area, sketch an east-west topographic profile across the middle of the photograph.

2 What geologic process is dominant in (a) the stream channels, (b) the valley slopes, and (c) the plateau surface?

3 What geologic process (i.e., downcutting of the stream channel, headward erosion, or slope retreat) will produce the most significant modifications in this landscape in the near future?

4 Compare and contrast the topography shown in Figures 8.1 and 8.2. What is the major difference between the two areas?

5 As erosion continues will the area in Figure 8.1 become more like that in Figure 8.2 or visa versa?

FIGURE 8.2: **Northern Rockies**

PROBLEMS

1 Examine several tributary systems and the large channels into which they flow. In a brief sentence or two, state the relationship between the size of the stream and the valley through which it flows.

2 What processes of stream erosion dominate in this area at the present time: headward erosion, downcutting, or slope retreat?

3 How will this area change as the processes of stream erosion continue?

- **a** Will the number of tributaries increase?
- **b** Will the length of the tributaries increase?
- **c** Will the divide migrate?
- **d** Will downcutting continue without more crustal uplift?
- **e** Will slope retreat become the dominant process?

FIGURE 8.3: **Kaaterskill, New York**

CI 20 ft 0 1 mile

PROBLEMS

1 What is the approximate gradient in feet per mile of the streams and creeks that flow to the southeast (Plattekill Creek and Kaaterskill Creek)?

2 What is the general gradient in feet per mile of Schoharie Creek and its major tributary, Gooseberry Creek?

3 On the basis of these observations, determine whether the divide between these two drainage systems will migrate to the northwest, southeast, or remain stationary. Explain the reason for your answer.

4 Study the drainage pattern of Kaaterskill Creek and Gooseberry Creek, and make a series of sketches showing how this drainage has changed in the recent geologic past.

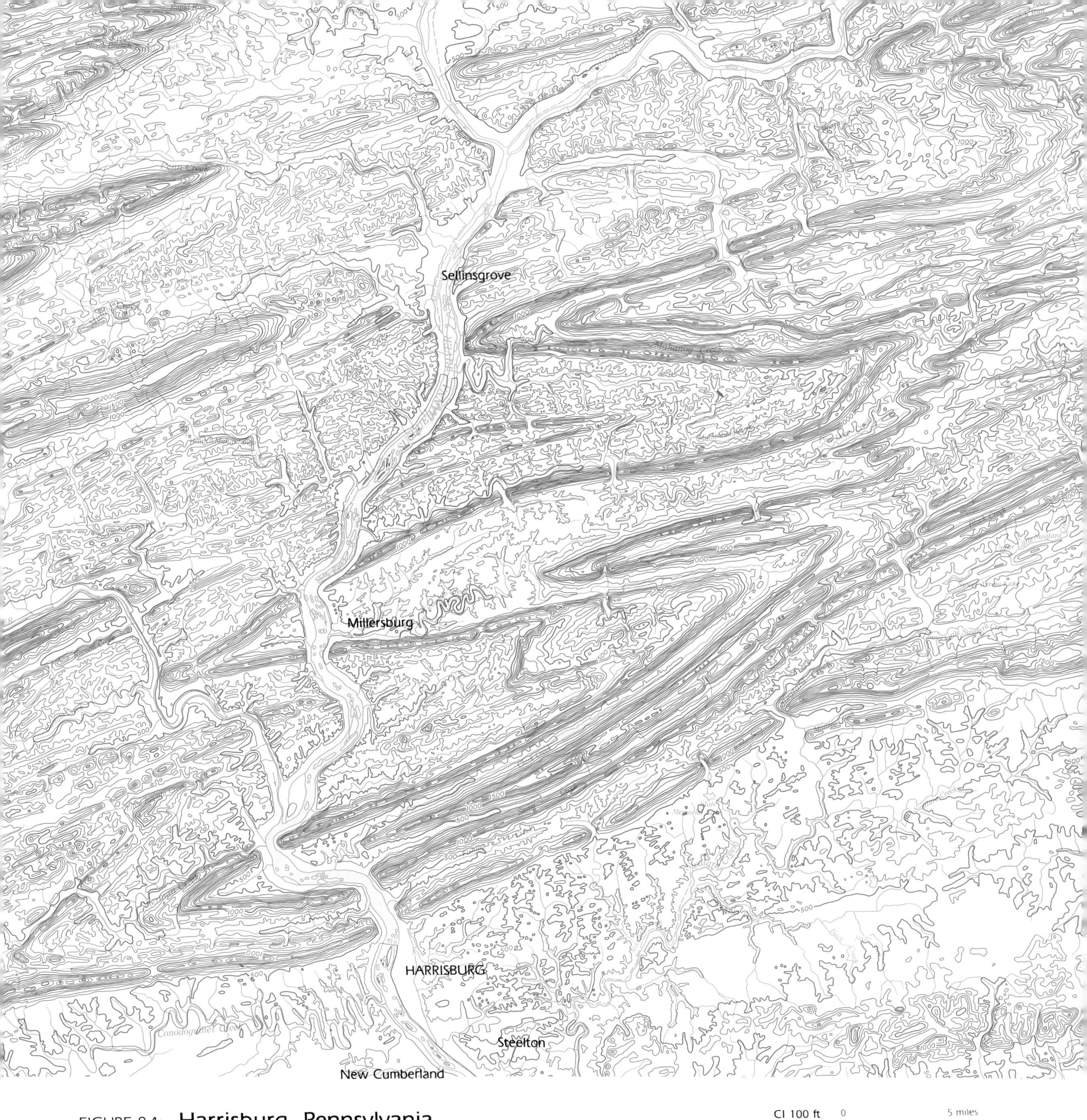

FIGURE 8.4: Harrisburg, Pennsylvania

CI 100 ft 0 5 miles

PROBLEMS

1 Compare this map with the images on pages 132 and 176. All are from the same province. What are the advantages of (a) a topographic map, (b) a landsat image and (c) a radar image.

2 Show by a heavy dashed line or in colored pencil the major tributaries of the Susquehanna River that developed by differential erosion along nonresistant strata. What type of drainage pattern is this?

3 Show by a heavy solid line or in colored pencil those streams that were superposed. What was the original drainage pattern of the Susquehanna River and its tributaries? What features on the map support your answer?

FIGURE 8.5: Canyonlands, Utah

Shaded relief CI 80 ft 0 1 mile

PROBLEMS

1 What is the maximum local relief in the area?

2 What is the minimum uplift indicated by the entrenchment of the Green River?

3 Draw a topographic profile along line **A-A′.**

4 Explain the striking differences between the pattern of the Green River and that of its major tributaries.

5 What major changes have occurred in the course of the Green River since its entrenchment?

6 What changes would you expect to occur in the future?

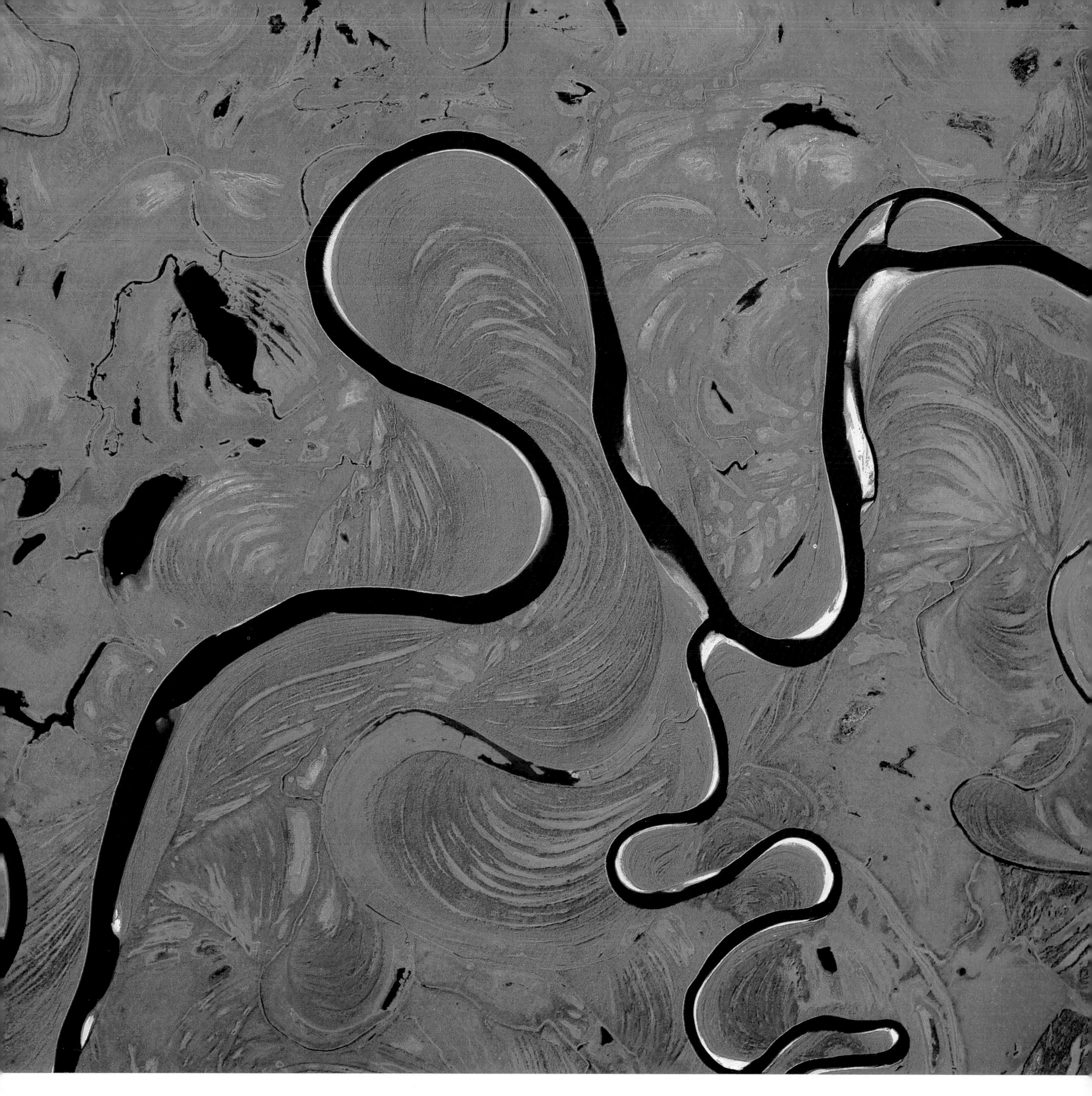

FIGURE 8.6: **River System—Alaska**

High-altitude infrared color photo 0 1 mile

PROBLEMS

1 Map and label the point bars formed by the major river. Indicate their relative age (1 = oldest, 2 = next oldest, and so forth). How many times has the river changed its course?

2 Express as a percentage the amount of shortening of the present course of the major river as compared with the river's course before the cutoff of the large meander in the central part of the photograph.

3 What changes would occur in the river gradient as a result of this shortening of the river course when a meander bend is abandoned?

4 What are the most significant geologic processes operating in a meandering river?

FIGURE 8.7: Lower Mississippi River

Landsat image 0 20 miles

PROBLEMS

1 Identify this area on the physiographic map, page 77. Locate on the Landsat image the areas where the course of the Mississippi River has changed significantly in recent years. Describe the process by which these changes occurred.

2 Estimate the width of the river's flood plain.

3 What landform dominates the flood plain of the Mississippi River?

4 What landforms dominate the upland adjacent to the floodplain? (reddish areas that appear rough)

5 Study the oxbow lakes along the Mississippi River, and make a series of sketches that show how an oxbow lake changes with time.

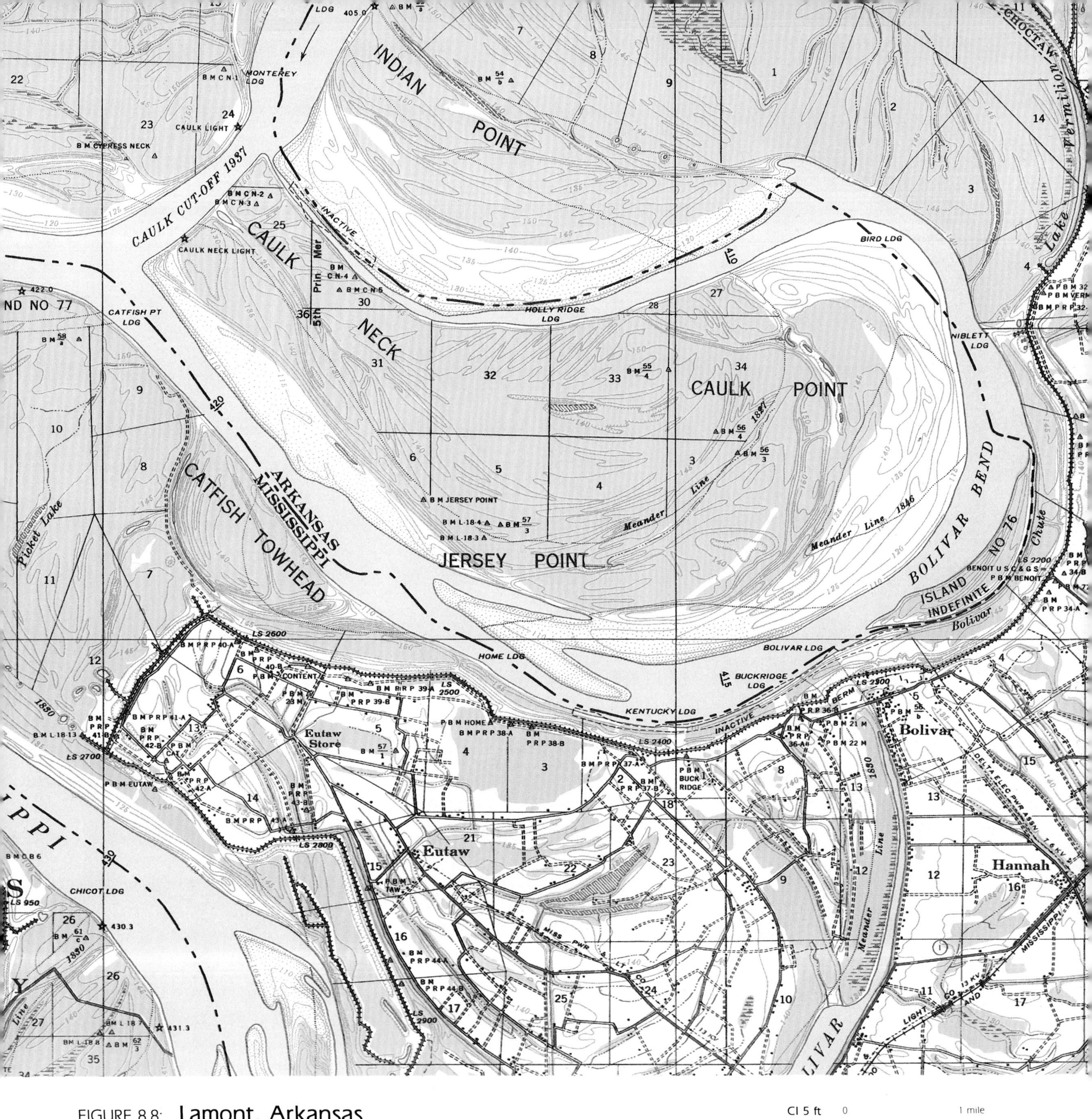

FIGURE 8.8: Lamont, Arkansas

CI 5 ft 0 1 mile

PROBLEMS

1 What major change occurred in this area in 1937?

2 What major changes occurred between 1827 and 1937?

3 What is the maximum local relief of the natural levees in this area?

4 Color the oldest meander scars green and the youngest yellow.

5 If the Mississippi River is 40 ft deep in this area, estimate the volume of sediment deposited on a point bar each year. Note the position of the inlands line in 1827, 1846, and the position at the time of cut-off in 1937.

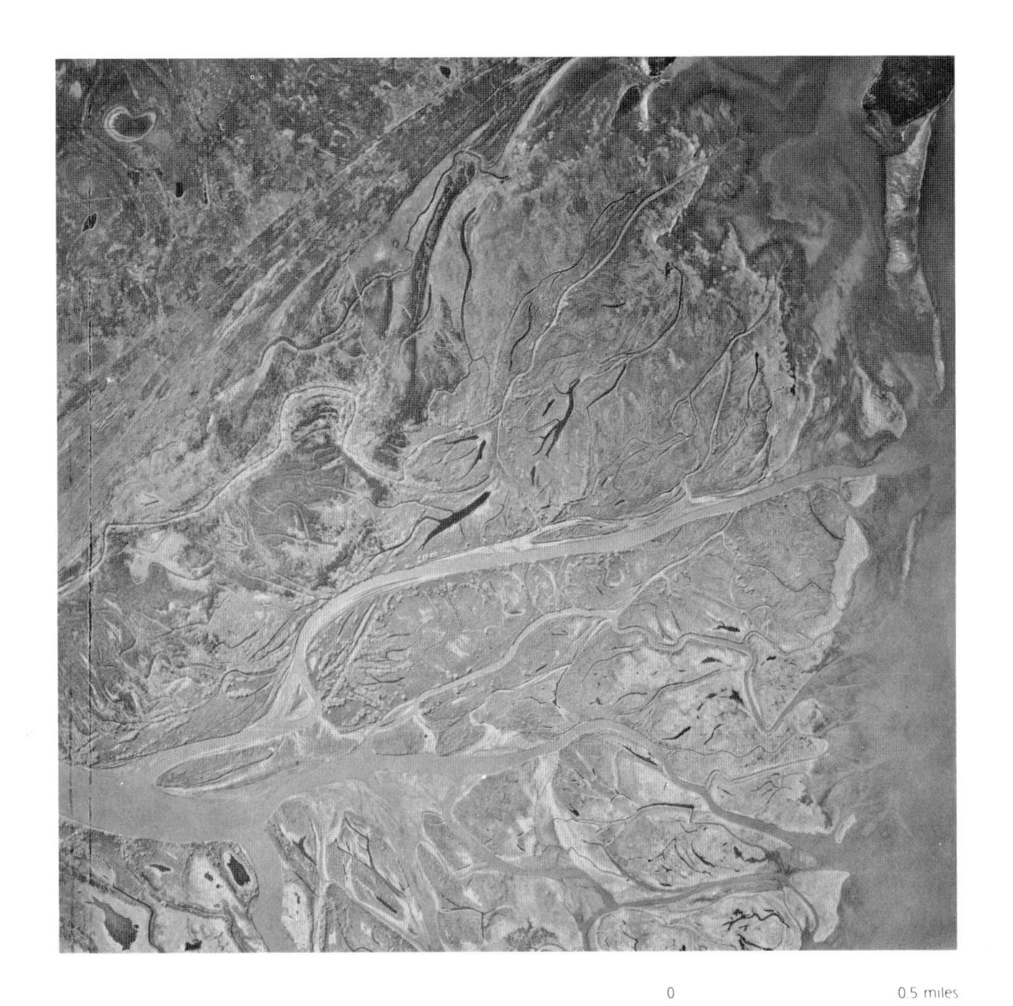

FIGURE 8.9: Delta—Canada

PROBLEMS

1 Color the main distributary channels blue and the splay channels yellow. Briefly explain why a stream splits into distributaries.

2 Is this delta dominated by fluvial, wave, or tidal action? Support your answer with evidence.

3 Note the small subdeltas built by the small stream in the lagoon in the northern part of the area. Why are there no similar deltas in the small streams that empty into the open sea in the southern part of the area?

FIGURE 8.10: Splays—Canada

PROBLEMS

1 Several splays have formed in this area. Label them according to relative age (1 = oldest, 2 = next oldest, and so forth).

2 How is a splay modified after the crevasse in the natural levee is sealed?

3 How does a crevasse originate in a natural levee?

4 Why is the entire river *not* diverted through a break or crevasse in the levee?

5 Explain the mechanism by which the stream channels in a splay split into distributaries.

FIGURE 8.11: **Mississippi Delta**

PROBLEMS

1 Study the morphology of the Mississippi Delta. Note especially the main stream channels and the areas where the streams split into distributaries. Label the main subdeltas according to relative age (1 = oldest, 2 = next oldest, and so forth).

2 What geologic processes operate on a subdelta after the main river channel changes to a new course?

3 What processes can you identify that are presently operating on the delta? (Label those areas *W* where wave processes are active, and those areas *F* where fluvial processes dominate.)

4 Why are there no beaches or barrier bars near the present subdelta area, whereas beaches and bars are common features to the west?

FIGURE 8.12: Alluvial Fans—California

PROBLEMS

1 What evidence do you find in this area that substantiates the assumption that streams cut the valleys through which they flow?

2 Sediment of different ages can be recognized on the fans by gradations of tone, texture, and relative elevation. Map these sediments according to their relative age (1 = oldest sediments, 2 = next oldest, and so forth).

3 Why does the drainage on the fan split into numerous distributaries?

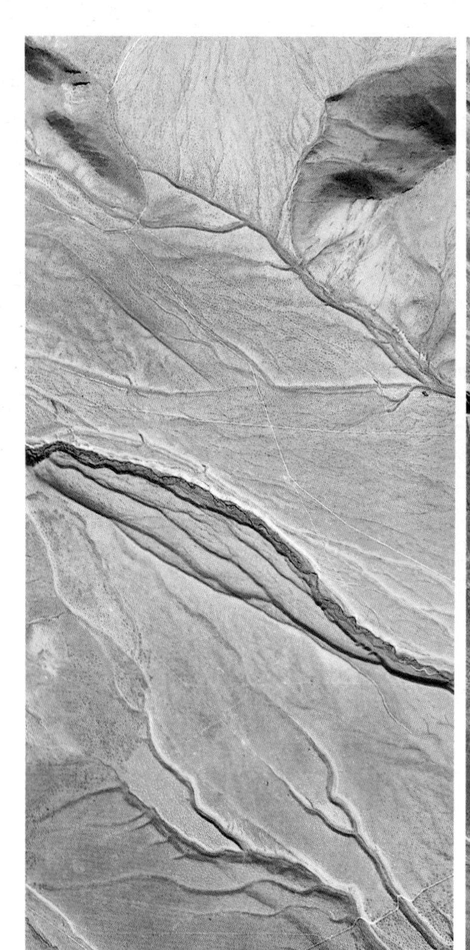
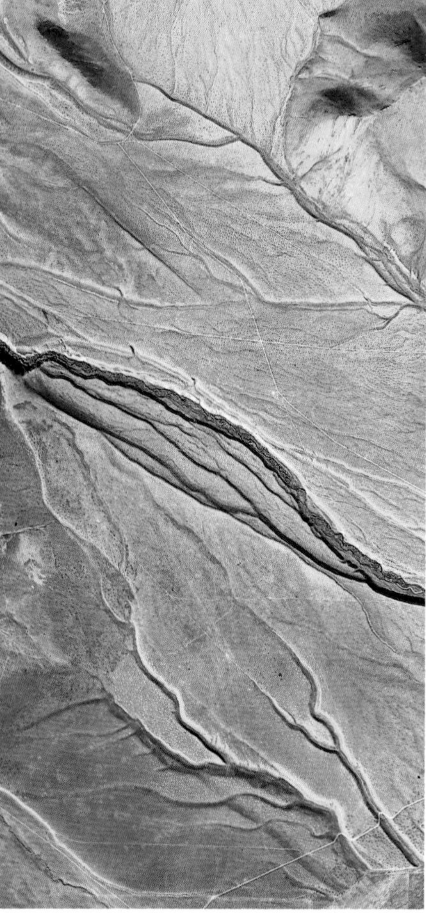

0 0.5 miles

FIGURE 8.13: Stream Terraces—Montana

PROBLEMS

1 What processes were involved in the deposition of the sediment on which the terraces are cut?

2 Color the stream terraces according to their relative age (yellow = oldest, green = next oldest, and blue = youngest). What geologic events do the terraces record?

3 Draw an idealized cross-section across the stream valley and show the morphology of the terraces and their relationship to the bedrock below.

FIGURE 8.14: Death Valley, California

Landsat image 0 30 miles

PROBLEMS

1 Study the image carefully. How many fans (large or small) can you see?

2 What factors govern the size of an alluvial fan (e.g., mountain height, climate, size of the drainage system)?

3 Draw a line showing the location of the fault along which the mountain was uplifted. Is there any evidence that recent movement has displaced some of the fans?

4 How would erosion and deposition in this area change if the climate became more humid?

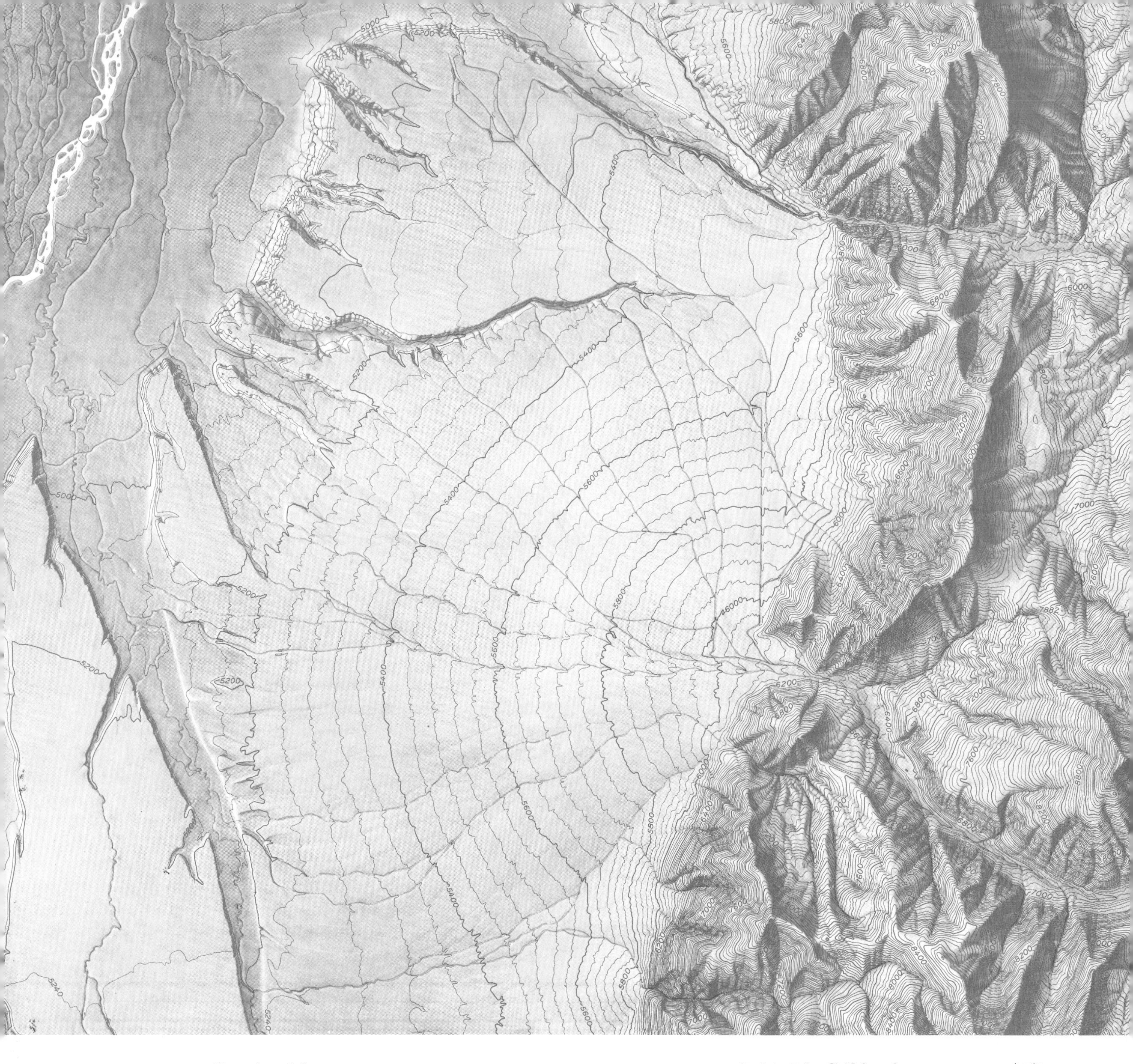

FIGURE 8.15: **Ennis, Montana**

Shaded relief CI 40 ft 0 1 mile

PROBLEMS

1 What is the gradient in feet per mile of the surface of the alluvial fans in this area?

2 What is the gradient in feet per mile of the valley floor beyond the fans?

3 What geologic processes formed the surface of the valley floor beyond the fans?

4 What evidence indicates that this area has recently been uplifted?

5 Estimate the minimum amount of uplift that has occurred in recent time.

6 What geologic structure occurs along the mountain front?

7 What evidence indicates that this area is in an arid to semi-arid climate?

FIGURE 8.16: Salt Lake City, Utah

CI 200 ft 0 5 miles

PROBLEMS

1 Study the contour lines in the Salt Lake valley and emphasize them with a brown pencil where necessary for clarity. On the basis of the contour lines and the drainage system coming from the mountains, trace the path of the drainageways as they flowed across the valley prior to urbanization.

2 This area has experienced serious flooding during the last few years from the melting of high snow packs and rapid spring runoff. Show on the map those areas of the city that would be most susceptible to flooding.

3 One of the flood control measures used was to sandbag the streets to create channels in which runoff flowed. Show on the map with a blue line those streets that would serve best as temporary channels for flood runoff.

4 Outline those areas in red that could be considered at risk from landslides and mudflows.

5 Explain how a geologic map of the area might help you to answer problem 4 better.

Exercise 9

GROUND WATER

OBJECTIVE

To recognize the distinctive landforms produced by ground water and to understand how landscapes are formed by solution activity.

MAIN CONCEPT

The solution activity of ground water is a significant agent of erosion in regions underlain by soluble rocks, such as limestone, gypsum, and rock salt. Where solution activity dominates, it commonly develops a distinctive landscape, called *karst topography*, which is characterized by sinkholes, solution valleys, and disappearing streams.

SUPPORTING IDEAS

Karst topography evolves through a series of stages until the soluble rock is completely removed.

DISCUSSION

KARST TOPOGRAPHY

Karst topography typically develops in humid areas where horizontal or gently dipping limestone beds are exposed at the surface. Sinkholes and solution valleys are the dominant landforms, and much of the drainage is underground. As a result, the region as a whole lacks a well-integrated surface drainage system. Tributary streams are few and are generally very short. Many minor streams appear suddenly as springs in blind valleys, flow for a short distance only, and then disappear into sinkholes. Only major streams flow in a defined, open valley.

An idealized sequence of stages in the evolution of a karst topography is shown in the diagrams in Figure 9.1. In the early stage (Figure 9.1A), solution activity forms a system of underground caverns that enlarge until, eventually, the cavern roof collapses, thus producing a sinkhole. As the sinkholes increase in number and size, some merge to form solution valleys. When solution valleys become numerous and interconnect, the area is considered to have reached the intermediate stage of development (Figure 9.1B). A considerable part of the original land surface has been destroyed, and maximum local relief is attained. Continued solution activity ultimately erodes the area down to the base of the limestone unit. Finally, only scattered, rounded hills and knolls remain (Figure 9.1C). Further erosion by solution activity can occur only if more limestone units exist below the surface.

A. EARLY STAGE: (1) Surface is nearly flat with a few small, scattered sinkhole depressions. Subterranean caverns are numerous. (2) Throughout the early stage, sinkholes become more abundant and increase in size.

B. MIDDLE STAGE: (1) Individual sinkholes enlarge and merge to form solution valleys with irregular branching outlines. (2) Much of the original surface is destroyed. There are many springs and disappearing streams. (3) Maximum relief, although not great, is achieved. Differences in elevation between rim and floor of a sinkhole rarely exceed 200 to 300 ft.

C. LATE STAGE: (1) Solution activity has reduced the area to the base of the limestone unit. (2) Hills formed as erosional remnants are few, widely scattered, and generally reduced to low, conical knolls.

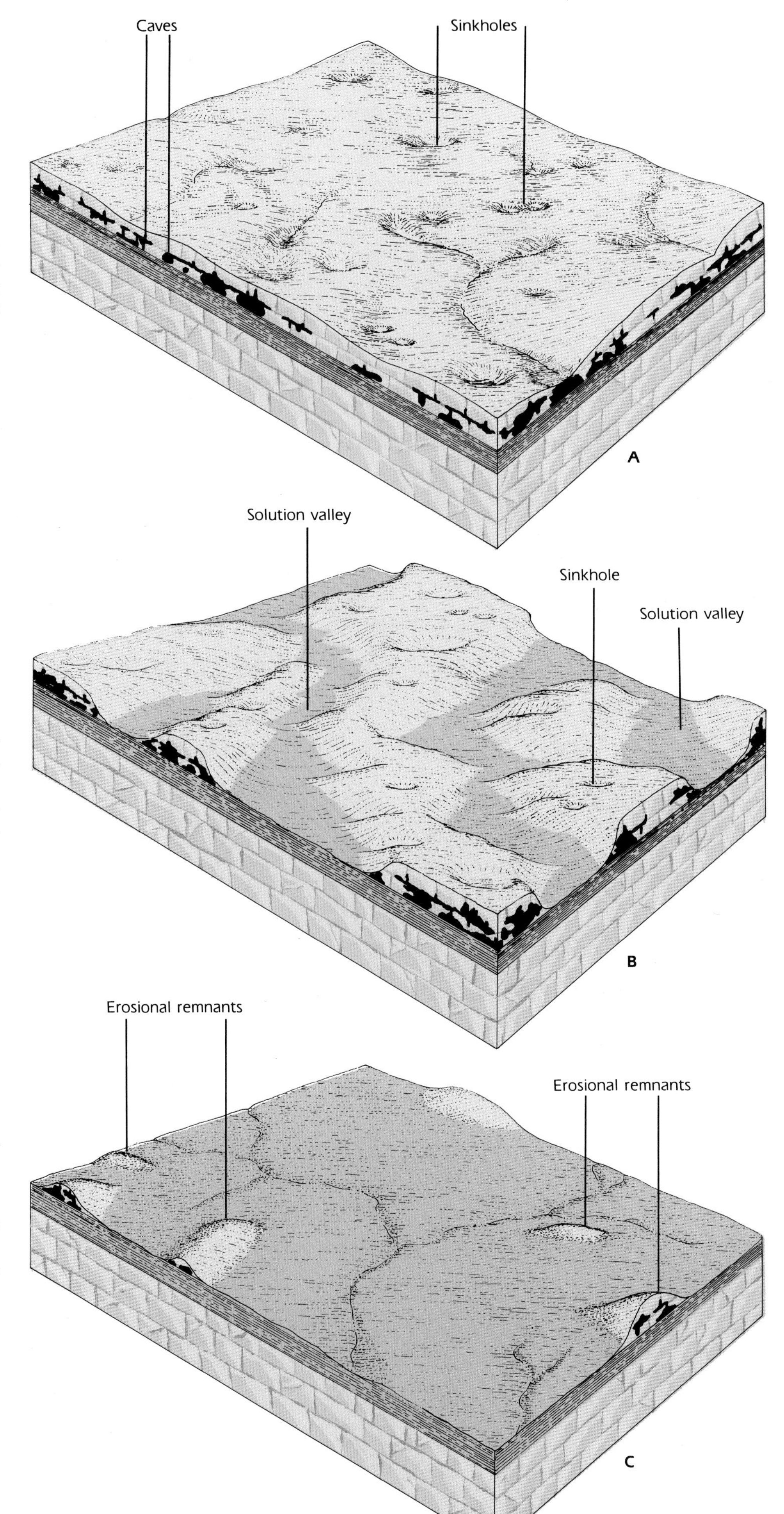

FIGURE 9.1

IDEALIZED EVOLUTION OF A KARST TOPOGRAPHY.

Stereogram color infrared photo 0 1 mile

FIGURE 9.2: Mammoth Cave, Kentucky

PROBLEMS

1 Describe the surface drainage in the area.

2 Is there evidence to indicate that the development of sinkholes is an ongoing process, like the development of oxbow lakes, or did most of these sinkholes form during one particular period of time?

3 How does a sinkhole change with time?

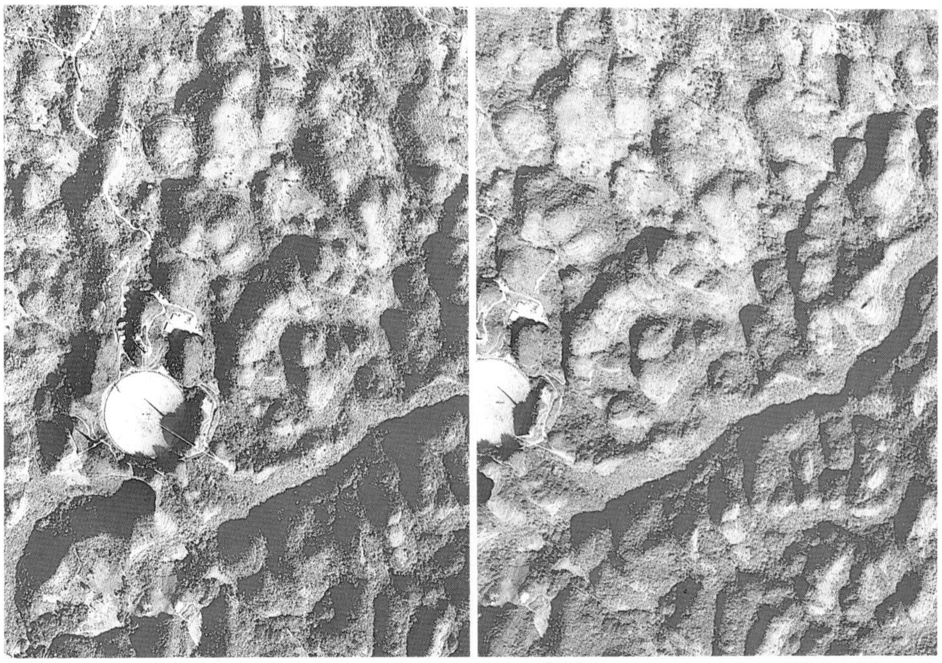

Stereogram 0 1 mile

FIGURE 9.3: Manati, Puerto Rico

The bedrock in Manati, Puerto Rico, is nearly horizontal and consists of pure, dense limestone. Sinkholes in this region are as much as 150 ft deep, and hills are up to 300 ft high.

PROBLEMS

1 List the evidence indicating that this topography was *not* produced by stream erosion.

2 Explain the origin of the circular and elongate hills in this region.

3 Why are there only a few sinkholes in the area? Where do sinkholes occur?

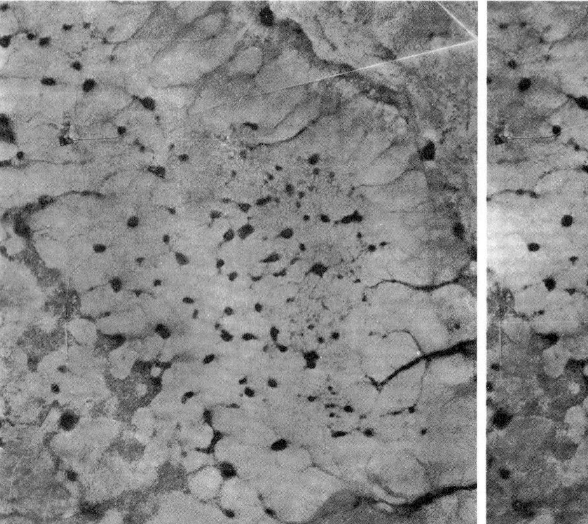

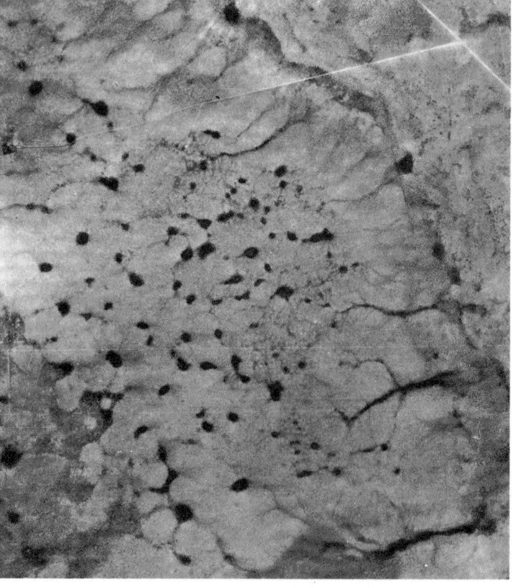

Stereogram 0 1 mile

FIGURE 9.4: West Texas

PROBLEMS

1 Compare the sinkholes in this area with those in Puerto Rico (Figure 9.3) and with those in the Mammoth Cave area (Figure 9.2). Consider features such as number, size, shape, and depth.

2 If all three areas are underlain by limestone, why is the topography of each area so different?

3 Trace the drainage system in this area. How does it differ from the drainage in the area shown in Figure 8.2?

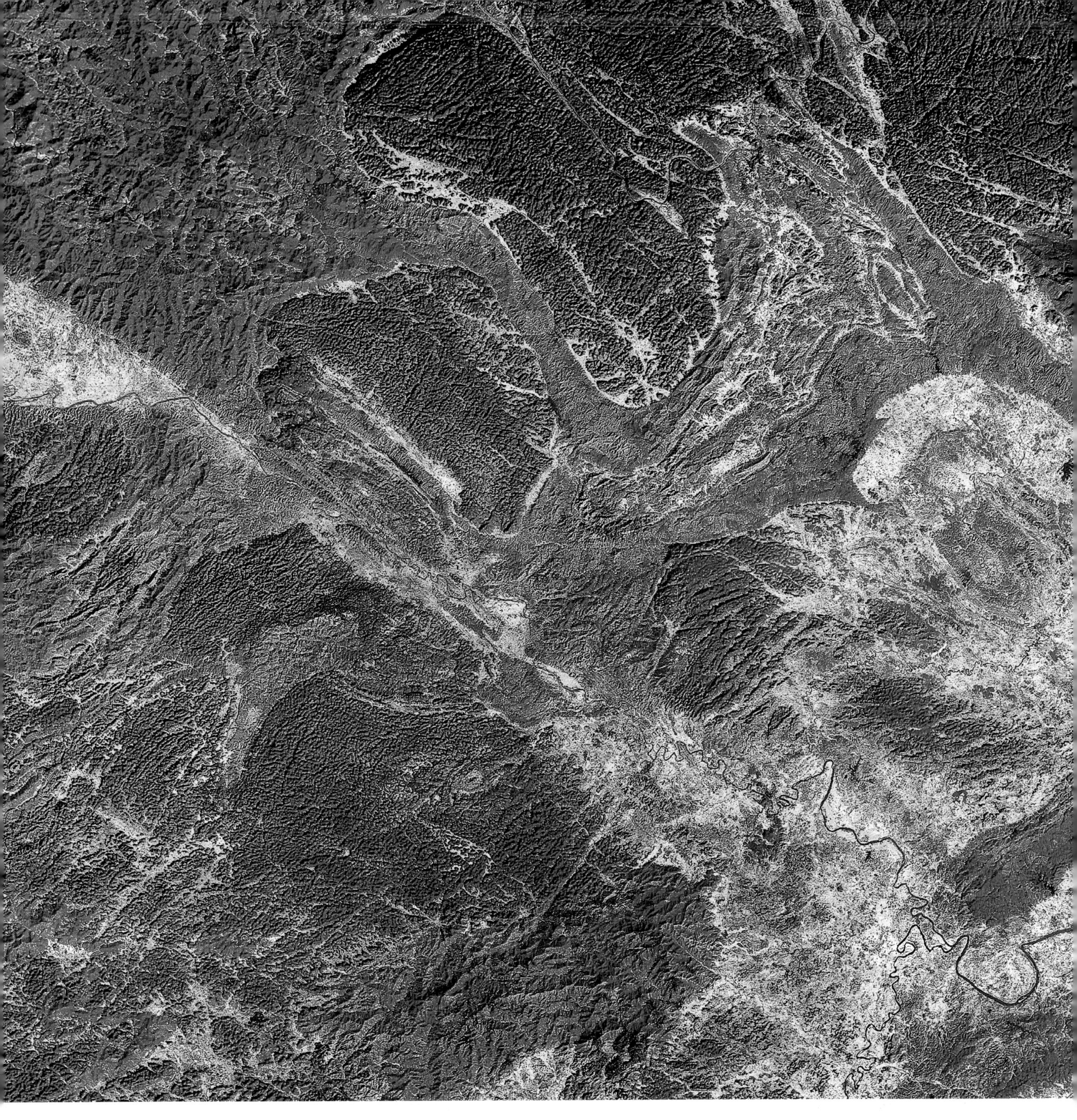

FIGURE 9.5: Nanming, China

Enhanced Landsat image 0 5 miles

This image shows the Nanming karst area of China. Although local details are not apparent on satellite imagery, the regional texture of the surface is exceptionally well expressed in this Landsat image.

PROBLEMS

1 What karst landforms dominate the area (i.e., sinkholes, solution valleys, or erosional remnants)?

2 What evidence indicates that a fracture system in the bedrock has influenced the karst topography of this area?

3 Would the karst landforms in this area be more similar to those of Kentucky, Puerto Rico, Florida, or Texas? Explain the basis for your answer.

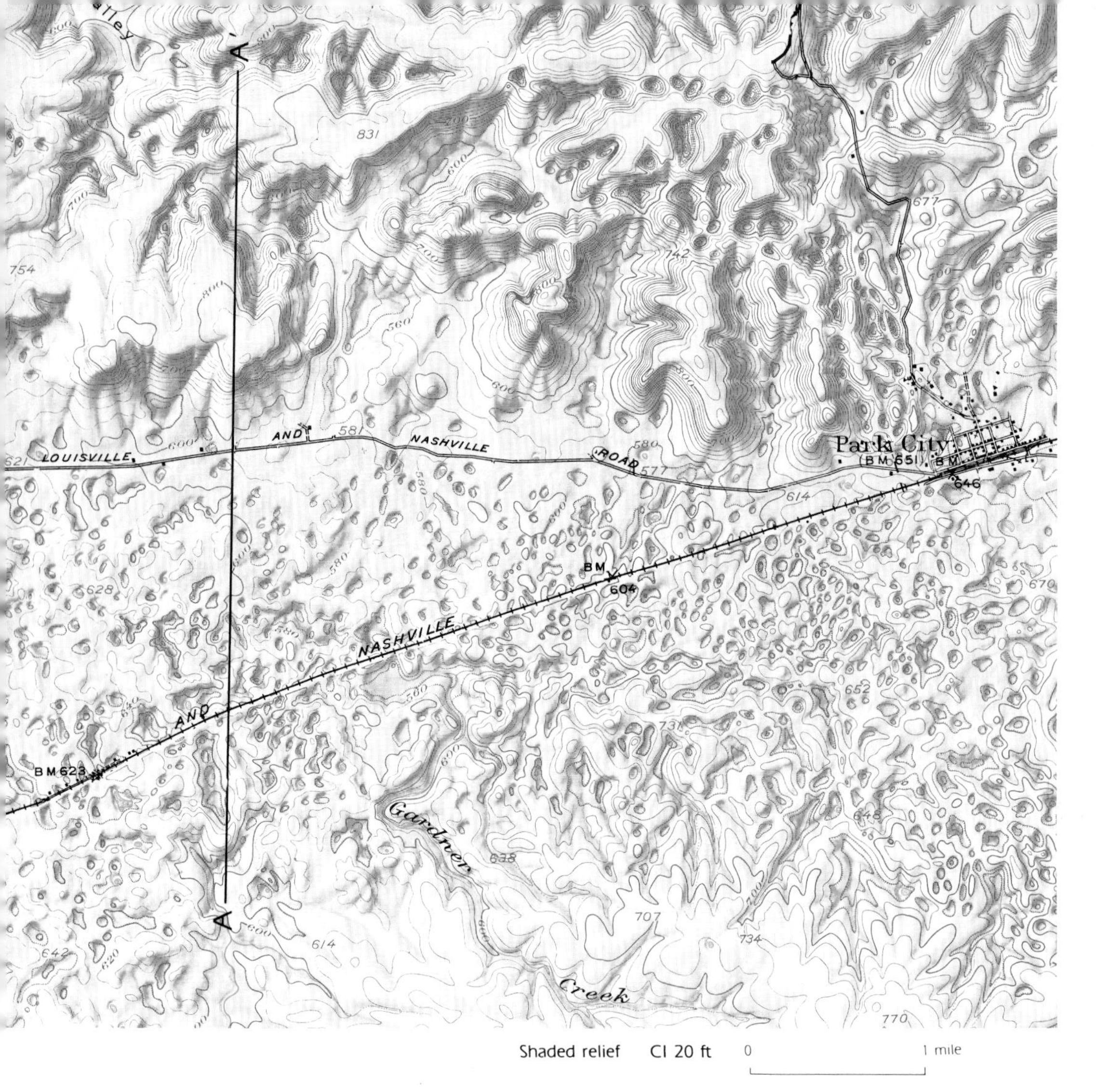

FIGURE 9.6: Mammoth Cave, Kentucky

The pockmarked surface of the Mammoth Cave area of Kentucky is a classic example of karst topography. As you study this map, remember that closed depressions are indicated by circular or elliptical contours with hachures pointing downslope.

PROBLEMS

1 Study the map carefully and outline the large solution valleys.

2 What happens to the water flowing in Gardner Creek and Little Sinking Creek?

3 Draw a topographic profile along **A-A′**. The plateau to the north is underlain by a sandstone unit roughly 100 ft thick. The bedrock is a limestone formation about 200 ft thick. The entire sequence of rock in this area dips 4 to 5 degrees to the north. With this information, sketch a geologic cross section showing the dipping beds on the topographic profile.

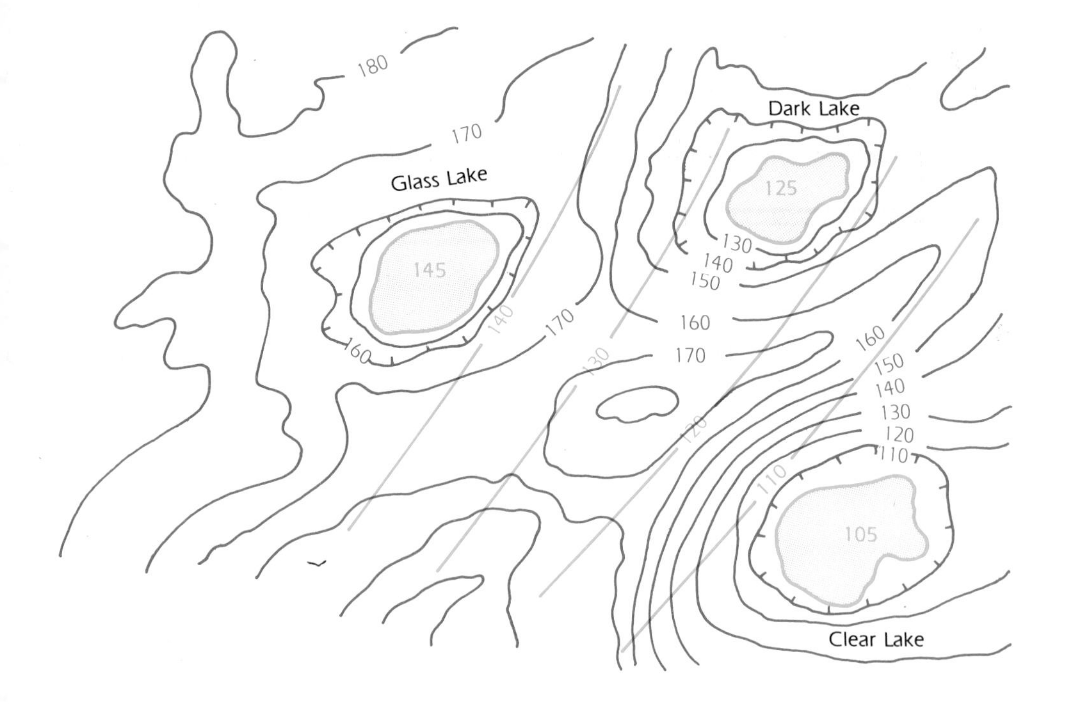

FIGURE 9.7: Contouring a Water Table

The elevation of the lakes in a region provides important information about the ground-water conditions. The surface of each lake is essentially the surface of the water table. The lakes are thus control points for the elevation of the water table, so we can construct a generalized contour map showing the configuration of the water table from these data. For example, on the map in Figure 9.7, the elevation of Dark Lake is between 120 and 130 ft, the elevation of Glass Lake is between 140 and 150 ft, and the elevation of Clear Lake is between 100 and 110 ft. For convenience, assume that the lake levels are 125, 145, and 105 ft, respectively. The water table can be contoured by applying the same principles used in contouring the land surface. The 140 and 130 ft contour lines would be located between Dark Lake and Glass Lake, and the 120 and 110 ft contour lines would be located between Clear Lake and Dark Lake. We thus observe that the water table slopes to the east.

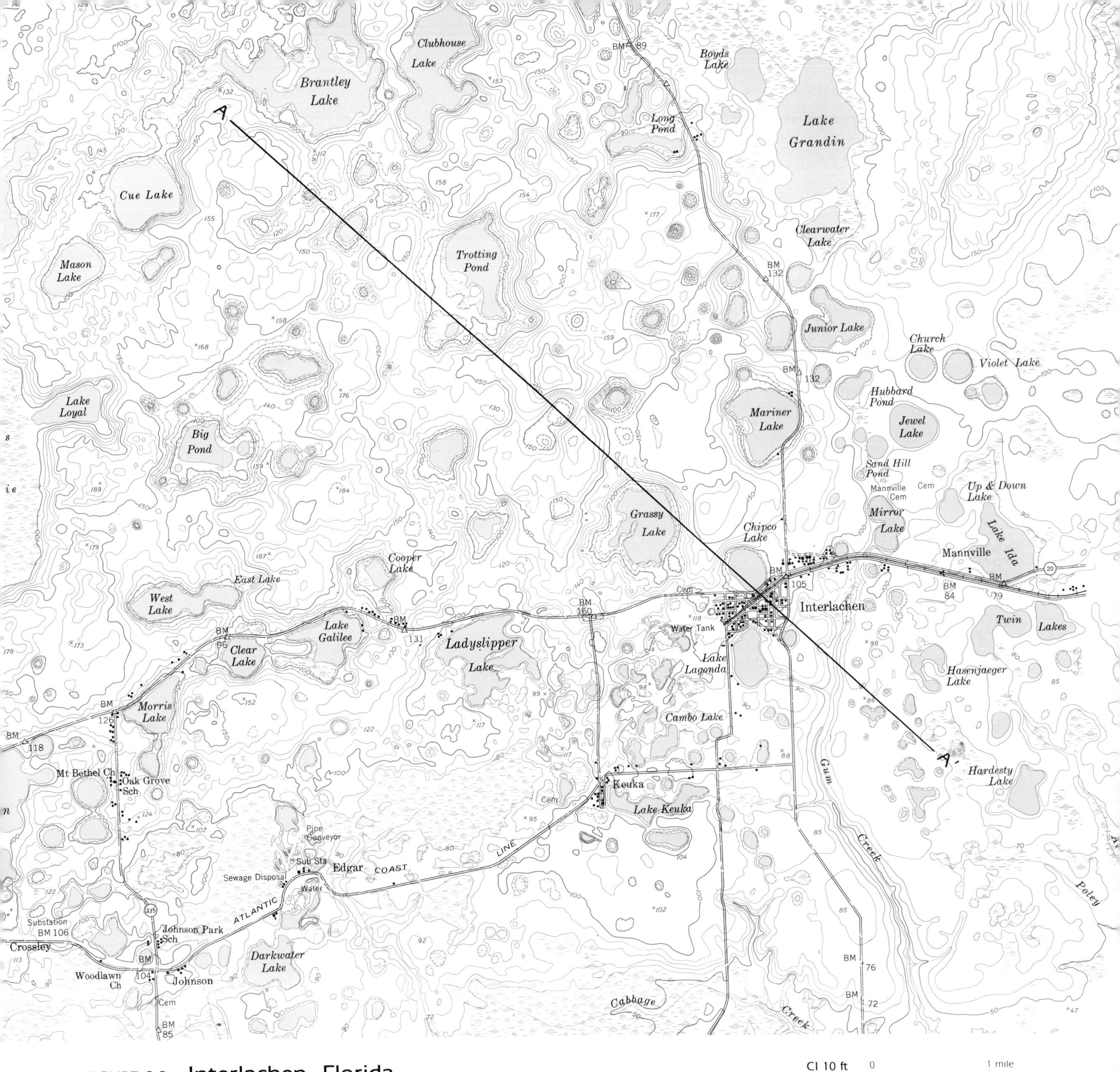

FIGURE 9.8: **Interlachen, Florida**

CI 10 ft 0 1 mile

PROBLEMS

1 Determine the elevation of most of the large lakes in this area, and construct a contour map of the surface of the water table. (Study Figure 9.7 and the discussion of how to contour the surface of a water table.) Use CI = 10′

2 Construct a topographic profile along line **A-A′**. Draw the water table with a blue line. Be sure to show how Gum Creek influences the water table.

3 Indicate the general direction of ground-water movement on the map with a series of arrows.

4 How deep would you have to drill to obtain water if the well site were located at the road intersection 1 mi west of Interlachen?

5 Many farms, small industries, and urban centers dump all of their untreated liquid waste into the subsurface. What happens to this waste? In what direction would the waste contaminants move in this area?

6 What natural hazards (i.e., floods, landslides, subsidence, earthquakes, or erosion) have the most geologic significance in terms of influencing construction work in this area?

FIGURE 9.9: Southern Florida

In recent years, the rapid urban development of the Dade County area of Florida has posed a serious threat to the environment, despite the efforts of some residents to maintain environmental quality. Most of the land surface in South Dade County is less than 10 ft above high-tide level, and the highest areas are only 25 ft above sea level. Water is the primary environmental concern in this area. For this reason, Dade County and its adjacent counties are part of a regional water control system developed to conserve and protect both the surface water and ground water, and to control freshwater and saltwater flooding. The water control measures include management of the drainage canals as well as water conservation and management of storage areas.

The major drainageways shown on this Landsat image include patterns of water movement in the Everglades Swamp and in canals, and modifications of existing drainageways. The canals form a system designed to reduce flood damage from storms and tides and to conserve fresh water.

A limestone sequence underlying the area contains two aquifers. The upper one, the Biscayne Aquifer, extends from the surface to as deep as 120 ft below ground and is the source of fresh water for the area. The other aquifer, the Floridian Aquifer, is more than 800 ft below the surface, and contains only brackish and salt water.

Water problems now being faced by South Dade County are saltwater intrusion into the Biscayne Aquifer, contamination of the aquifer by recharge from canal and surface waters, and depletion of the aquifer. The demand for water to serve the residents of the community has lowered the water table in the Everglades and threatens the natural flora and fauna there.

In an area of such low-lying topography, drainage problems are serious as well. Flooding is always a potential hazard. Drainage canals have helped to control the flooding, but the canals affect adversely the ecosystem of the Everglades by diverting the swamp's freshwater source. To prevent drainage problems in urban areas, building codes that require a minimum elevation for lots and streets have been established. The consequent storm runoff, which results from urbanization, increases the flooding from high tides caused by hurricanes.

South Dade County faces the environmental problems that plague all urban areas, plus the problems created by the county's particular physical characteristics. Changes are occurring rapidly as the natural forests and the Everglades are being converted for agricultural use, and eventually to commercial and residential development.

PROBLEMS

1 Show with a series of arrows the original surface drainage patterns in southern Florida.

2 What effect has the construction of canals had on this drainage?

3 What areas would be most susceptible to damage from hurricanes and heavy rainfall?

4 What effect would the Keys have on hurricanes approaching from offshore?

5 What particular problems need to be considered before building on a limestone terrain?

6 Why has most of the urbanization of this area been concentrated in its present location?

7 How would you expect the pattern of urbanization to appear 20 years hence?

8 What would be the best sites for the disposal of solid waste? Why?

9 What problems are likely to result from excessive pumping of ground water?

Enhanced Landsat image

0 10 miles

Exercise 10

VALLEY GLACIATION

OBJECTIVE

To recognize the types of landforms developed by valley glaciers and to understand the processes responsible for their development.

MAIN CONCEPT

Valley or alpine glaciers are systems of moving ice that flow down preexisting stream valleys. As the ice moves, it erodes the landscape, creating deep U-shaped valleys, sharp interstream divides, cirques, horns, and hanging valleys. Sediment deposited near the end of the glacier typically forms terminal moraines or is reworked by meltwater and deposited as outwash.

SUPPORTING IDEAS

1 During a glacial epoch, many thousands of miles of ice flow through systems of valley glaciers and erode the former stream valleys by abrasion and glacial plucking.

2 Frost action and mass movement along the valley walls deliver large amounts of rock debris to the glacier. The debris is then transported as lateral and medial moraines.

3 Deposition occurs at the end of the glacier, where the rate of melting exceeds the rate of ice flow. Terminal moraines and outwash plains are major depositional features.

DISCUSSION

The series of diagrams in Figure 10.1 illustrates how valley glaciers modify a landscape previously sculptured by running water. Before glaciation (Figure 10.1A), the topography is characterized by V-shaped stream valleys and rounded hills. When one looks up the major valleys, spurs (or ridges) appear to overlap as the stream curves and bends. Valley size is proportional to the size of the stream that flows through it. At stream junctions, tributaries join the major streams without waterfalls or rapids.

During glaciation (Figure 10.1B), snowfields form in the high mountain ranges, and glacier systems expand down the major stream valleys. Glacial erosion by abrasion and plucking, together with frost action, tend to produce sharp, angular landforms. Bowl-shaped depressions, called *cirques,* develop at the glacier head and erode headward. Where several cirques merge, a sharp, angular peak, called a *horn,* is produced.

When the glacial period ends and the ice melts (Figure 10.1C), the topography is strikingly different from the preglacial landscape. Divides between glacial valleys develop sharp ridges called *arêtes.* Valleys that have been glaciated are typically straight and U-shaped in cross section. Major glacial valleys are deep, and spurs are truncated. Glacial tributaries form hanging valleys where they enter the larger glacial valleys. Debris deposited at the end of the glacier forms *moraines,* and meltwaters rework glacial sediment to form an *outwash plain.*

A. PREGLACIATION: Topography before glaciation is characterized by V-shaped stream valleys, overlapping spurs, and rounded hills.

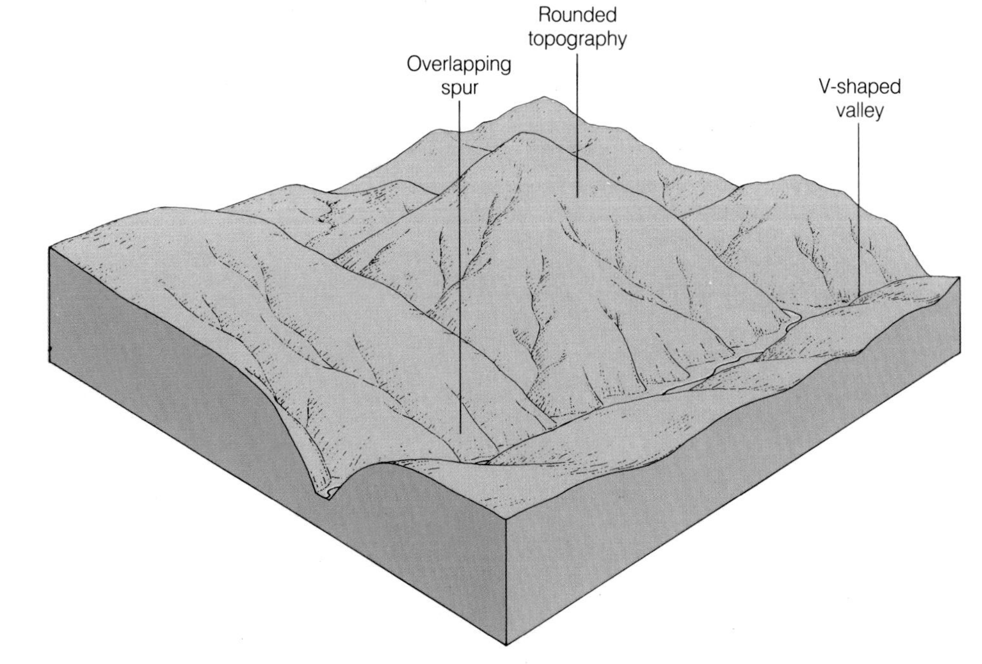

B. GLACIATION: Valley glaciers develop from snowfields in the high peaks and expand down stream valleys. Major glaciers thus have a network of tributaries that follows the drainage system.

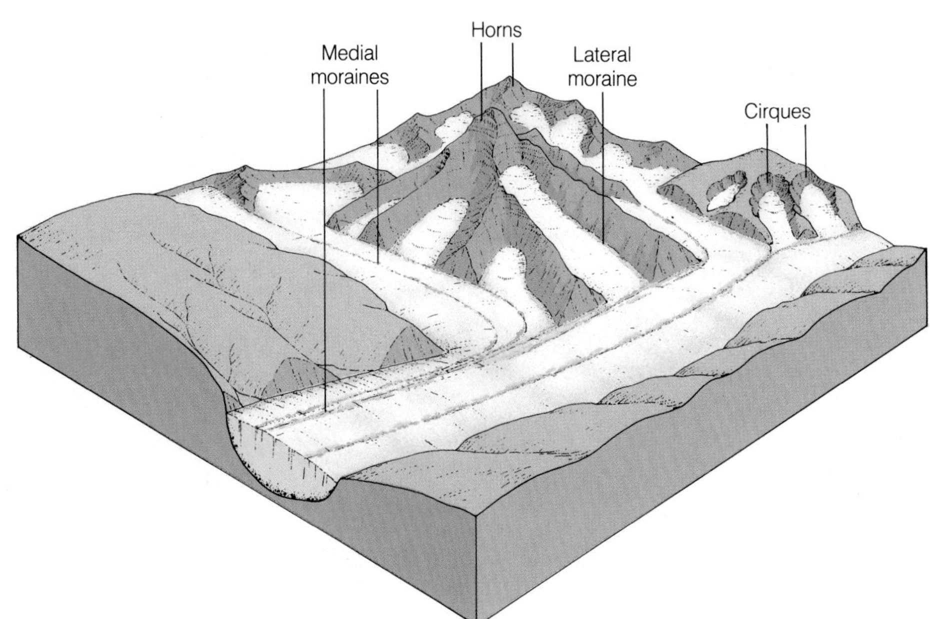

C. POSTGLACIATION: Broad, deep, U-shaped valleys are the most characteristic landform developed by valley glaciation. Cirques, horns, and arêtes are glacial features that create spectacular scenery in the highlands. Hanging valleys, often with high waterfalls, occur where tributaries enter the main valley.

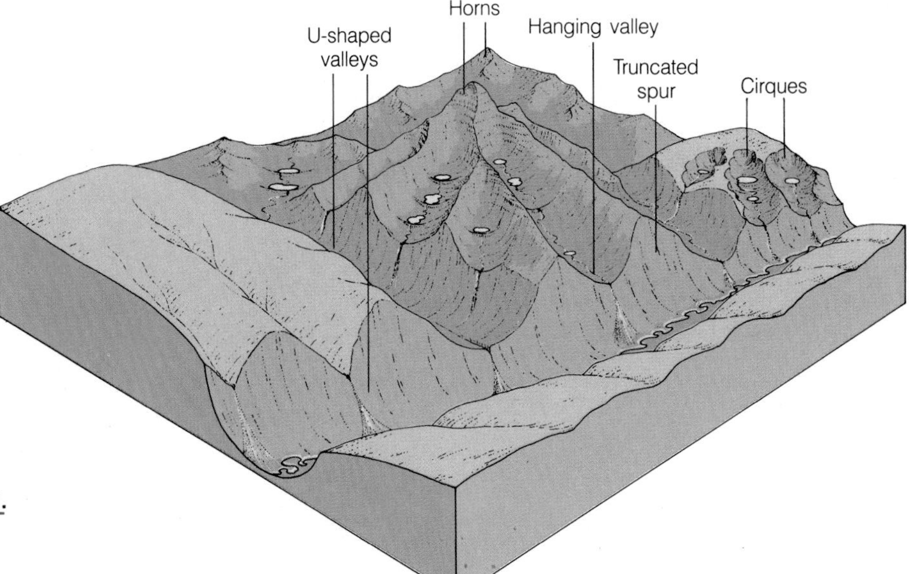

FIGURE 10.1

LANDFORMS DEVELOPED BY VALLEY GLACIATION.

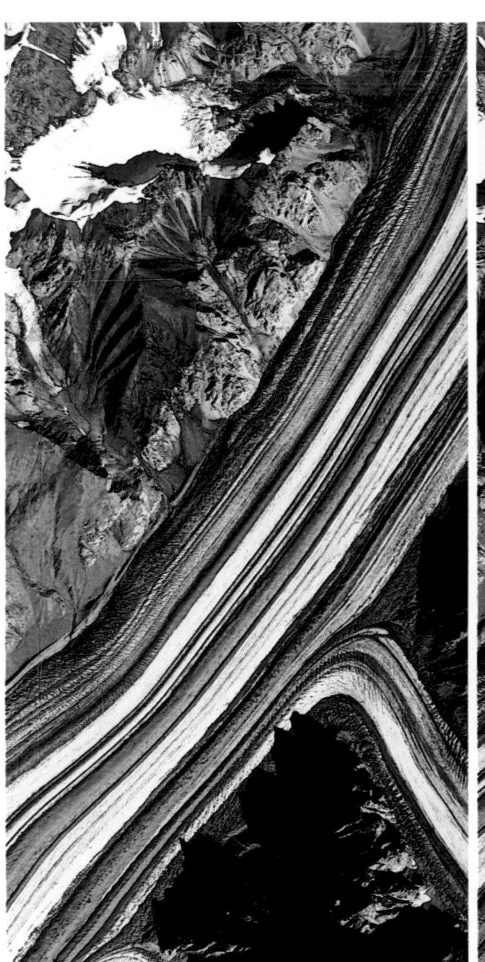

Color infrared photo 0 1 mile

FIGURE 10.2: Valley Glaciers—Alaska

PROBLEMS

1 What evidence in this photograph indicates that lateral and medial moraines are derived largely from mass movement on the valley walls, rather than from glacial plucking?

2 What evidence indicates that the various tributary glaciers flow at different rates?

3 Explain the difference in color and texture of the medial moraines.

4 What evidence shown in these photos indicates that glacial ice flows?

Color infrared photo 0 1 mile

FIGURE 10.3: Valley Glaciation—Alaska

PROBLEMS

1 Several different terminal moraines are distinguishable by differences in form, texture, and growth of vegetation (red color). Outline the major terminal moraines and explain briefly the glacial history they record.

2 Explain why the end of this glacier is completely covered with moraine sediments.

3 How can you distinguish between older and younger lateral moraines in this area?

4 What is the origin of the small lakes near the bottom of the photographs?

FIGURE 10.4: Cordova, Alaska

CI 100 ft 0 1 mile

PROBLEMS

1 Study the landforms and label the following features: (a) outwash plain, (b) arêtes, (c) recessional moraines, (d) medial moraines, (e) hanging valleys, (f) horns, (g) braided streams, (h) cirques, and (i) ice falls.

2 Draw arrows on the map to show the flow direction of the tributaries to the Scott Glacier.

3 Draw a topographic profile across the valley and approximately 1 mi south of the end of Scott Glacier. How does a glacial valley differ from a stream valley?

4 Can we reasonably assume that a glacier modifies the valley through which it flows? What field evidence can you cite to support your answer?

5 Draw a longitudinal profile up Scott Valley to the front of the glacier. If the gradient were projected up the valley under the glacier, what would be the approximate thickness of the ice at the 1800-ft contour line near the middle of Scott Glacier?

FIGURE 10.5: Glacial Topography—Alaska

Color infrared photo 0 1 mile

During the Ice Age, extensive snowfields and alpine glaciers were prevalent in the high mountain ranges of the world. Many of these glaciers have now disappeared, but they have left a clear depositional and erosional record of their former activity. Careful study of glaciated landscapes permits us to outline the areas of former glaciers and to interpret the Pleistocene history of the regions.

PROBLEMS

1 Study the numerous hanging valleys, the linear grooves and striations on the valley walls, and changes in the morphology of the valley walls from the floor to the tops of the spurs. On the basis of these features, draw the upper level of the glacier system that once existed in this area.

2 Label all glacial landforms.

3 Explain why the main stream meanders along the valley floor.

4 How has the topography of this area been modified since the glaciers receded?

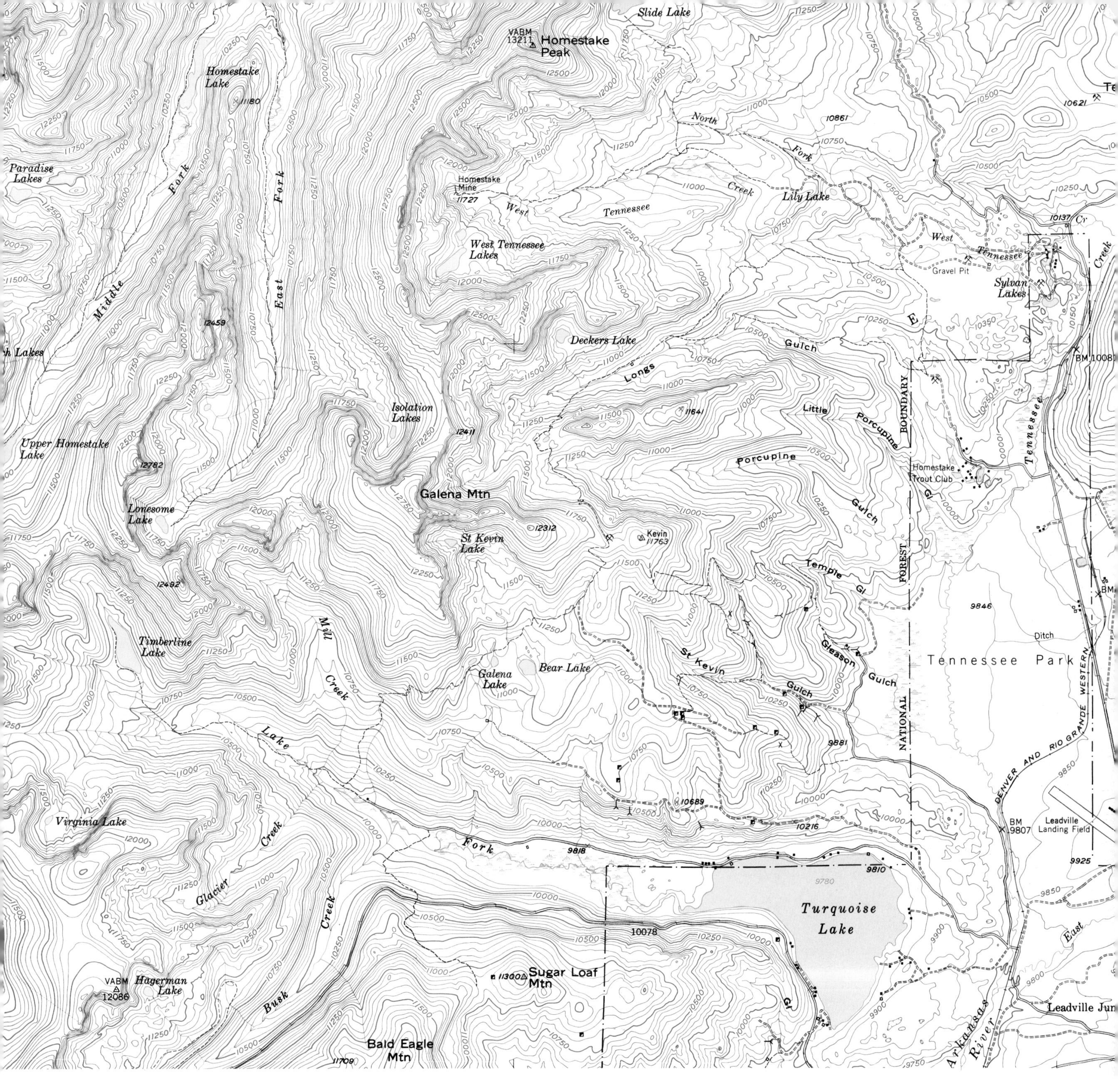

FIGURE 10.6: Holy Cross, Colorado

CI 50 ft 0 1 mile

PROBLEMS

1 Study carefully the landforms shown on the map, and color those areas in blue that you believe were formerly covered by glaciers. (A good approach is to first outline the major cirques, then trace the margins of the U-shaped valleys downstream.)

2 Study the landforms near the lower end of the major valleys. Locate the moraines, and color them green.

3 Compare the glacial features in this area with those in Figure 10.5. List the glacial features that are common to both areas, then list those that are unique to the Holy Cross area.

4 What is the origin of the small lakes located at the head of many of the major valleys?

5 Why have swamps and lakes developed in the central part of many of the major valleys?

Exercise 11

CONTINENTAL GLACIATION

OBJECTIVE

To recognize the major types of landforms created by continental glaciers and to understand the processes responsible for their development.

MAIN CONCEPT

A continental glacier is a system of flowing ice, up to 2 mi thick, that covers large parts of a continent. As the ice within the glacier moves, it erodes the soil and bedrock, and transports the erosional debris toward the margins of the ice, where the debris is deposited in a variety of landforms.

SUPPORTING IDEAS

1 Much of the sediment load of continental glaciers is deposited at the ice margins to form end moraines.

2 Drumlins, eskers, and ground moraines form beneath the ice.

3 Meltwater from the glacier may rework large amounts of glacial debris and deposit it in an outwash plain in front of the ice sheet.

DISCUSSION

Continental glaciers originate in polar regions where precipitation is sufficient to build up and maintain a thick body of ice. Their margins are commonly irregular or lobate, because the advancing ice front moves more rapidly into lowlands than over higher terrain. The thickness of a continental glacier commonly exceeds 10,000 ft but is rarely more than 15,000 ft. An important effect of continental glaciation is that the weight of the ice depresses the earth's crust, so the land surrounding the ice sheet slopes toward the glacier. This sloping of the terrain toward the ice sheet may produce lakes along the ice margins or permit an arm of the sea to invade temporarily and fill the depression.

The idealized diagrams in Figure 11.1 illustrate the major landforms developed by continental glaciation and the processes by which they are produced. A number of significant landforms result from the geologic processes operating at the margins of continental glaciers, for it is here that the flowing ice leaves the glacial system by melting and evaporation (Figure 11.1A). Much of the sediment carried by the glacier is deposited along the ice margins as moraine, or is reworked by meltwater and deposited as outwash. The sediment deposited at the ice margins follows the lobate outline of the glacier to form terminal moraine, recessional moraine, and interlobate moraine. Lakes that form in the depression next to the ice sheet margins receive sediment from the meltwater streams. This sediment accumulates as deltas or may be dispersed over the lake bottom. Elsewhere, meltwater issues from tunnels beneath the ice, and forms braided streams that deposit stratified fluvial sand and gravel on the outwash plain. Blocks of ice broken off from the glacier may be partly buried by sediment as the glacier begins to recede. Moraine and outwash sediment may be reshaped by subsequent advances of the glacier to produce streamlined hills, called *drumlins.*

When the glacier retreats, ridges of moraine mark the former positions of the ice margins, and well-sorted, stratified clay and silt mark the sites of former lakes (Figure 11.1B). The outwash plain is characterized by deposits of fluvial sand and gravel. Depressions, known as *kettles,* form where partly buried ice blocks have melted. Sinuous ridges, called *eskers,* result from sediment deposited on the floor of former ice tunnels, and numerous lakes fill depressions in the ground moraine. Preglacial drainage has been greatly modified or obliterated, so the postglacial drainage is completely deranged.

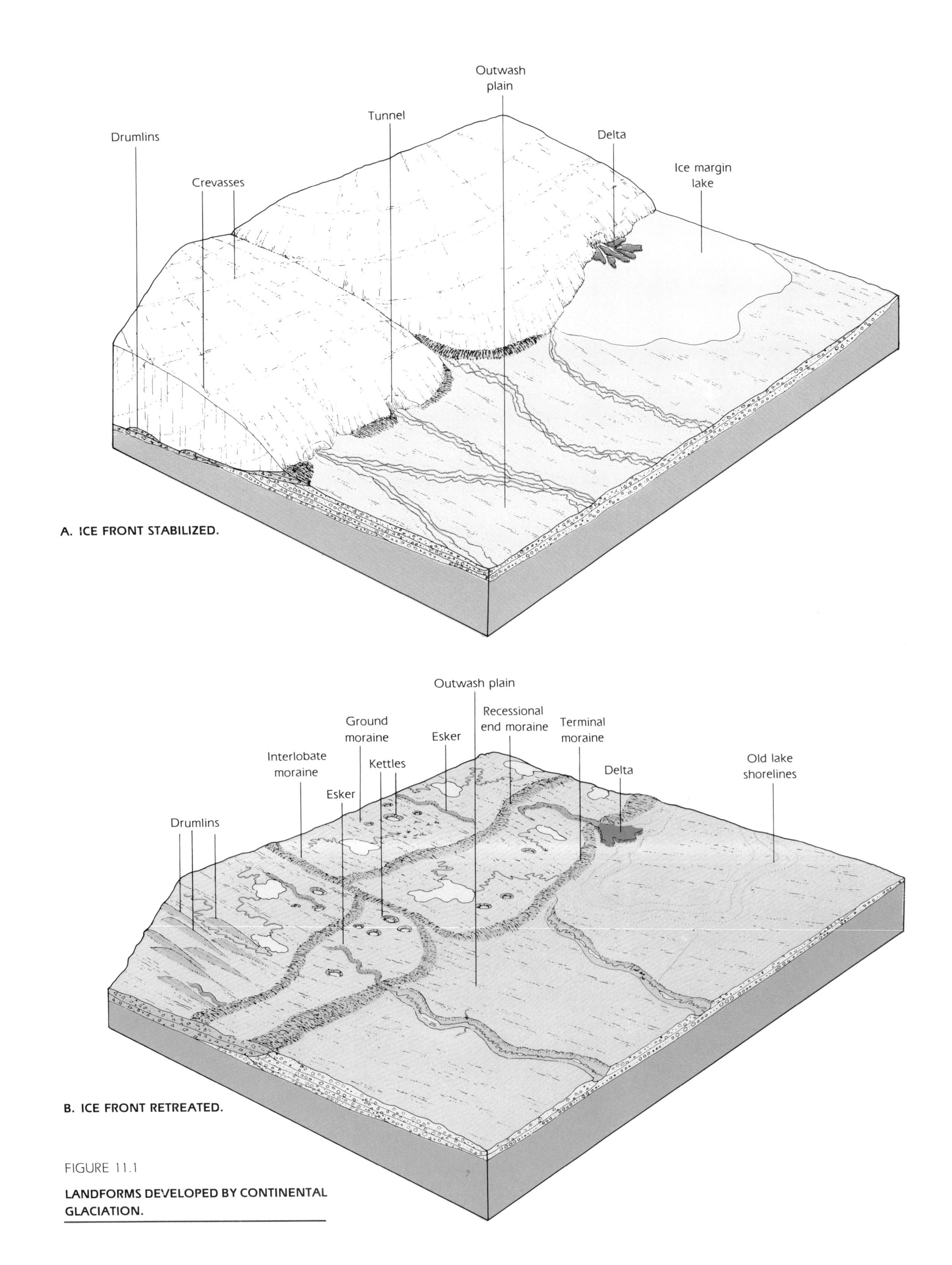

FIGURE 11.1

LANDFORMS DEVELOPED BY CONTINENTAL GLACIATION.

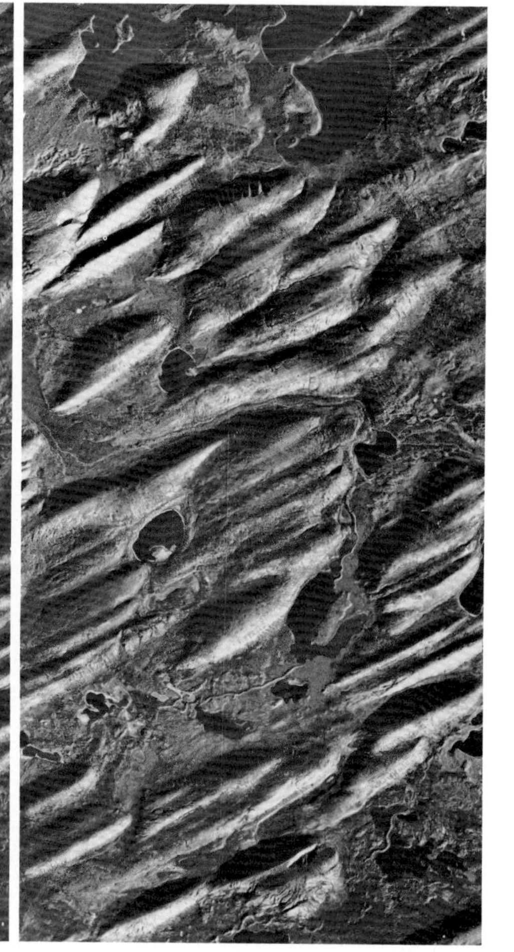

FIGURE 11.2: Drumlins—Canada

PROBLEMS

1 Show by arrows the direction of glacier movement in this area.

2 Consider the shape, orientation, and grouping of the drumlins shown here. Cite evidence to support the conclusion that drumlins result from glaciation and not from some other geologic process, such as wind action.

3 To what extent has stream erosion modified the drumlins?

4 Locate the low, sinuous ridges trending roughly parallel to the drumlins, and color them yellow. Explain how these features originated.

5 Many cities use sanitary landfills to dispose of solid waste. What problems might arise if this were the method of waste disposal in a drumlin area?

0 1 mile

FIGURE 11.3: Glacial Landforms—North Dakota

PROBLEMS

1 Draw a line along the boundary between the end moraine and ground moraine. What topographic features characterize each area?

2 What is the origin of the lakes in the northeast part of the area?

3 Draw several arrows on the photograph to show the direction of ice movement. Explain how you reached this conclusion.

4 Why is the drainage so poorly developed in this area?

FIGURE 11.4: **Jackson, Michigan**

PROBLEMS

1 Locate this area on the map on page 113. Study the landforms in this area and identify (a) eskers, (b) ground moraine, and (c) terminal moraine. Draw the approximate boundaries between these features. Color the ground moraine light green, the terminal moraine dark green, and the eskers orange.

2 The poorly developed drainage and the extensive swamp and marsh are typical of areas where continental glaciation has occurred. To emphasize the lack of integration in this drainage system, use a dark blue pencil to trace the drainage of the Grand River and its tributaries. How does this drainage pattern compare with the drainage patterns of the river systems studied in Exercise 8?

3 Is there any evidence from the Jackson drainage pattern that the landforms are primarily depositional and that they have been only slightly modified by erosion?

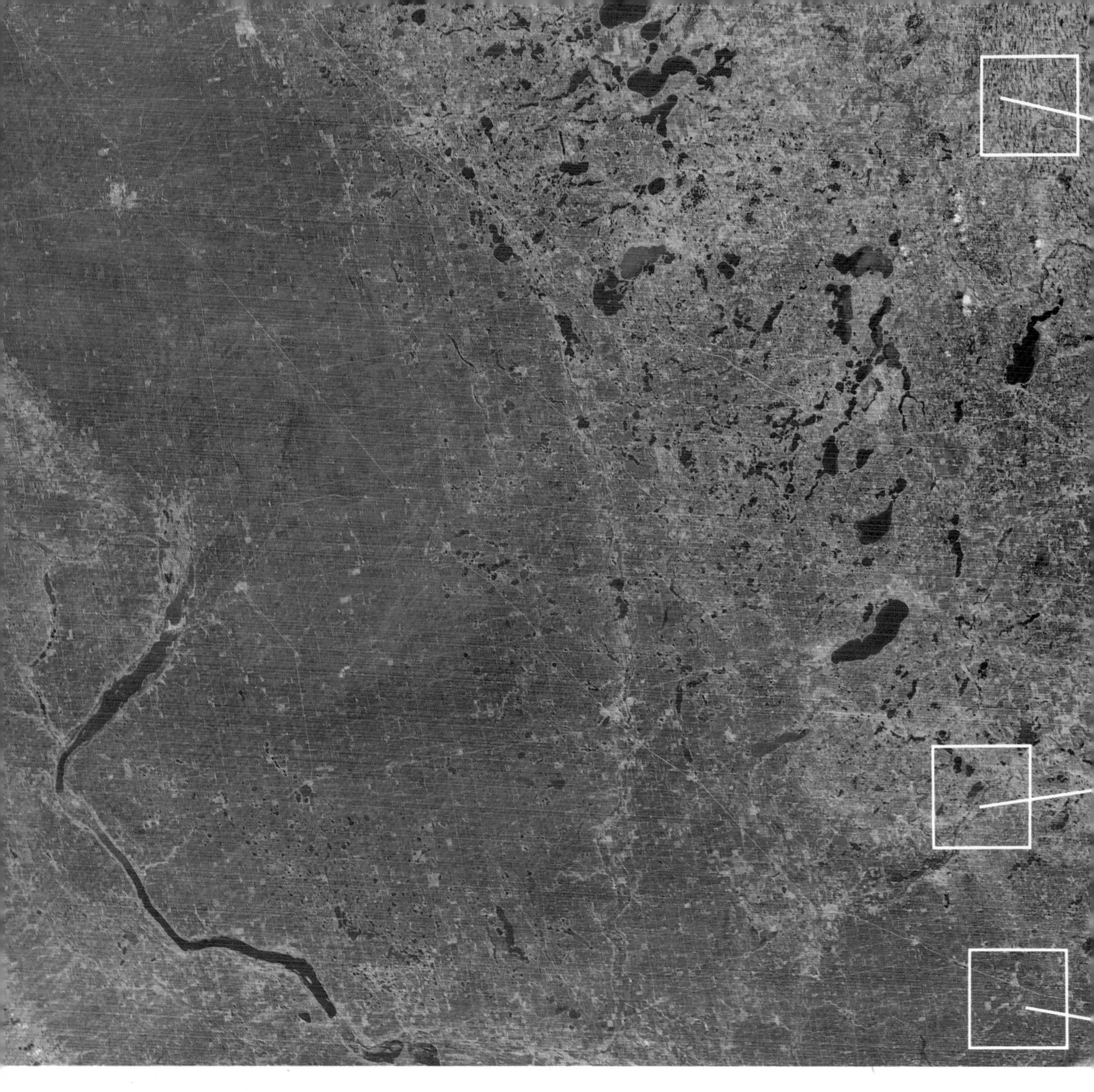

FIGURE 11.5: Glacial Topography—North Dakota

A

Landsat image 0 10 miles

PROBLEMS

1 On the Landsat image (Figure 11.5A), draw a line along the contacts between lake sediment and end moraine, and between end moraine and ground moraine. What topographic features characterize each area?

2 What is the origin of the numerous small lakes in the upper central part of the image?

3 What is the nature and origin of the linear fabric in the extreme northeast corner of the Landsat image? (Study also the high-altitude, color infrared photograph in Figure 11.5B).

4 Describe the topography shown in the high-altitude photograph in Figure 11.5C. How is this general topography apparent on the Landsat image (Figure 11.5A).

Color infrared photos

0 1 mile

B

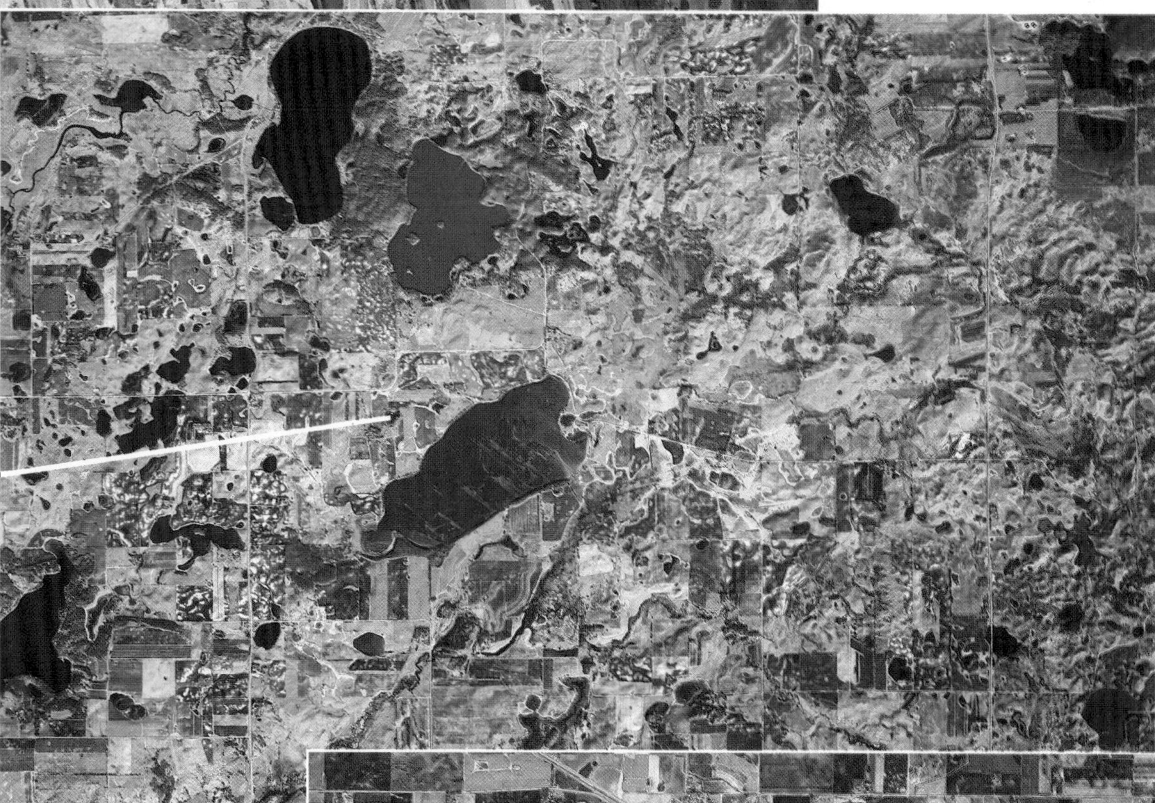

0 1 mile

C

5 Describe the topography shown in Figure 11.5D. How does it differ from that shown in Figure 11.5C?

6 On the basis of the glacial landforms shown in Figure 11.5A, construct a generalized paleogeographic map outlining the former position of the ice front. Use arrows to indicate the direction of ice movement.

D

0 1 mile

FIGURE 11.6: Glacial Features—Great Lakes Area

PROBLEMS

1 Study the patterns of the major glacial features (i.e., outwash, lakes, and moraines), and note the contact of older and younger deposits that are associated with them. On the basis of these relationships, determine the relative ages of the following: ground moraine or drift, eskers, outwash sediments, and lake sediments.

2 How many glacial stillstands are represented by the end moraines in Ohio and Indiana southwest of Lake Erie? How many stillstands are represented by the moraines south of Lake Michigan?

3 Was the glacial front advancing or retreating when these moraines formed?

4 Compare and contrast the glacial features formed by the lobes of ice that advanced down the areas of Lake Erie, Lake Michigan, and Green Bay into eastern Wisconsin.

5 Note the breaks in many of the end moraines. What is the probable cause of these breaks?

6 In the northeastern Indiana area, study the pattern of outwash sediments (yellow) in the vicinity of the moraines near Union City, Mississinewa, Salomie, Wabash, and Fort Wayne. What does this pattern indicate about the direction of the drainage during glacial times?

7 What effect did the end moraines have on the location of ancient glacial lakes such as lakes Maumee, Whittlesey, Glenwood, and Calumet (indicated by red lines on the map as lacustrine sediments and structures)? These ancient lakes extended beyond the present-day shores of Lake Michigan and Lake Erie.

SYMBOLS

Strandlines

Striation direction

Streamline features

0 25 miles

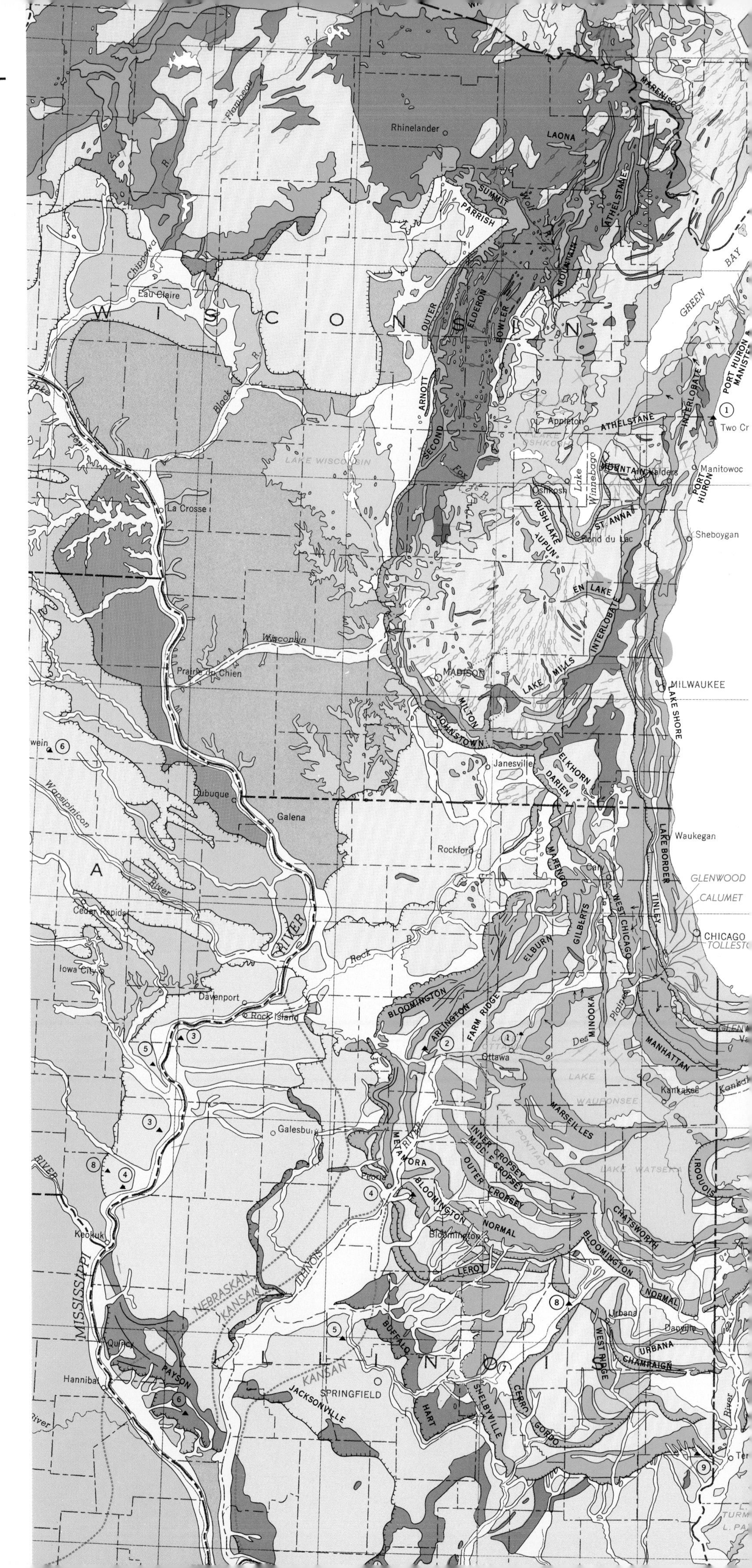

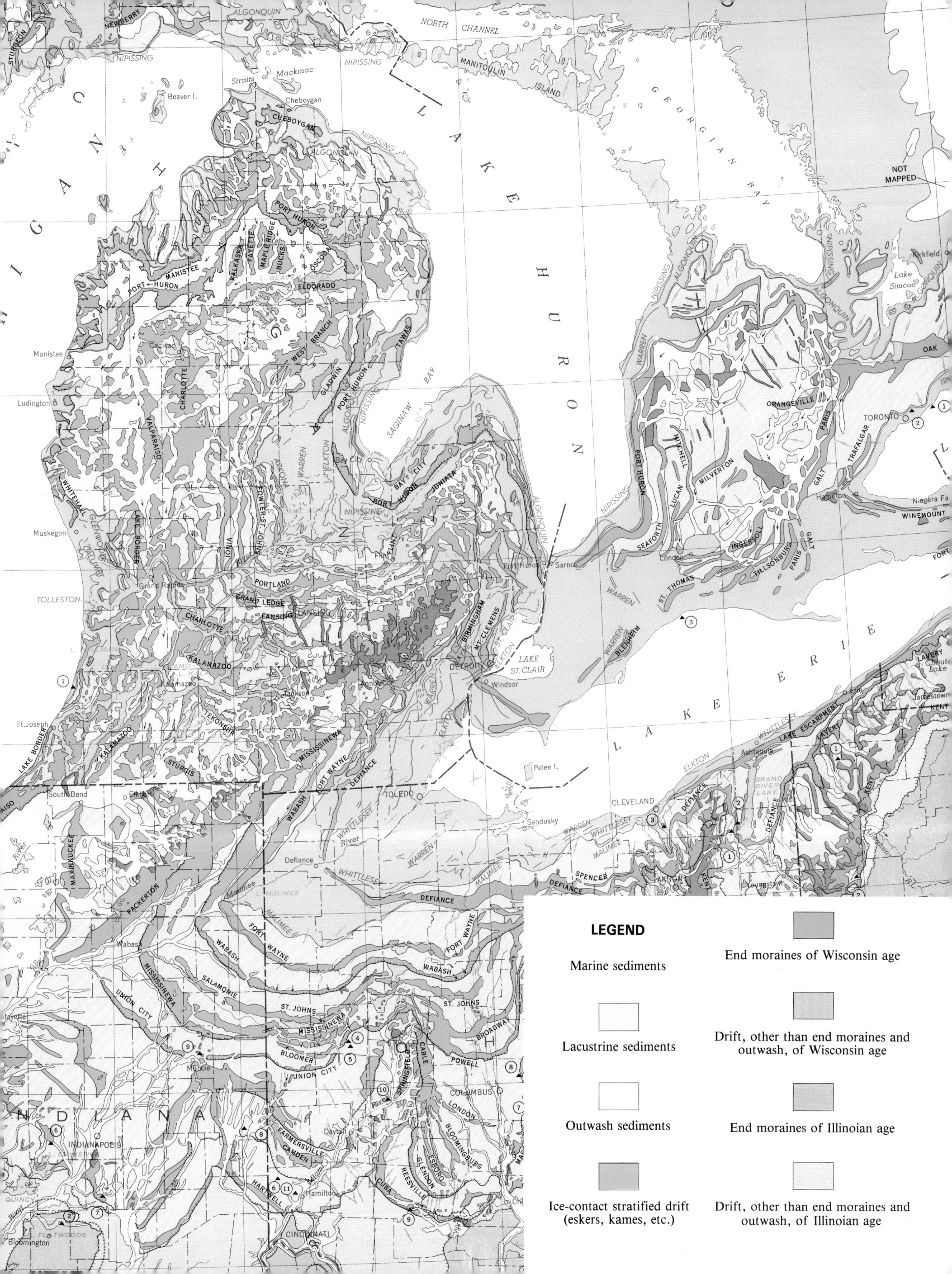
LEGEND
Marine sediments
Lacustrine sediments
Outwash sediments
Ice-contact stratified drift (eskers, kames, etc.)
End moraines of Wisconsin age
Drift, other than end moraines and outwash, of Wisconsin age
End moraines of Illinoian age
Drift, other than end moraines and outwash, of Illinoian age
NORTH CHANNEL
MANITOULIN ISLAND
GEORGIAN BAY
LAKE HURON
SAGINAW BAY
LAKE ERIE
LAKE ST. CLAIR
Straits of Mackinac
Beaver I.
Cheboygan
Manistee
Ludington
Muskegon
Grand Rapids
Kalamazoo
St. Joseph
South Bend
Bay City
Flint
Lansing
Jackson
Ann Arbor
Port Huron
Sarnia
London
Hamilton
Toronto
Niagara Falls
Kirkfield
Lake Simcoe
DETROIT
Windsor
Pelee I.
TOLEDO
Sandusky
CLEVELAND
Ashtabula
Erie
Jamestown
Defiance
Wabash
Muncie
INDIANAPOLIS
COLUMBUS
Dayton
Hamilton
CINCINNATI
Bloomington
INDIANA
NOT MAPPED
NIPISSING
ALGONQUIN
WARREN
WHITTLESEY
MAUMEE
ELKTON
TOLLESTON
PORT HURON
MANISTEE
KALKASKA
FAYETTE
MAPLE RIDGE
BUCKS
OSCODA
ELDORADO
WEST BRANCH
TAWAS
GLADWIN
CHARLOTTE
VALPARAISO
WHITEHALL
LAKE BORDER
IONIA
FOWLER-ST. JOHNS
PORTLAND
GRAND LEDGE
LANSING
KALAMAZOO
TEKONSHA
STURGIS
MISSISSINEWA
FORT WAYNE
DEFIANCE
WABASH
BIRMINGHAM
MT. CLEMENS
JUNIATA
BAY CITY
MAXINKUCKEE
PACKERTON
SALAMONIE
UNION CITY
ST. JOHNS
BROADWAY
BLOOMER
POWELL
SPRINGFIELD
CABLE
LONDON
BLOOMINGBURG
FARMERSVILLE
CAMDEN
HARTWELL
CUBA
REESVILLE
GLENDON
ESBORO
SPENCER
KENT
LAVERY
LAKE ESCARPMENT
GRAND RIVER LAKE
SEAFORTH
LUCAN
MITCHELL
MILVERTON
ORANGEVILLE
PARIS
GALT
TRAFALGAR
INGERSOLL
TILLSONBURG
ST. THOMAS
BLENHEIM
WINEMOUNT
OAK

Exercise 12

SHORELINE PROCESSES

OBJECTIVE

To recognize the major types of coasts and to understand the processes involved in their development.

MAIN CONCEPT

The configuration of a shoreline changes as a result of erosion and deposition until it reaches a state of equilibrium.

SUPPORTING IDEAS

1 Wave refraction is a fundamental process by which energy is concentrated on headlands and dispersed across bays.

2 Longshore drift is one of the most important processes of sediment transport along a coast.

3 Erosion along a coast tends to develop sea cliffs by the undercutting action of waves and longshore currents. As the cliff recedes, a wave-cut platform develops.

4 Most coasts can be classified on the basis of the geologic processes that were most significant in determining their configuration.

5 The worldwide rise in sea level associated with the melting of the continental glaciers drowned many coasts. The configuration of our present-day coasts may therefore be due to a variety of geologic processes that operated on the land before sea level rose, and *not* to the marine processes operating today.

DISCUSSION

Wind-generated waves provide most of the energy for shoreline processes. The energy moves out from the storm area where the waves are generated. As the waves approach a headland, they are commonly bent or *refracted* so their crest line tends to become parallel to the coastline. Wave refraction occurs along an irregular coast because, in shallow water, part of the wave begins to "drag bottom" and slow down, while the remainder of the wave in deeper water moves forward at normal velocity. As a result, wave refraction concentrates energy on headlands and disperses it in bays. One way to visualize the concentration and dispersal of energy resulting from wave refraction is to draw equally spaced rays (lines perpendicular to wave crests) and to note how the spacing changes as the waves are refracted. Energy will be concentrated in zones where the rays are close together and dispersed where the rays are farther apart.

Longshore drift is generated where waves advance obliquely to the shore. As the waves strike the shore, water and sediment moved by the breakers are transported obliquely up the beach in the direction of wave advancement. When the energy is spent, water and sediment move as backwash directly down the beach, perpendicular to the shore. The sediment thus moves down the beach in a zigzag path.

As a result of wave erosion and deposition, the configuration of a coast evolves until energy is distributed evenly along the coast so that neither large-scale erosion nor deposition is occurring. Such a coast is called a *shoreline of equilibrium* and is characterized by a smooth, gently curving beach. Few present-day coasts approach this state of equilibrium, however, because most shorelines all over the world have been affected greatly by the rise in sea level associated with the melting of the glaciers during the last 30,000 years. Virtually all coastal areas have been submerged, and river valleys have become embayed or drowned. The configuration of many coasts, therefore, is not the result of marine processes at all. Rather, shoreline configuration is the result of the terrestrial geologic processes that operated in the area before sea level rose. Some coasts, however, do show evidence of significant change since present-day sea level was established, because waves and currents have rapidly modified the preexisting topography.

A meaningful classification of coasts is based on the geologic processes that are responsible for the configuration of the shoreline. Two principal subdivisions are recognized: (1) *primary coasts,* formed by terrestrial processes, and (2) *secondary coasts,* formed by marine processes. Various subdivisions of these are shown in Figure 12.1.

PRIMARY COASTS (CONFIGURATION DUE TO TERRESTRIAL PROCESSES).

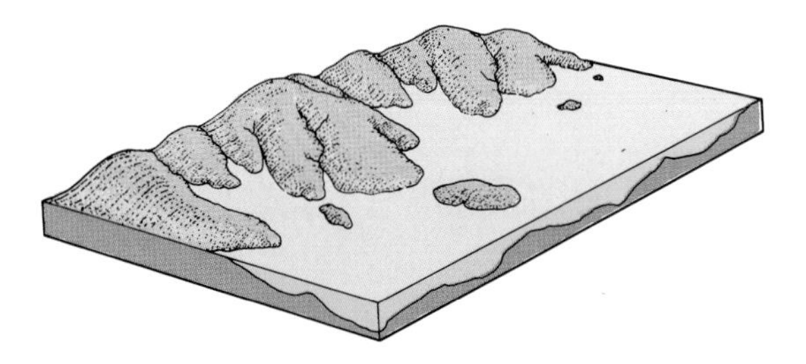

A. STREAM EROSION: The configuration of the landscape was developed by stream erosion when the area was not covered by the sea. A subsequent rise in sea level, resulting from the melting of Pleistocene glaciers, drowned the river valleys.

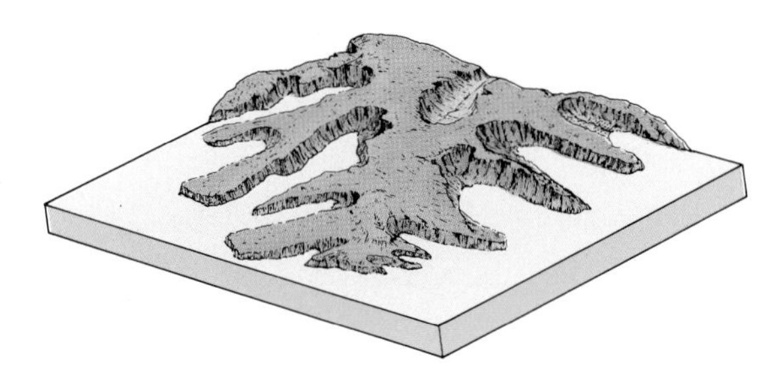

B. GLACIAL EROSION: When glaciers melt and sea level rises, long arms of the sea extend many miles up the deep glacially formed valleys. These drowned glacial valleys, called fiords, are not modified greatly by wave action.

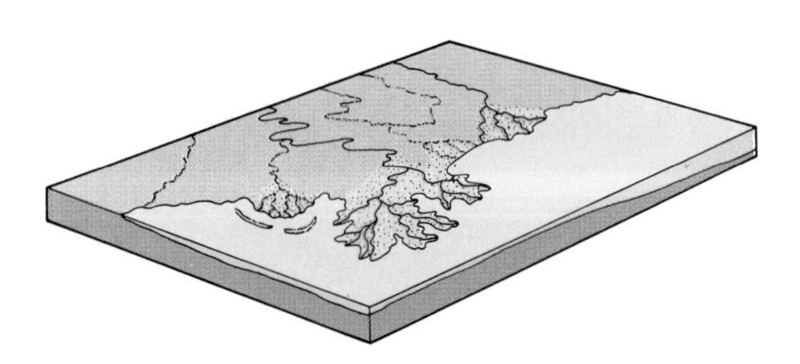

C. STREAM DEPOSITION: Where a major river enters the sea, it commonly deposits more sediment than waves and currents can carry away, so new coastal land is added in the form of a delta. Deltaic coasts develop best in protected bays, where wave action is minimal.

SECONDARY COASTS (CONFIGURATION DUE TO MARINE PROCESSES).

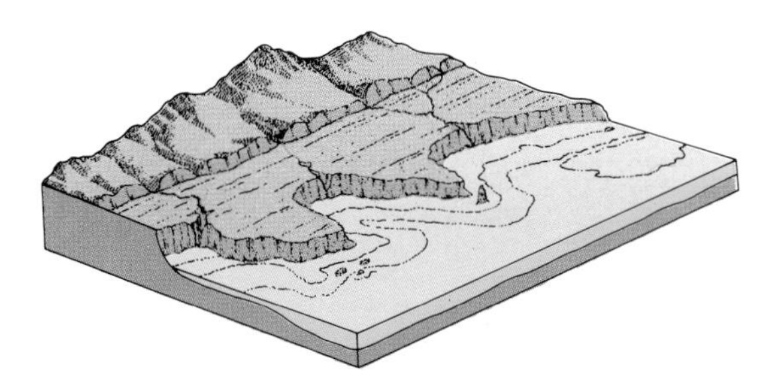

D. WAVE EROSION: In areas where the rock is relatively uniform in composition, wave erosion forms straight sea cliffs. If different rock types are present, wave erosion produces bays in the softer material and leaves the resistant rock projecting into the sea as rocky points.

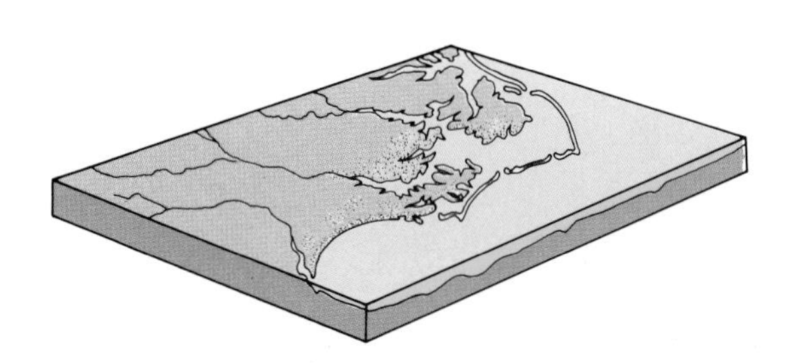

E. MARINE DEPOSITION: Coasts built by sediment deposited by waves and currents are readily recognizable by beaches, barrier islands, spits, and bars.

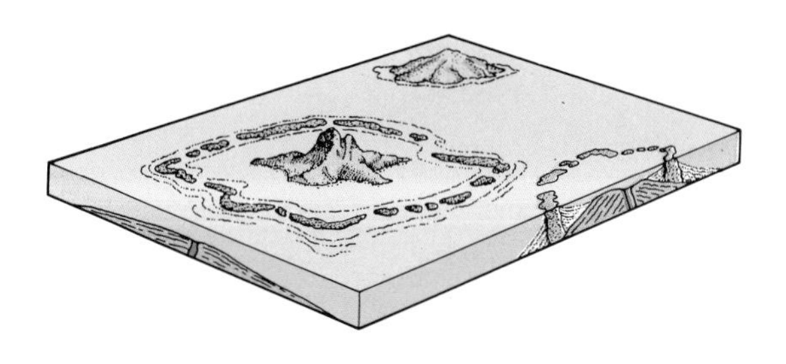

F. ORGANICALLY BUILT COASTS: Configuration of some shorelines is controlled by growth of organisms such as coral reefs and mangrove trees.

FIGURE 12.1

MAJOR TYPES OF SHORELINES.

FIGURE 12.2: Coast—Southern California

0 0.5 miles

PROBLEMS

1 Study this area and label all of the coastal landforms that you can identify.

2 Assuming that the present wave pattern is normal show with arrows the prevailing pattern of longshore drift.

3 Note the five jetties (long, narrow seawalls) constructed from headland and positioned up, along the main beach. How do these structures affect longshore drift?

4 Show how the energy between points *A*, *B*, *C*, and *D* will be distributed along the coast as a result of wave refraction. (A good approach to solving this problem is to draw lines along the crests of the major waves to emphasize their regional pattern, then draw rays from points *A*, *B*, *C*, and *D* toward the shore.)

5 Show the position of the ancient shoreline that existed during the time when the wide wave-cut platform (on which the air strip was built) was formed.

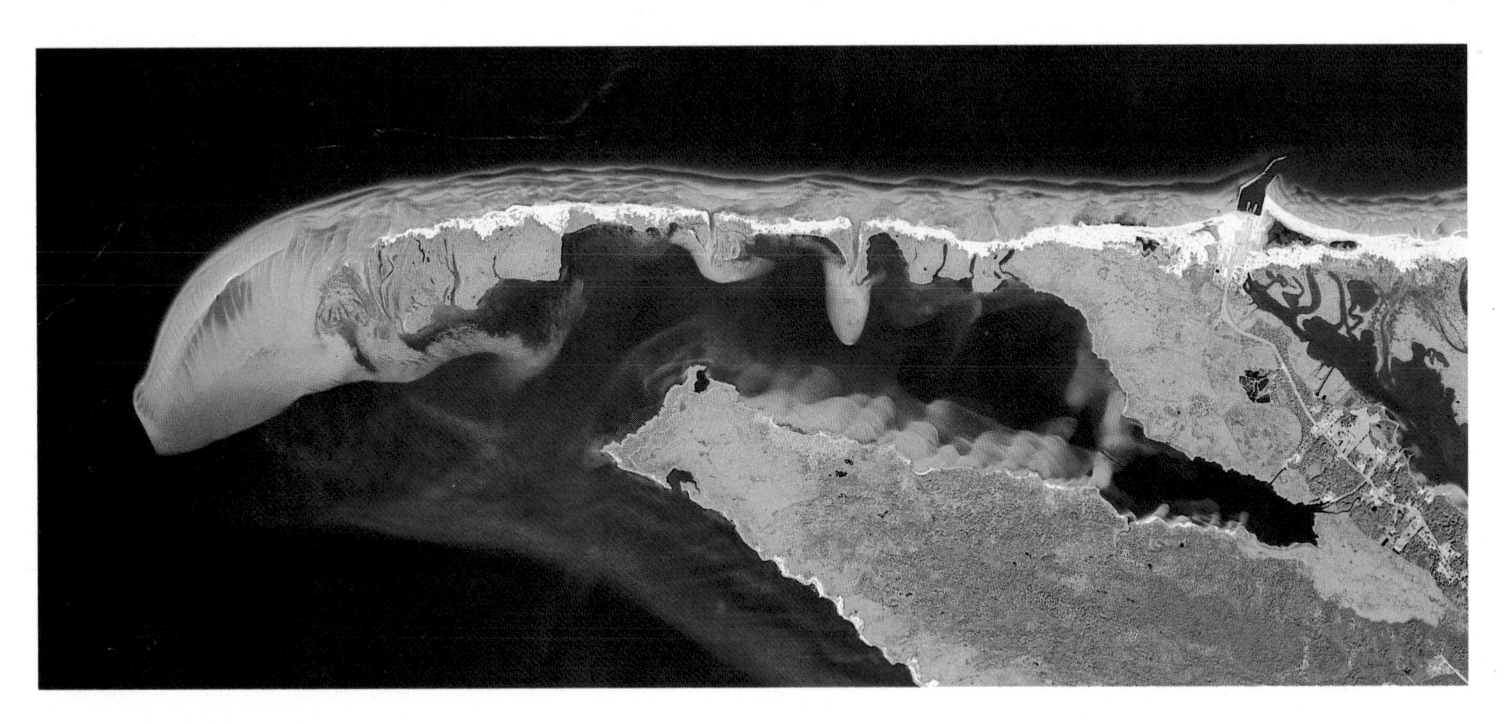

FIGURE 12.3: Spit—North Carolina

Color infrared photo 0 2 miles

PROBLEMS

1 Show the orientation of the dominant wave pattern along this coast.

2 What evidence suggests that longshore drift occurs underwater as well as on the beach?

3 Note that the beach is broken by numerous inlets connecting the bay with the open ocean. What processes are probably responsible for these features?

4 Assuming that present processes continue, show by a series of lines how this coast will evolve.

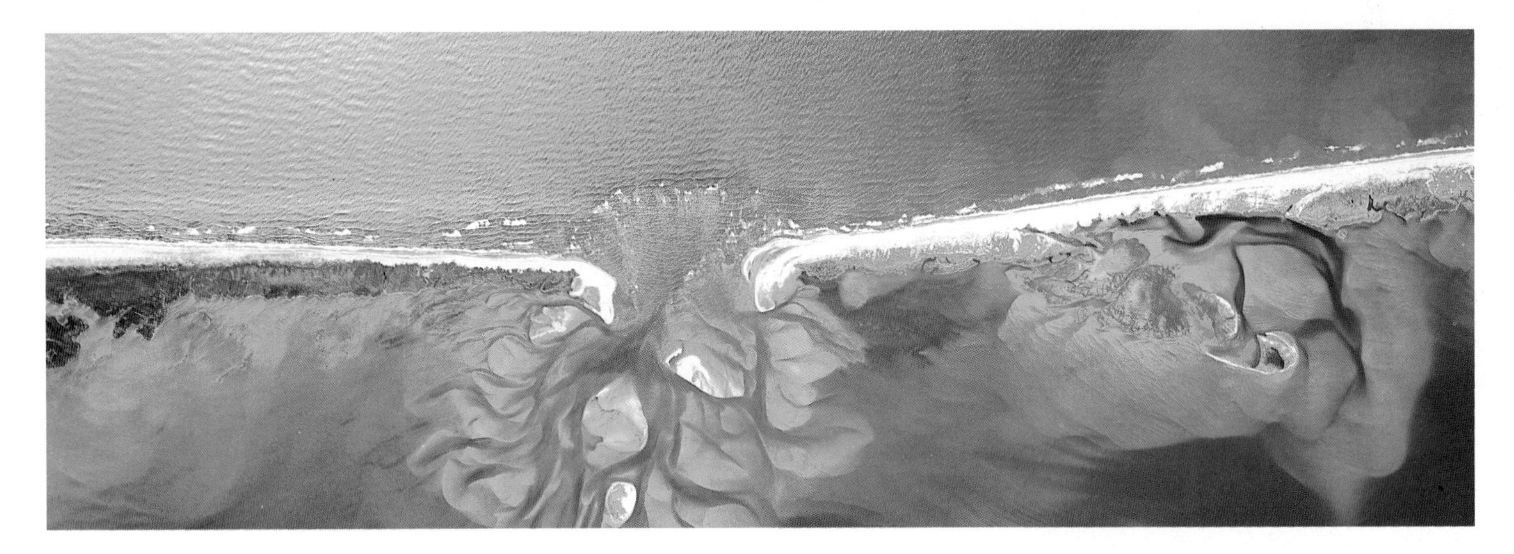

FIGURE 12.4: Coastal Features—North Carolina

Color infrared photo 0 0.5 miles

PROBLEMS

1 The major landforms in this area result from longshore currents, which move sediment along the coast, and from tidal currents, which move sediment in and out of the tidal inlet. Study the shape of the various coastal features, and show by a series of arrows the direction of sediment transport.

2 Why is the tidal delta larger in the lagoon than in the open ocean?

FIGURE 12.5: Newfoundland Coast, Canada

PROBLEMS

1 Study the coast and label the following features

a headlands, **b** sea stacks, **c** wave-cut terrace

d tidal flat, **e** tidal channels, **f** sea cliff.

2 Make a sketch map showing how this coast will appear when it approaches more closely a state of equilibrium.

3 Erosion along a coast produces a variety of landforms, which are soon modified or destroyed. Color the erosional feature that you believe will be the end product of wave erosion on this coast.

4 List all of the landforms that you recognize as the product of tides.

FIGURE 12.6: Coastal Features—Quebec, Canada

0 0.5 miles

PROBLEMS

1 Label the following features and explain their origin: (a) baymouth bar, (b) tidal inlet, (c) tidal delta, (d) tidal channels, (e) lagoon, (f) tidal marsh, (g) bayhead delta, (h) wave-cut platform, and (i) delta.

2 What is the source of the sediment that is being deposited both seaward and landward of the inlet between the spits?

3 How will this coast change with time?

4 Compare this area with that in Figure 12.5. Which area represents the more advanced stage of coastal development?

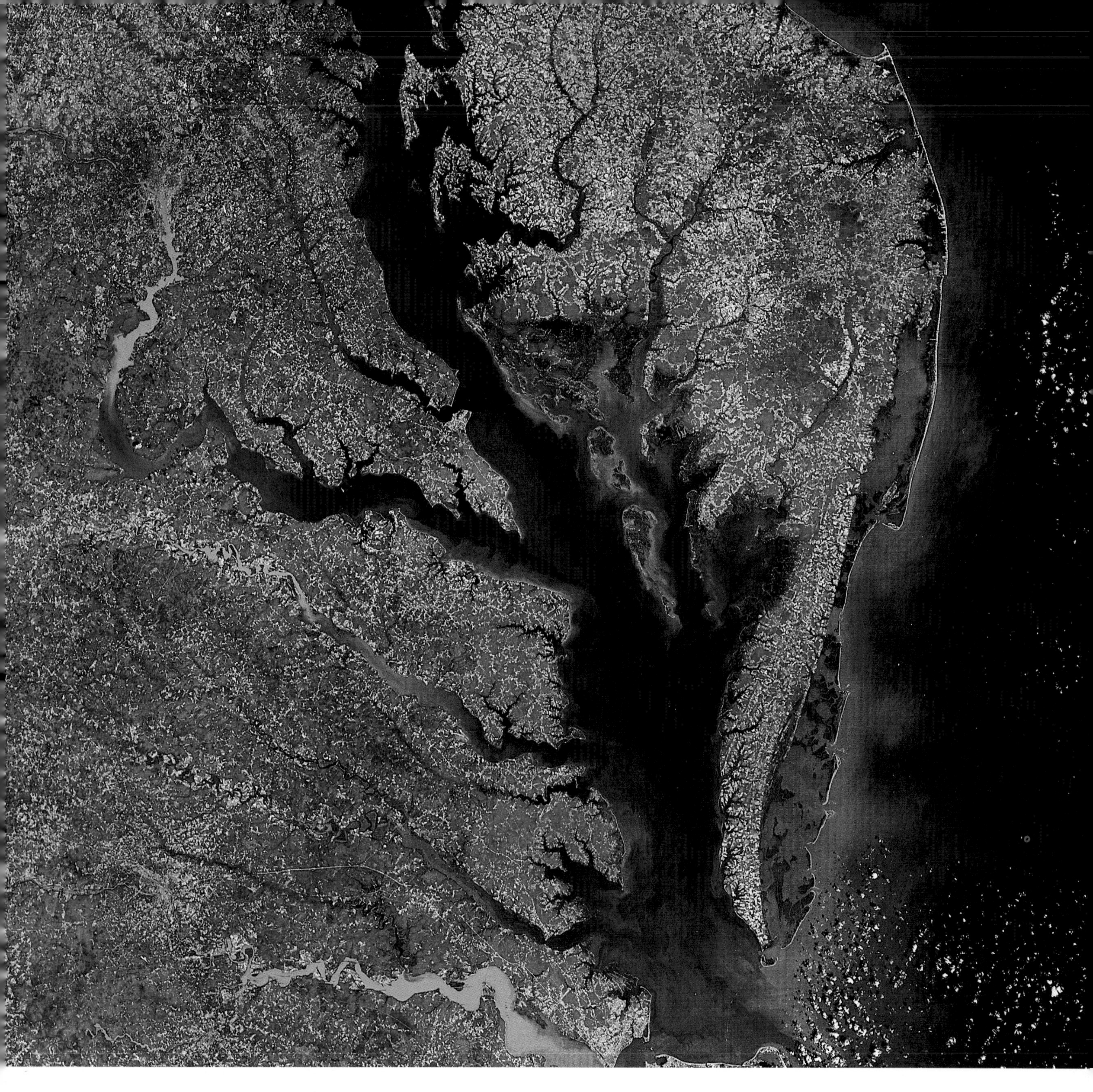

FIGURE 12.7: Chesapeake Bay

Enhanced Landsat Mosaic 0 25 miles

PROBLEMS

1 Explain the origin of the coastline shown in this image.

2 What causes the water in some areas to appear light blue, whereas in other areas the water appears dark blue?

3 What is the origin of the long, narrow bays? How will they be modified with time?

4 Make a series of sketches showing how the Chesapeake Bay area will appear in the future as marine processes and sedimentation in the estuary continue.

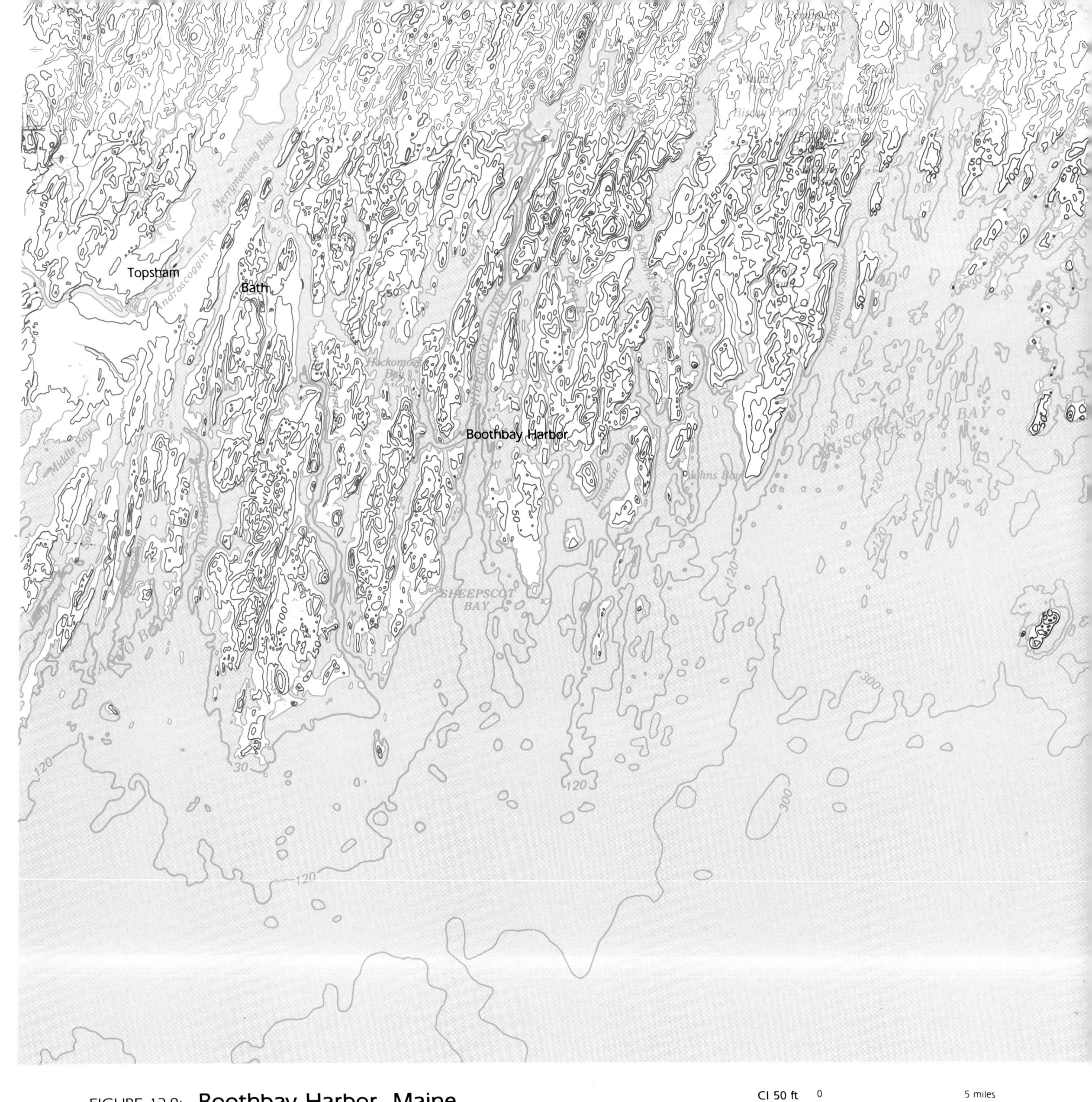

FIGURE 12.8: **Boothbay Harbor, Maine**

CI 50 ft 0 5 miles

PROBLEMS

1 Study the land contours and submarine topography. What evidence is there of a recent rise in sea level?

2 On the basis of the area's topography (both submarine and surface), what has been the minimum rise in sea level?

3 To what degree have marine erosion and deposition modified this topography?

4 Was the effect of the glaciation that occurred in this general area primarily erosional or depositional? What field evidence would you seek to substantiate your answer?

FIGURE 12.9: San Pedro, California

CI 20 ft 0 0.5 miles

PROBLEMS

1 Construct a topographic profile along line **A-A′**, and label all of the depositional and erosional features that you can identify.

2 As erosion of the land continues, the wave-cut terrace will become wider. What effect will a broad, shallow terrace have on the ability of waves to undercut the sea cliff?

3 Identify all of the wave-cut terraces shown on this map, and color each a different color. What do these terraces indicate about the tectonic history of the area?

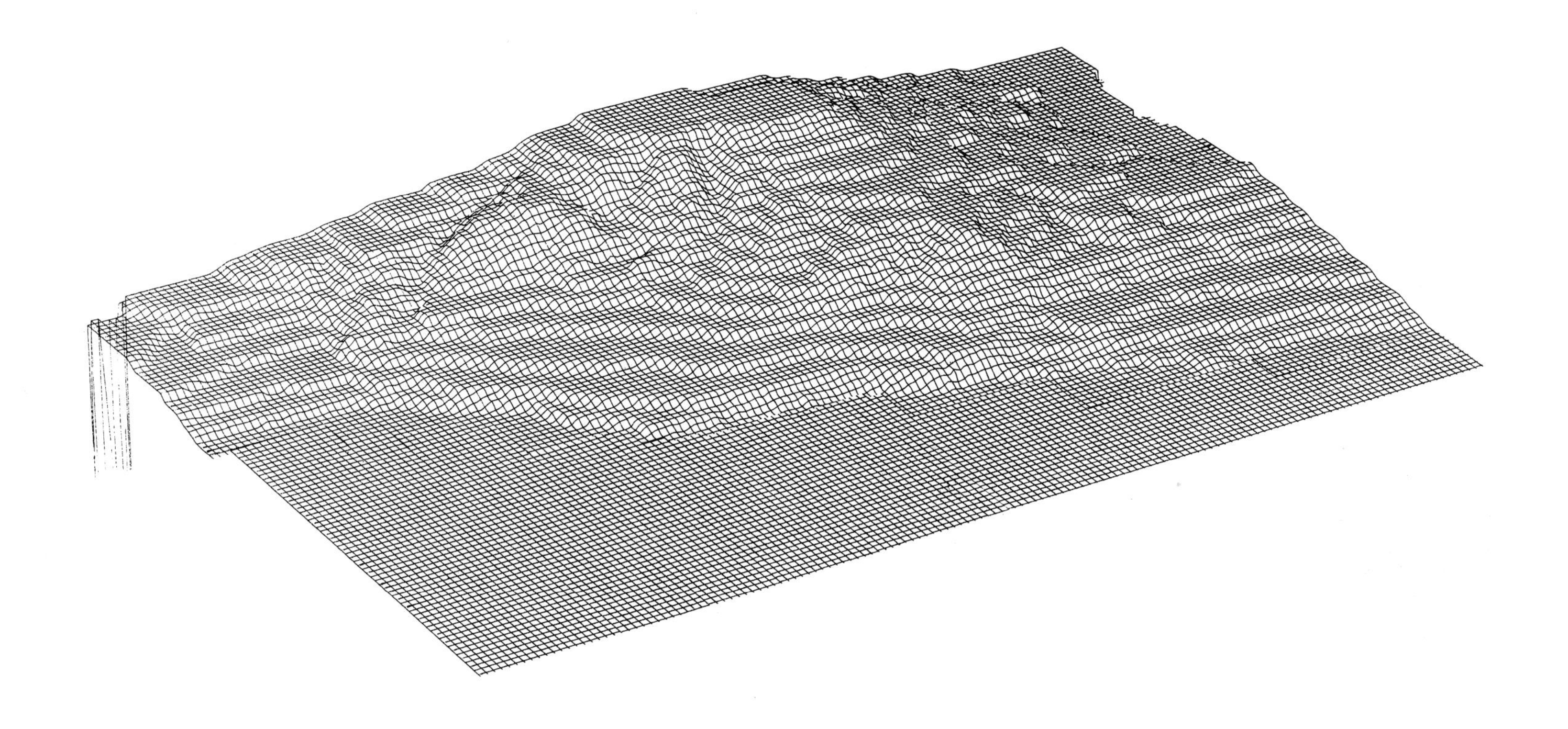

FIGURE 12.10: San Pedro Area, California

Computer-generated block diagram

PROBLEMS

1 This diagram was generated by having a computer draw numerous closely spaced profiles across the region shown in Figure 12.9. The vertical scale has been exaggerated. How many marine terraces can you identify on this model?

2 Compare your answer to problem 1 with your answer to problem 3 concerning Figure 12.9. What are the advantages of computer-generated models compared with standard topographic maps?

3 What are some of the limitations of computer-generated cartographic data?

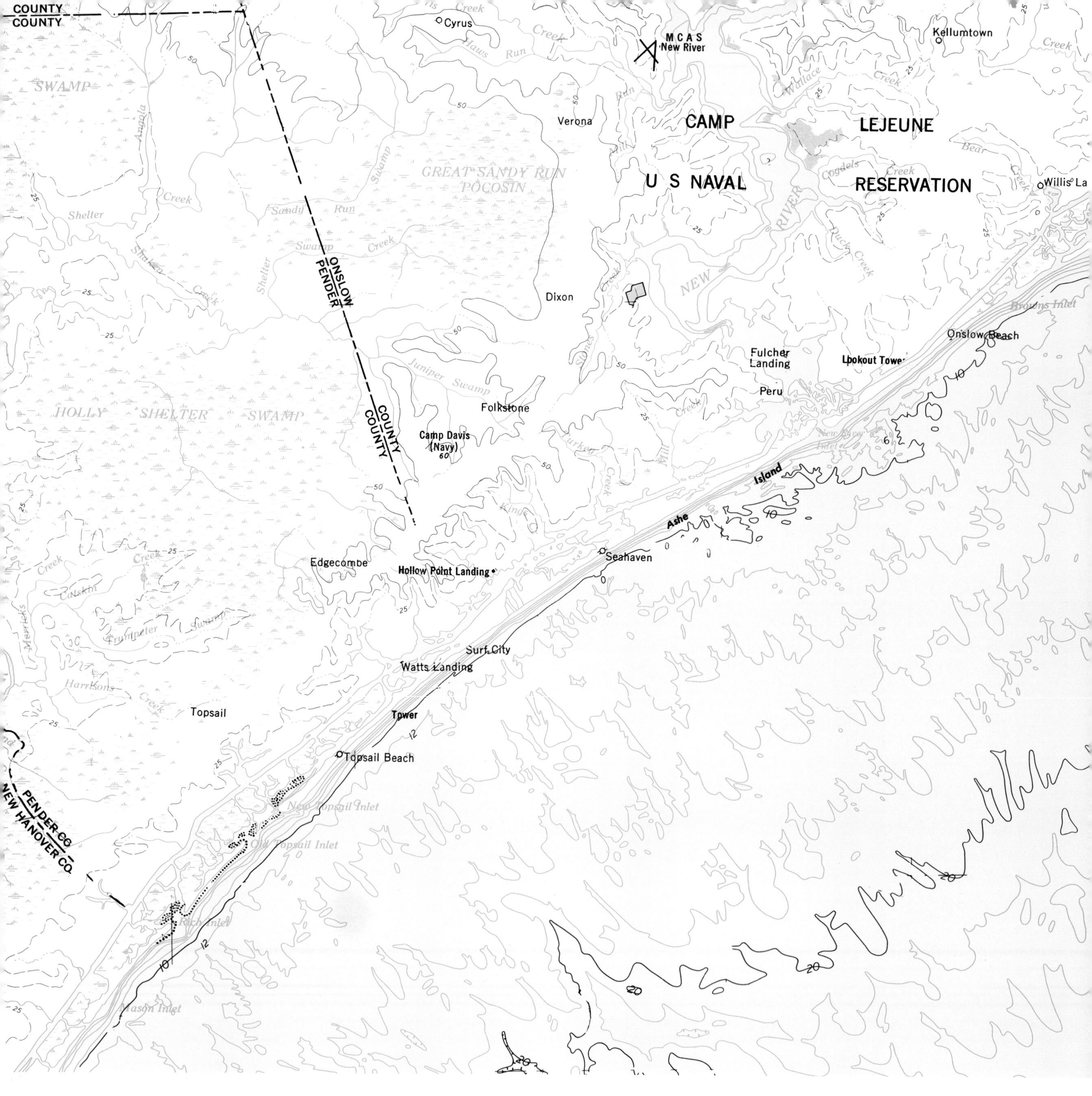

FIGURE 12.11: **Beaufort, North Carolina**

CI 50 ft   0 ——— 5 miles

PROBLEMS

1 Explain the origin and evolution of the large barrier island, or spit, that runs north and south on the east side of the map.

2 Are the lagoons presently being filled with sediment? What evidence can you cite to support your answer?

3 Explain the origin of the swamps in this area?

4 How will this coastal area change if sea level remains constant?

5 How will this coast change if sea level gradually rises?

6 What are the principal environmental hazards that you might expect in this coastal region?

Exercise 13

EOLIAN PROCESSES

OBJECTIVE

To recognize the major landforms produced by eolian activity and to understand the factors that control their development.

MAIN CONCEPT

Wind is capable of transporting large quantities of sand, which, when deposited, form distinctive types of dunes.

SUPPORTING IDEAS

1 The variety of dune types results from variations in sand supply, variations in wind velocity, variations in wind direction, and the different characteristics of the surface over which the sand moves.

2 Dunes may be stabilized by the growth of vegetation.

3 The wind does not produce large-scale erosional features in most desert regions, but it can form deflation basins, which are significant features in some areas.

DISCUSSION

FORMATION OF SAND DUNES

Wind is an effective geologic agent, capable of transporting loose sand and dust. The most significant landforms created by the wind are sand dunes, which can assume a variety of fascinating shapes and patterns. Wind-formed sand dunes cover vast areas of the earth's surface. In the great deserts of the world, dunes dominate the landscape. They are also significant features along many coasts, in local areas of semiarid regions, and in polar regions that have a sand supply sufficient for dune formation. The most common dune varieties are as follows:

1 *Transverse dunes* develop where wind direction is constant and the supply of sand is large.

2 *Barchan dunes* form where the supply of sand is limited and wind direction is constant.

3 *Longitudinal dunes* result from two slightly different wind directions and are elongate in the direction of the prevailing wind.

4 *Star dunes* have a high central point from which three or four ridges extend like the spokes of a wheel. They are believed to result from winds blowing in three or more directions.

5 *Parabolic dunes* develop along the coast from strong onshore winds.

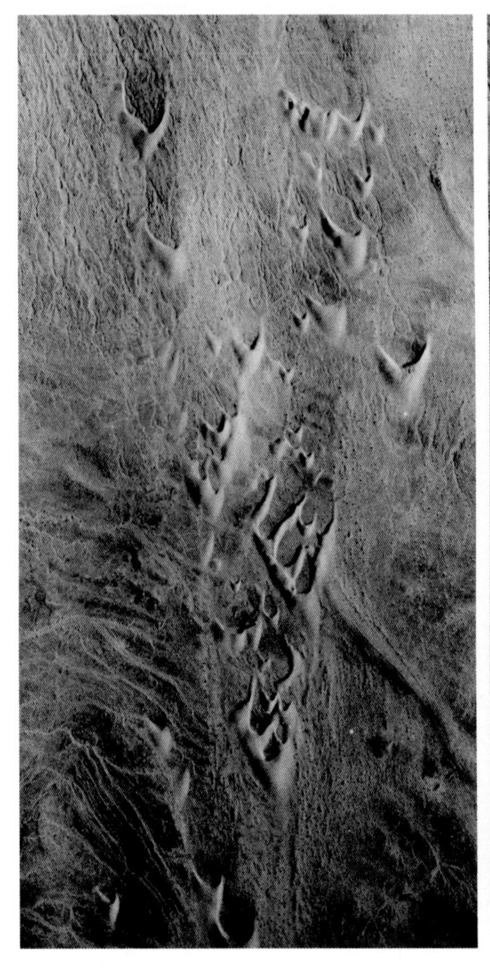
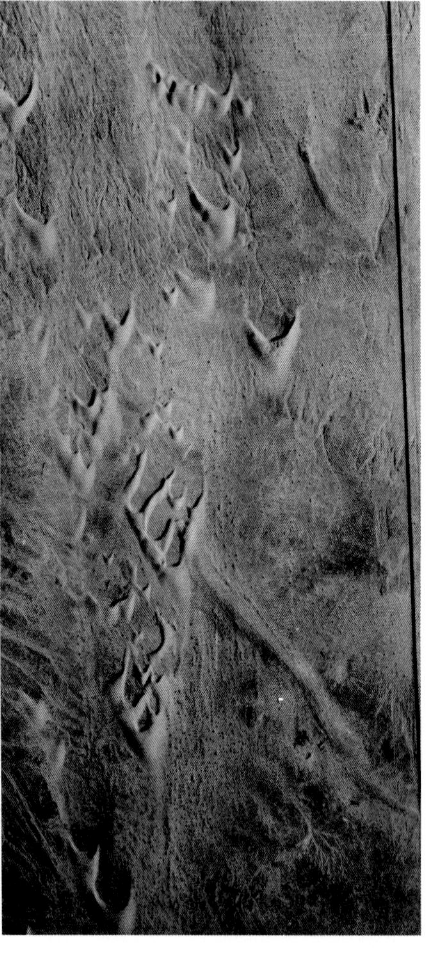

FIGURE 13.1: Sand Dunes—California

PROBLEMS

1 List the types of dunes that can be identified in the area of stereoview.

2 What factors determine the type of dune that forms?

3 By tracing on the photograph, make a map showing the area covered by each dune type. Show the general pattern of the dune crests, and indicate the direction of the prevailing wind.

4 Note that in several areas, one dune type grades into another. After carefully studying the photograph, explain briefly how and why this transition takes place.

5 What is the most probable source of sand in a desert area such as this: (a) wind abrasion of the general surface, (b) mass movement and slope retreat of local valley walls or escarpments, or (c) alluvium transported by a drainage system?

FIGURE 13.2: Coastal Dunes—Oregon

PROBLEMS

1 What type of sand dune is most common in this area?

2 Make a sketch map showing the major dune fields according to their relative age. How many generations of dune systems can you identify?

3 Explain the origin of the elongate ridges and furrows just inland from the beach.

4 Sediment is being transported by three systems: eolian, beach, and stream. Show by arrows the direction of sediment transport in each of these systems.

5 Which system of sediment transport dominates in this area: fluvial or beach?

FIGURE 13.3: Sand Hills—Nebraska

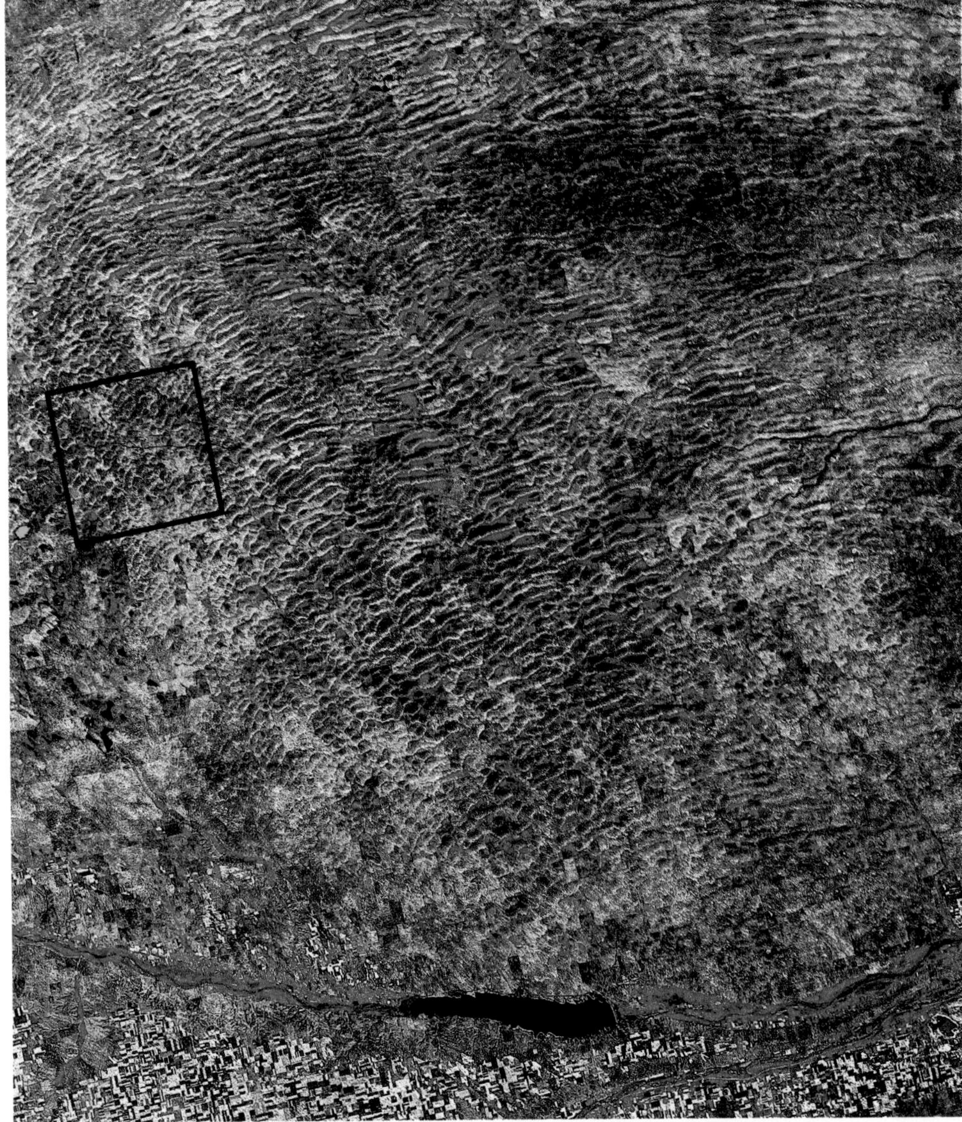

PROBLEMS

1 On the image, draw a short arrow in each square inch to indicate the wind direction during dune development. Was the wind direction constant or variable?

2 How have the dunes affected the drainage in this area?

3 Are the dunes older or younger than the North Platte River (the large east-flowing stream near the bottom of the image)?

4 Are the dunes older or younger than the small streams near the eastern border of the image?

5 How will this area change in the future?

Landsat image

FIGURE 13.4: Sand Hills—Nebraska

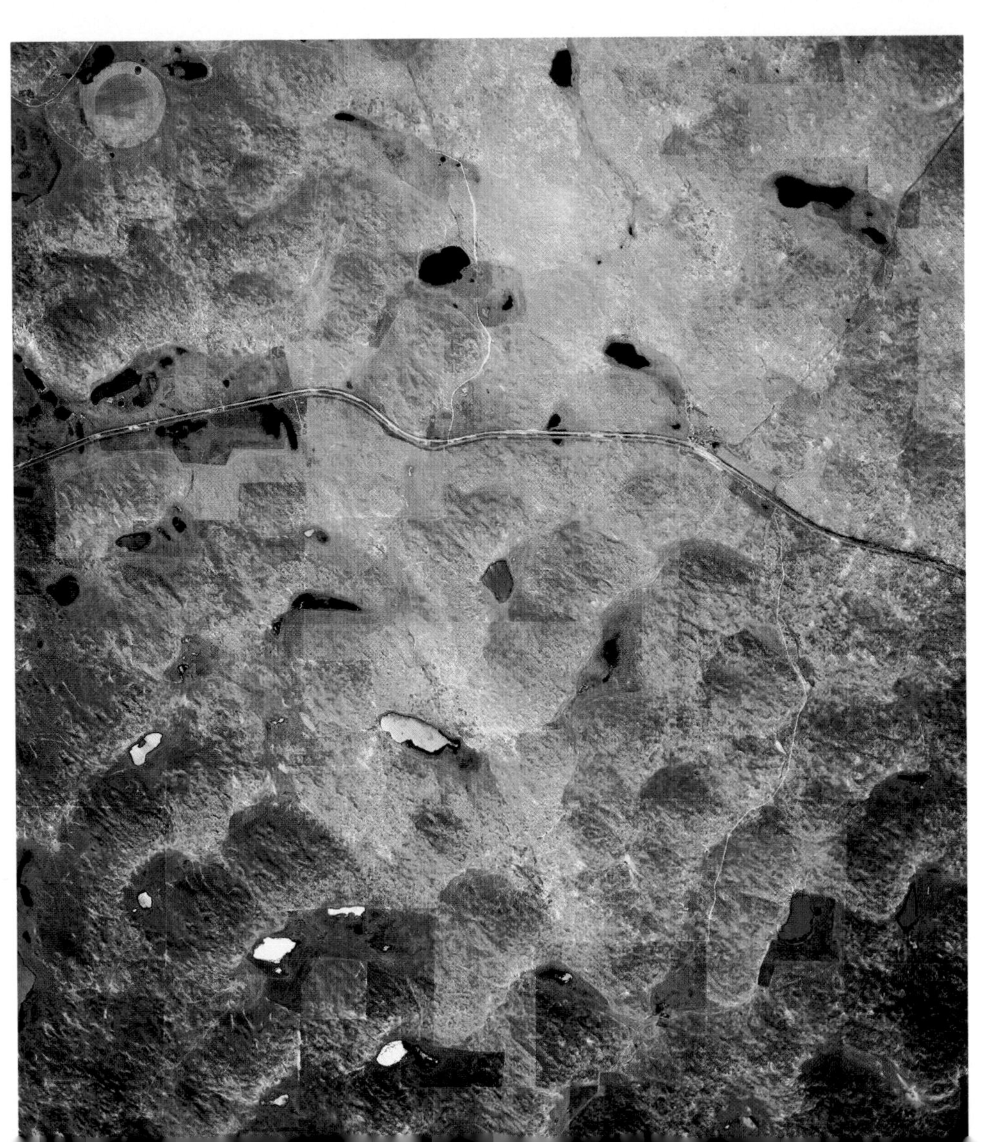

PROBLEMS

1 This is a high altitude infrared photograph of the area outlined on the landsat image in Figure 13.3 above. Study the high-altitude aerial photograph, and show by a series of arrows the wind direction in this area at the time when the dunes were developing.

2 Identify the type of dune that occurs in this area?

3 Explain the origin of the lakes.

4 How have the geologic processes that developed this topography been modified in the last several thousand years?

5 How will this area change in the future?

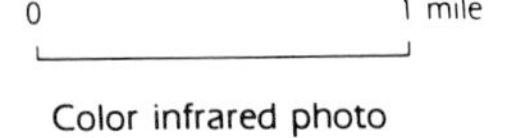

Color infrared photo

FIGURE 13.5: Sand Seas—North Africa

Enhanced Landsat image 0 5 miles

PROBLEMS

1 How many dune types can you recognize in this area?

2 Plot the prevailing wind direction with a series of arrows. What causes the desert conditions in North Africa?

3 How does the origin of this dune field compare with that of the sand hills of Nebraska?

4 What evidence indicates that the landscape now buried beneath the "sea of sand" was eroded originally by running water?

5 In which directions were the streams flowing?

6 How has the topography of the mountainous landscape influenced the types of dunes and their distribution?

Exercise 14

PLATE TECTONICS

OBJECTIVE

To become aware of the location and topographic expression of the major tectonic features of the earth and to understand how these features are related to the plate tectonic theory.

MAIN CONCEPT

The theory of plate tectonics explains the origin and evolution of the earth's structural features as the result of a system of moving lithospheric plates. This motion involves sea-floor spreading, continental drift, and subduction, and results in the creation of new oceanic crust, the growth of continents, volcanism, earthquakes, and mountain building.

SUPPORTING IDEAS

1 Plate boundaries are the most dynamic zones of the earth's lithosphere.

2 The major components of all continents are (a) shields, (b) stable platforms, and (c) folded mountain belts.

3 Plate motion is indicated by most major structural and topographic features.

4 Many local characteristics (e.g., island arcs, mountain belts, volcanic activity, and features on the ocean floor) can be explained by specific conditions of the tectonic setting.

DISCUSSION

The theory of plate tectonics grew out of our newly acquired abilities to study the geology of the ocean floors, measure rock magnetism, and study the earth's seismicity on a global scale. The vast amount of data generated by these studies have been recorded on a variety of maps. In this way, the data may be synthesized, analyzed, and interpreted. One of the most useful maps for this purpose is the physiographic map of the world that accompanies this manual. It shows in relief the surface features of both the ocean floors and the continents. This exercise is involved largely with the study of this map—plotting geologic data and interpreting the tectonic relationships of the earth's features on a global scale. To do this effectively, let us first study the characteristics of the major structural features of the continents as shown on Landsat imagery.

FIGURE 14.1: Canadian Shield

Landsat image 0 20 miles

PROBLEMS

1 Study the Landsat image and determine the general shape of the granitic bodies in this part of the shield? What type of igneous intrusions are they?

2 Study the structural trends (commonly expressed by linear lakes) of the metamorphic rocks. What evidence can you find that these rocks have been intensely deformed by horizontal stresses in the crust?

3 What geologic features indicate that the rocks in the shield are the roots of ancient mountains?

4 Why might you conclude that great volumes of rock have been eroded to form the shield?

5 Draw an idealized cross-section showing the general characteristics of the structure and rock types of this area. Is this area typical of shields in general? See Figure 17.19.

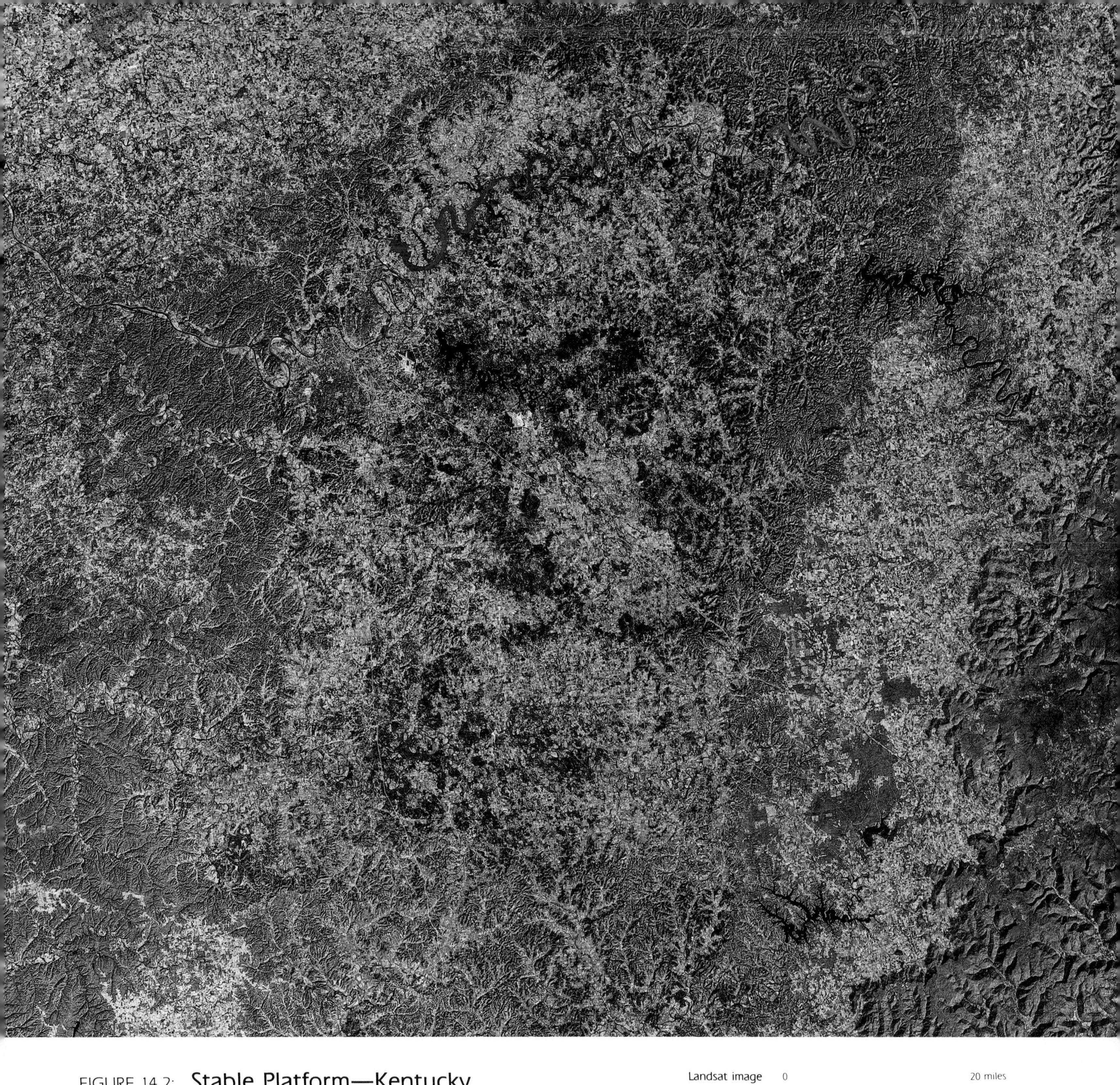

FIGURE 14.2: Stable Platform—Kentucky

Landsat image 0 20 miles

PROBLEMS

1 How does the deformation of the stable platform differ from that of mountain belts?

2 Compare the landforms shown on the Landsat image of a shield (Figure 14.1) with the landforms shown on this image of the stable platform. What is the major difference between the two areas?

3 Why are the rocks of the stable platform predominantly shallow-marine sediments?

4 The Nashville dome is typical of the broad structural upwarps of a stable platform. The crest of the dome has been eroded, so the outcropping rock formations display an elliptical pattern (see Figure 17.7). One formation is expressed on this Landsat image in dark brown and has a distinctive texture produced by a fine drainage network. Trace the boundaries of this formation on the image. Next, draw a generalized cross section across the dome to show the broad configuration of the structure and the surface expression of the rock formations.

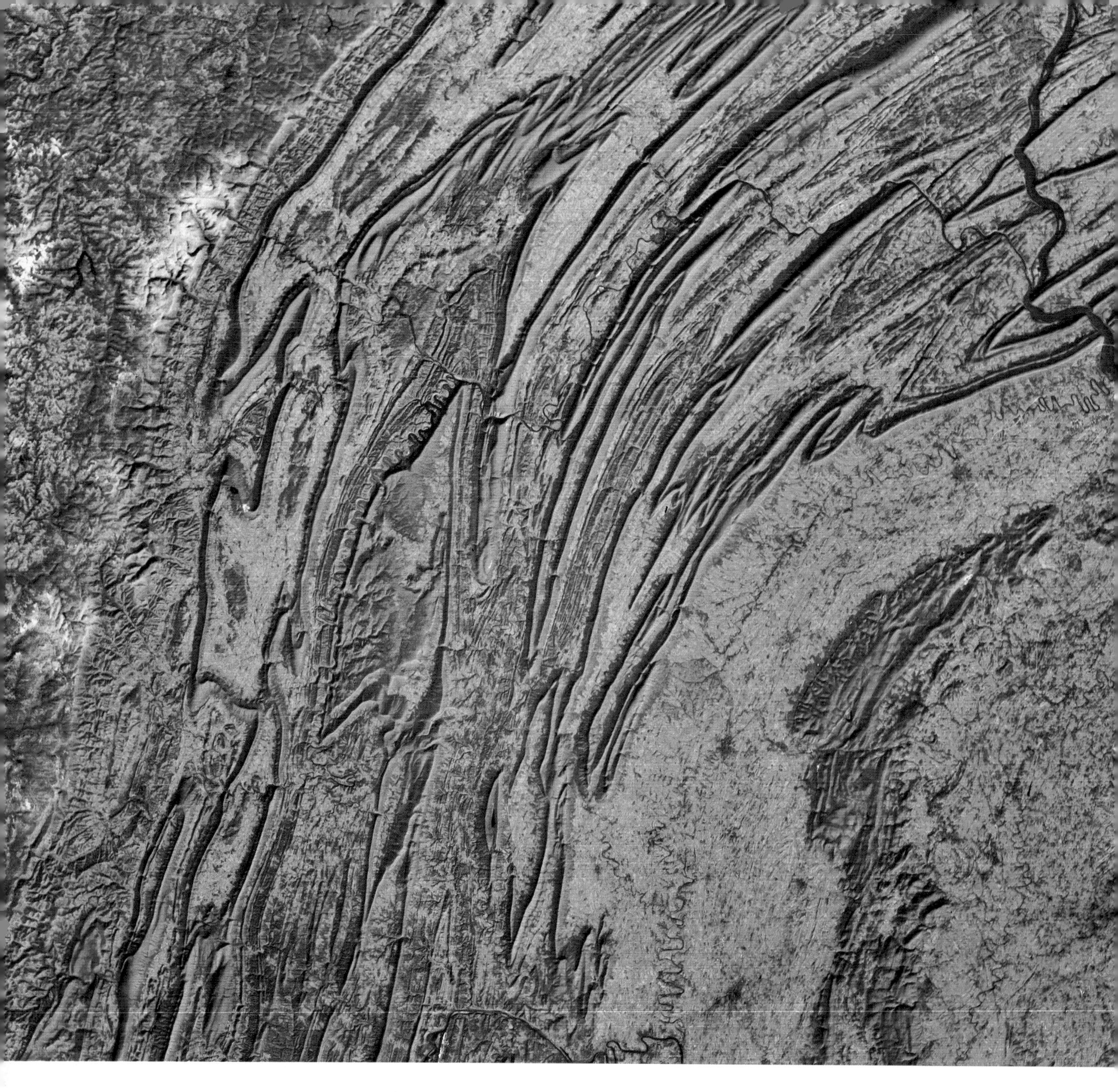

FIGURE 14.3: **Folded Appalachian Mountains**

Landsat image 0 20 miles

In this area, the entire sequence of sedimentary rocks (nearly 40,000 ft thick) has been deformed into a series of tight folds. Erosion has sculpted the resistant formations into prominent ridges that zigzag across the landscape. The low, intervening valleys are eroded in nonresistant shale and limestone formations.

PROBLEMS

1 Study the Landsat image and note the configuration of the prominent ridges. Trace their extent across the area. Refer to the map patterns of plunging folds (see Figure 17.11, p. 160), and label the anticlines and synclines.

2 Draw an idealized cross section of the folded belt to show the style and degree of deformation. Show the folds below the surface with solid lines and show the parts of the folds that have been eroded away (above the surface) with dashed lines.

3 Indicate by arrows the direction of the stresses that produced this deformation.

FIGURE 14.4: Basin and Range—Nevada

Landsat image 0 20 miles

PROBLEMS

1 Study the Landsat image and draw an idealized cross section of this area. Show the style and degree of deformation. How do the mountains in this area differ from those in Figure 14.3?

2 What active geologic and geophysical phenomena would you expect to find in this area?

3 Show by arrows the direction of the stresses that produced this deformation.

4 How is this area likely to evolve if the present tectonic forces continue?

5 Label the following features: (a) alluveal fans, (b) bajadas, (c) playa lakes.

PALEOMAGNETISM

DISCUSSION

When a lava has cooled and solidified, the magnetic domains within the iron minerals contained in the lava align in the direction of the earth's magnetic field. The solidified lava thus preserves a record of the earth's magnetic field at the time when the rocks were formed. Scientists recognized as early as 1906 that the poles of the magnetic field preserved in some rocks were oriented in the opposite direction from the poles evident in other specimens, as if the earth's north and south magnetic poles had switched places. Subsequent studies showed that these magnetic reversals do indeed occur, and that the magnetic poles have been reversed many times during the geologic past. The effect of these reversals has been to create periods of "normal" magnetism (i.e., periods in which the polarity matched the present position of the north and south magnetic poles), and periods when the magnetic field was reversed.

To test the plate tectonic theory, scientists proposed that if new oceanic crust is created at the oceanic ridges by the extrusion of lava, then the rocks on the ocean floor adjacent to the ridges should preserve a record of the periods of normal and reversed magnetism. If we measure the strength of the earth's magnetic field across the ridge, a series of anomalies, or differences, should occur. Where rocks solidified during periods of "normal" magnetism, higher values should be found, because the induced magnetism adds to the present strength of the magnetic field. Lower values should be found in rocks solidified during reversed magnetism, because polarity of these rocks reduces the local effect of the earth's present magnetic field.

If we were to map the magnetic anomalies across the oceanic ridge, alternating bands of high and low magnetism should appear on both sides of the ridge, one side being the mirror image of the other. Such a pattern would provide strong evidence for sea-floor spreading and the theory of plate tectonics.

Figure 14.5 shows a map of a portion of the Atlantic Ocean and the magnetic measurements made by a research vessel as it crossed the ridge on four traverses. When the measurement curve on the traverse is above zero, the strength of the magnetic field is greater than normal (i.e., the paleomagnetism in the rocks is adding to the strength of the present magnetic field). Where the curve drops below zero, the strength of the magnetic field is less than normal because the paleomagnetism in the rocks has a reverse polarity and thus reduces the present strength of the magnetic field. The objective of this part of the exercise is to determine the zones of normal and reverse polarity along each track of the research vessel and to correlate the zones along this part of the mid-Atlantic ridge.

PROCEDURE

1 On each track, mark off the points where the magnetic curve intersects the line of zero field strength. Start at the ridge and work outward, both ways. These points are the divisions between zones that have normal and reverse polarity.

2 Connect the equivalent points on each track, and shade in the areas of each negative anomaly.

3 Color the bands of the positive anomalies as follows: youngest = red, next youngest = orange, next = light green, and oldest = dark green.

PROBLEMS

1 Describe briefly the patterns of the magnetic anomalies shown on the map that you have constructed. Do the patterns cross or parallel the oceanic ridge? Are the patterns on either side of the ridge similar or different?

2 Write a brief paragraph explaining how the pattern of magnetic anomalies on the sea floor supports the theory of plate tectonics.

3 Refer to the geomagnetic time scale in your text and determine the age of each band on the map.

4 Calculate the average rate of spreading of the floor of the Atlantic Ocean during the last 3 million years. (Use the map scale and the time duration of the rock units determined from the geomagnetic time scale in your text.)

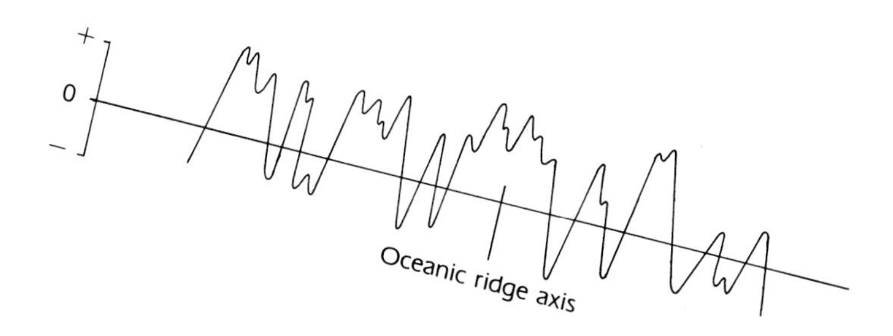

+
0
−
Oceanic ridge axis

+
0
−
Oceanic ridge axis

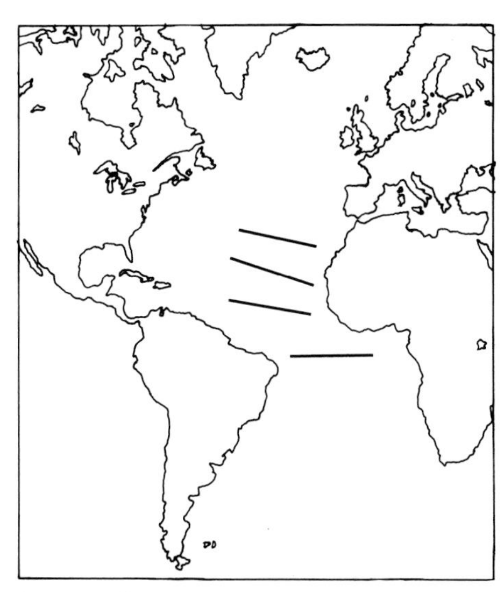

Location of traverses

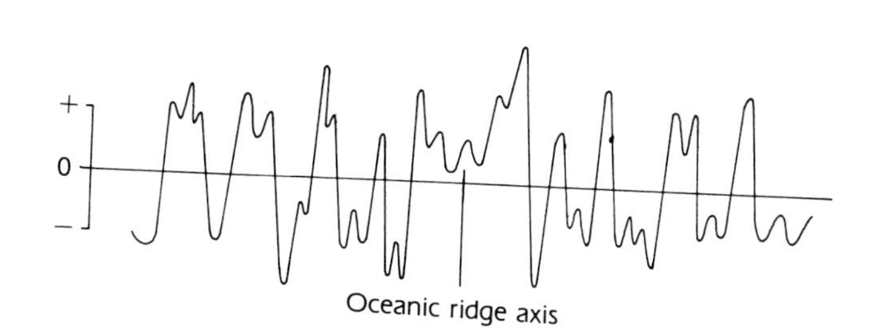

FIGURE 14.5

MAP OF PART OF THE ATLANTIC OCEAN BASIN showing four traverses across the basin and the magnetic measurements recorded on each traverse.

PLATE BOUNDARIES AND PLATE MOTION

PROBLEMS

1 Draw the major plate boundaries on the physiographic map of the world. A good procedure to follow is this:

a Outline the divergent plate boundaries and the major transform fault boundaries in red. *Work first on a piece of tracing paper, then when you are satisfied with your results, carefully transfer your work to the physiographic map.*

b Trace the extent of the major convergent plate boundaries, as indicated by oceanic trenches, in purple.

c Add the boundaries indicated by the presence of active, mobile mountain belts.

d Add the remaining boundaries that are less definite.

2 List the features shown on the physiographic map that would indicate the direction of plate motion.

3 On the basis of your answer to problem 2, show the direction of plate motion with a series of arrows. Use a different colored arrow for each type of feature that indicates plate movement.

4 Using the tectonic information that you have compiled on the physiographic map, draw an idealized cross section through the plate margins that occur near (a) South America, (b) Central America, (c) Japan, (d) the Himalaya Mountains, and (e) the Indonesian island arc.

5 On the basis of the tectonic information that you have plotted, explain briefly the origin of (a) the Red Sea, (b) the Black Sea, (c) the Mediterranean Sea, and (d) the Gulf of California.

6 Assuming that the present directions of plate motion continue in the future, what will happen to each of these seas or small ocean basins during the next few million years?

7 Study the Himalayan mountain belt on the northern border of India. This is the highest mountain range in the world. How is the existence of this mountain belt explained by plate tectonic theory?

8 Why are the Himalaya Mountains higher than the Andes Mountains?

9 What is the relationship between the islands of Indonesia, the Himalaya Mountains, and the Alps?

10 Fracture zones and transform faults characteristically cut across the oceanic ridge. Assuming that this is also true of the fracture zones in the northeastern Pacific Ocean, where is the ridge in that area? Explain what happened to obscure it.

11 If you were involved with the deep-sea drilling project, where would you suggest drilling to find the oldest sediments in the Pacific Ocean?

12 Where on the North American plate would you expect to find the thickest accumulation of marine sediments?

13 The average rate of plate motion is about 1.5 in. per year. Assuming that this rate of plate movement has been constant, how old is the Atlantic Ocean?

14 Explain the origin of the Cascade Mountains in Oregon and Washington.

15 Note the similarities and differences of these two tectonic settings: (a) the area between Florida and South America (Caribbean Sea) and (b) the area between South America and Antarctica (Scotia Sea). From your observations, do you conclude that these areas had similar or different origins?

16 Is the southern part of Central America most likely to be (a) a segment of continental shield, (b) part of a stable platform, (c) a former volcanic island arc, or (d) part of a mountain belt of folded sedimentary rocks?

17 Australia and New Guinea are part of the same continental block, and are separated from each other by a very shallow sea. What is the origin of the mountains of New Guinea?

EXPLANATION

ACTIVE PLATE BOUNDARIES

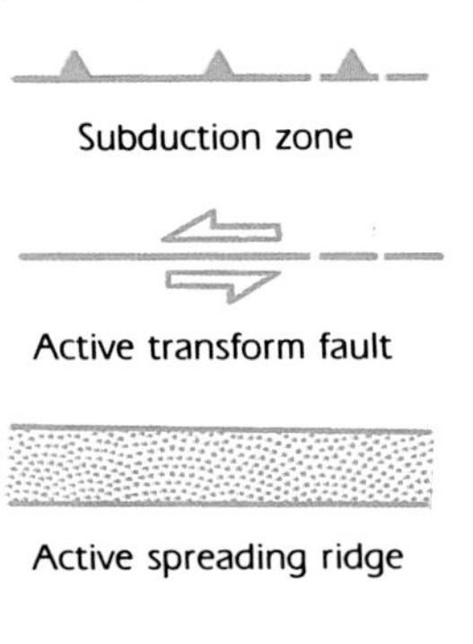

MAGNETIC LINEATIONS

2

3

Magnetic anomalies Labels show correlation with geomagnetic polarity time scale

INTRAPLATE STRUCTURES

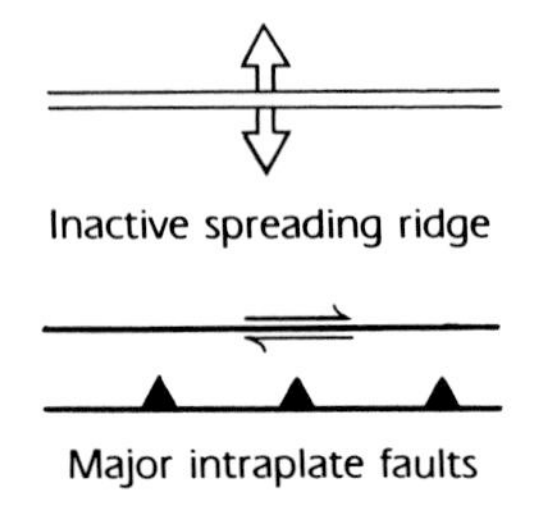

MOTION VECTORS

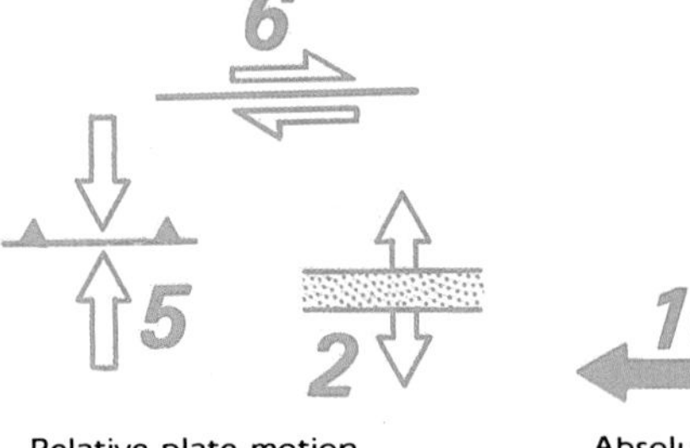

Relative plate motion In cm/yr

Absolute motion

VOLCANIC CENTERS

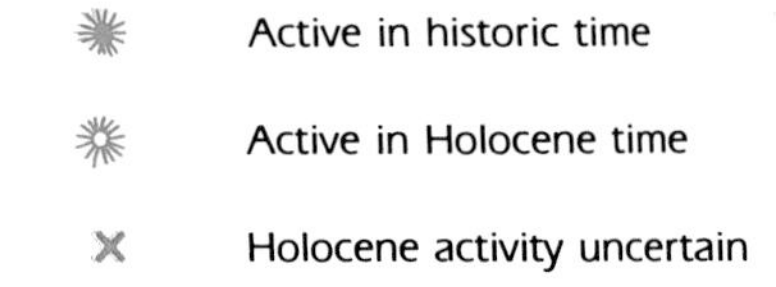

EARTHQUAKE EPICENTERS

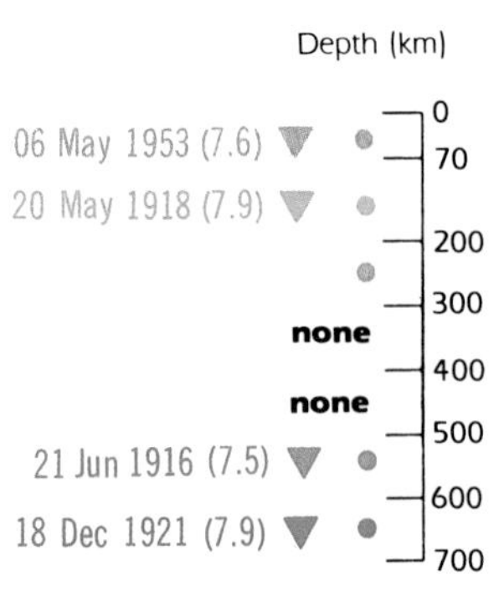

FIGURE 14.6: **Tectonic Map of Southwestern Pacific Ocean Basin**

0 250 miles

PROBLEMS

1 Study the explanation of this map on the facing page. Compare this map with the same area on the physiographic map of the world. How is the oceanic ridge expressed physiographically in this area?

2 How are the fracture zones expressed physiographically?

3 Where are the earthquakes concentrated in this part of the Pacific Ocean?

4 What is the range of focal depth for the earthquakes in this part of the Pacific Ocean?

5 What is the maximum rate of plate motion in this area? Where does it occur?

6 How many volcanoes have erupted in historic times in this part of South America?

7 Draw an idealized cross section along line **A-A′** and label the following features:

- **a** Major topographic features of the ocean floor and continental surface
- **b** Oceanic crust and continental crust
- **c** Magnetic anomalies and their ages
- **d** Earthquake foci and volcanoes

FIGURE 14.7: Tectonic Map of North America

PROBLEMS

1 Study the structural patterns of the shield. What evidence on the map indicates that the shield consists of the roots of a series of deformed mountain belts? Indicate on the map by arrows the direction of the compressive forces necessary to produce the structural trends in each major area of the shield.

2 If a mountain belt results from the collision of two lithospheric plates, how many major orogenic pulses are apparent in this shield?

3 How does the radiometric age of the rocks in the shield support the concept of continental growth by accretion?

4 Note the thickness of the sedimentary rock cover of the stable platform as indicated by contour lines. How thick is this rock cover throughout most of the platform?

5 Explain why the stable platform sedimentary rock cover was probably deposited in very shallow water rather than in the deep ocean.

6 How are the orientation and location of the mountain systems of North America explained by the theory of plate tectonics?

7 The bright red colors on the map represent granitic intrusions into young, folded mountain belts. Use the theory of plate tectonics to explain how the intrusions formed and how they are related to mountain building.

8 Explain the tectonic origin of Baja, California. If the present tectonic forces continue to be active in the western United States, what will happen to Baja and southern California?

9 Explain the origin of the volcanoes in the Cascade Mountains.

10 Note the submarine contours west of Central America. What tectonic features do they represent?

11 How does the theory of plate tectonics explain the concentration of volcanic activity along the west coast of Central America?

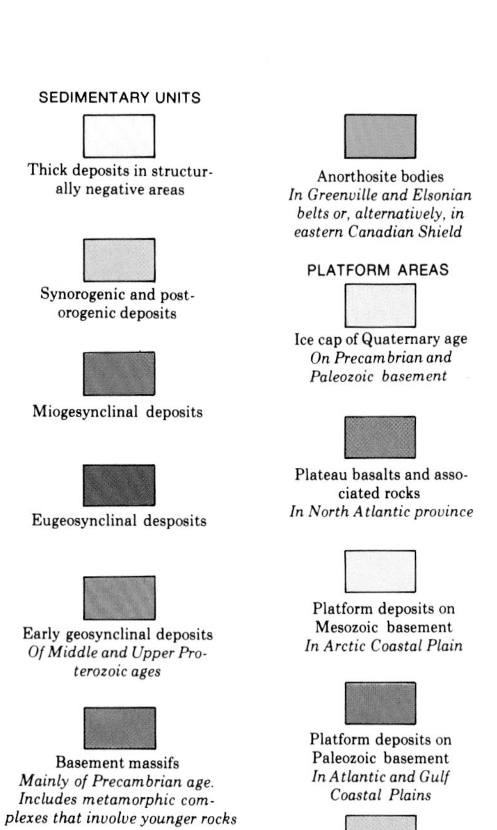

SEDIMENTARY UNITS

Thick deposits in structurally negative areas

Synorogenic and post-orogenic deposits

Miogesynclinal deposits

Eugeosynclinal desposits

Early geosynclinal deposits
Of Middle and Upper Proterozoic ages

Basement massifs
Mainly of Precambrian age. Includes metamorphic complexes that involve younger rocks

VOLCANIC AND PLUTONIC UNIT

Granitic plutons
Ages are generally within the span of the tectonic cycle of the foldbelts in which they lie

SPECIAL UNIT

Eugeosynclinal deposits of the Pacific border
Includes Franciscan Formation of California

PRECAMBRIAN FOLDBELTS

Dark colors show areas of paraschist and paragneiss derived from supracrustal rocks; light colors show areas of granite and orthogneiss of plutonic origin

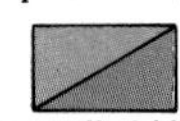

Greenville foldbelt
Deformed 880-1,000 m.y. ago

Rocks of the Hudsonian foldbelt
Overprinted by Elsonian event about 1,370 m.y. ago

Hudsonian foldbelts
Deformed 1,640-1,820 m.y. ago

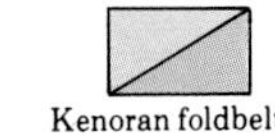

Kenoran foldbelts
Deformed 2,390-2,600 m.y. ago

Anorthosite bodies
In Greenville and Elsonian belts or, alternatively, in eastern Canadian Shield

PLATFORM AREAS

Ice cap of Quaternary age
On Precambrian and Paleozoic basement

Plateau basalts and associated rocks
In North Atlantic province

Platform deposits on Mesozoic basement
In Arctic Coastal Plain

Platform deposits on Paleozoic basement
In Atlantic and Gulf Coastal Plains

Platform deposits on Precambrian basement
In central-craton

Platform deposits within the Precambrian
Mainly in the Canadian Shield

STRUCTURAL SYMBOLS

Normal fault
Hachures on downthrown side

Transcurrent fault
Arrows show relative lateral movement

Thrust fault
Barbs on upthrown side

Subsea fault
Long dashes based on topographic and geophysical evidence; short dashes, based on geophysical evidence only

Axes of sea-floor spreading

Flexure
Arrows on depressed side

Salt domes and salt diapirs
In Gulf Coastal Plain and Gulf of Mexico

Volcano

+2000
0
2000

Contours on basement surfaces beneath platform areas
All contours are below sea level except where marked with plus symbols. Interval 2,000 meters

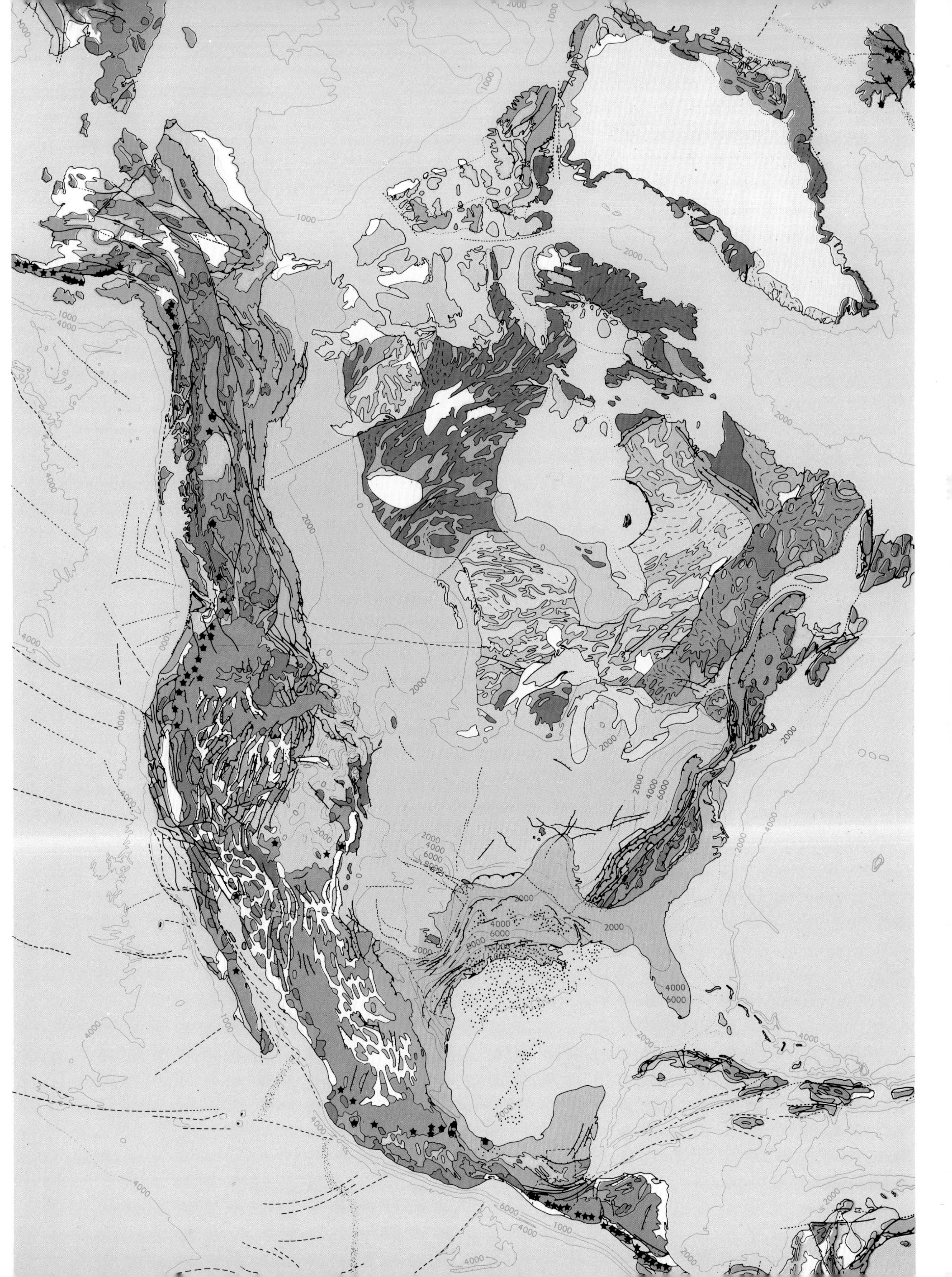

2000
1000
1000
1000
2000
1000
4000
2000
4000
2000
0
2000
4000
2000
4000
2000
4000
6000
2000
4000
6000
8000
2000
2000
4000
6000
8000
2000
4000
6000
1000
4000
2000
4000
4000
2000
6000
4000
1000
4000
2000

TECTONIC FRAMEWORK OF THE WORLD

1 This part of the exercise is to be done on the physiographic map accompanying this manual. To help you visualize clearly the major structural elements of the continents, color the shields, stable platforms, and folded mountain belts in very light tones of different color.

2 Would Greenland be considered part of the North American continental crust or a separate fragment of continental crust?

3 Is the southern part of Central America most likely to be (a) a segment of a shield, (b) an area of folded Paleozoic and Mesozoic strata, or (c) a volcanic arc connecting two continental masses?

4 Two large shields are exposed in South America: (a) the Brazilian Shield south of the Amazon River, and (b) the Guiana shield north of the Amazon. On the basis of the map and your knowledge of the continents, do you believe that the rocks of these shields are connected structurally beneath the Amazon basin, or are the shields separate and distinct structural units?

5 What is the origin of the series of elongate lakes in East Africa?

6 The Ethiopian highlands are composed of extensive flood basalts. In light of the major tectonic features of this general area, explain how the highlands originated?

7 North and South America, Australia, and India all have mountain belts. Why are there no major mountain belts along the east, south, and west coasts of Africa?

8 Why are Asia, Europe, and India considered to be three independent continental masses in the geologic sense?

9 What structural element is dominant in Russia: shield, stable platform, or folded mountain belt? (Before attempting to answer this question, study North America on the map and take note of the difference in patterns used to distinguish the shield, platform, and mountainous areas. Use this observation as you study Asia on the map.)

10 The Keweenawan Peninsula of northern Michigan, which extends out into Lake Superior, is composed of more than 60,000 ft of the rock type shown in Figure 3.15 of this lab manual. A narrow subsurface zone of this rock can be traced to eastern Kansas. How could this deposit have originated in the center of North America?

11 Africa is a relatively high continent; large areas of the shield are more than 6000 ft above sea level. On most other continents, the shield is near sea level. How might this phenomenon be explained in light of Africa's unique relationship to the oceanic ridge system?

12 Most of the stable platform and shield of Australia are relatively low, generally less than 1000 ft above sea level. How might this be explained in light of Australia's relationship to the surrounding plate boundaries?

Exercise 15

SEISMOLOGY

OBJECTIVE

To understand how the study of seismic waves is used to determine the internal structure of the earth.

MAIN CONCEPT

Changes in direction and variations in velocity with which seismic waves pass through the earth are evidence of distinct discontinuities in internal structure and give us a basis for interpreting the internal structure of the earth.

SUPPORTING IDEAS

1 Seismic reflection profiles provide an effective way to study relatively shallow subsurface structure and stratigraphy.

2 The shadow zone of both the *P* and *S* waves indicates the size of the earth's core, and provides important information about the state of matter (solid or liquid) in the core and mantle.

DISCUSSION

The exploration of the subsurface structure of the Earth by monitoring seismic waves has been possible since the 1920s. With the recent advent of solid state electronic components, the development of magnetic tape records, and the use of computers, seismic exploration has reached a high level of sophistication. The seismic records in Figure 15.1 are an example of the type of seismic data that geologists commonly analyze. Each vertical line is a record of energy received at a recording station. When the energy is from a reflective horizon, such as a resistant sandstone formation, it is recorded as a positive or negative segment of the curve (deviation to the right or left, respectively). Positive reflections, which occur when seismic energy encounters a higher velocity layer below a lower velocity layer, are generally shaded black for emphasis. This is quite clear in Figure 15.1A, which shows small-scale detail from a recording. When a series of seismic records are made along a line, they show the configuration of the rock units in the subsurface and can be interpreted in much the same way as geologic cross sections (Figures 15.1B–15.1D). (There are, however, a number of geophysical limitations that must be considered by the geologist, but for our purposes, we will interpret the seismic sections as we would interpret a geologic cross section.)

A. A RECORD FROM A SINGLE RECORDING STATION shows variations in wave velocity. The positive segments of the curve (deviations to the right) result from strata that transmit seismic waves at a high velocity. These segments are rendered black for emphasis.

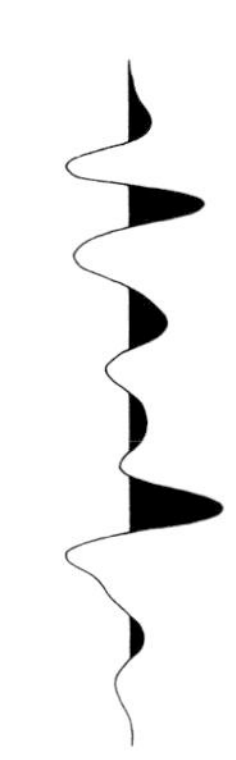

B. A NUMBER OF RECORDING STATIONS POSITIONED IN A LINE produced this seismic record. The positive segments of the record outline the boundaries between rock units that have different seismic characteristics (for example, a unit of sandstone and a unit of shale).

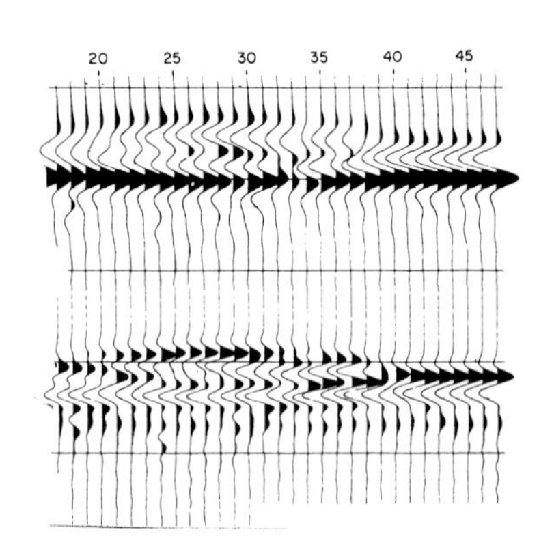

C. A LINE OF SEISMIC STATIONS may record a profile of the major rock bodies.

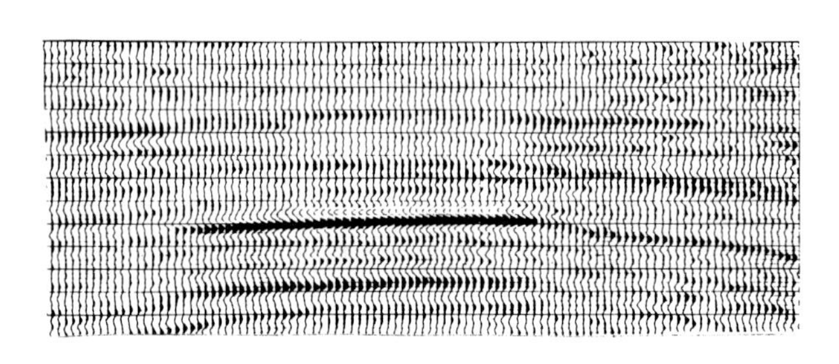

D. A LONGER LINE OF SEISMIC STATIONS may record the subsurface structure of the rock units.

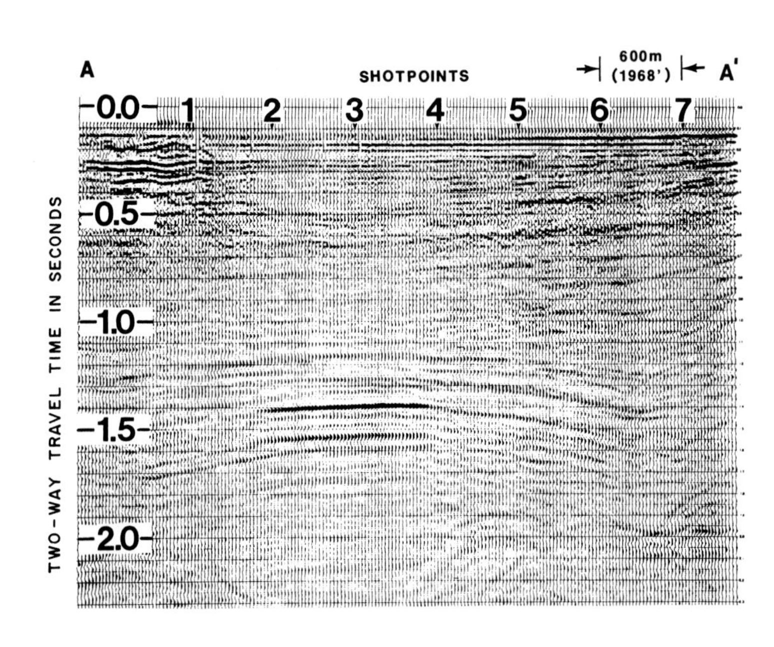

FIGURE 15.1 **EXAMPLES OF SEISMIC RECORDS.**

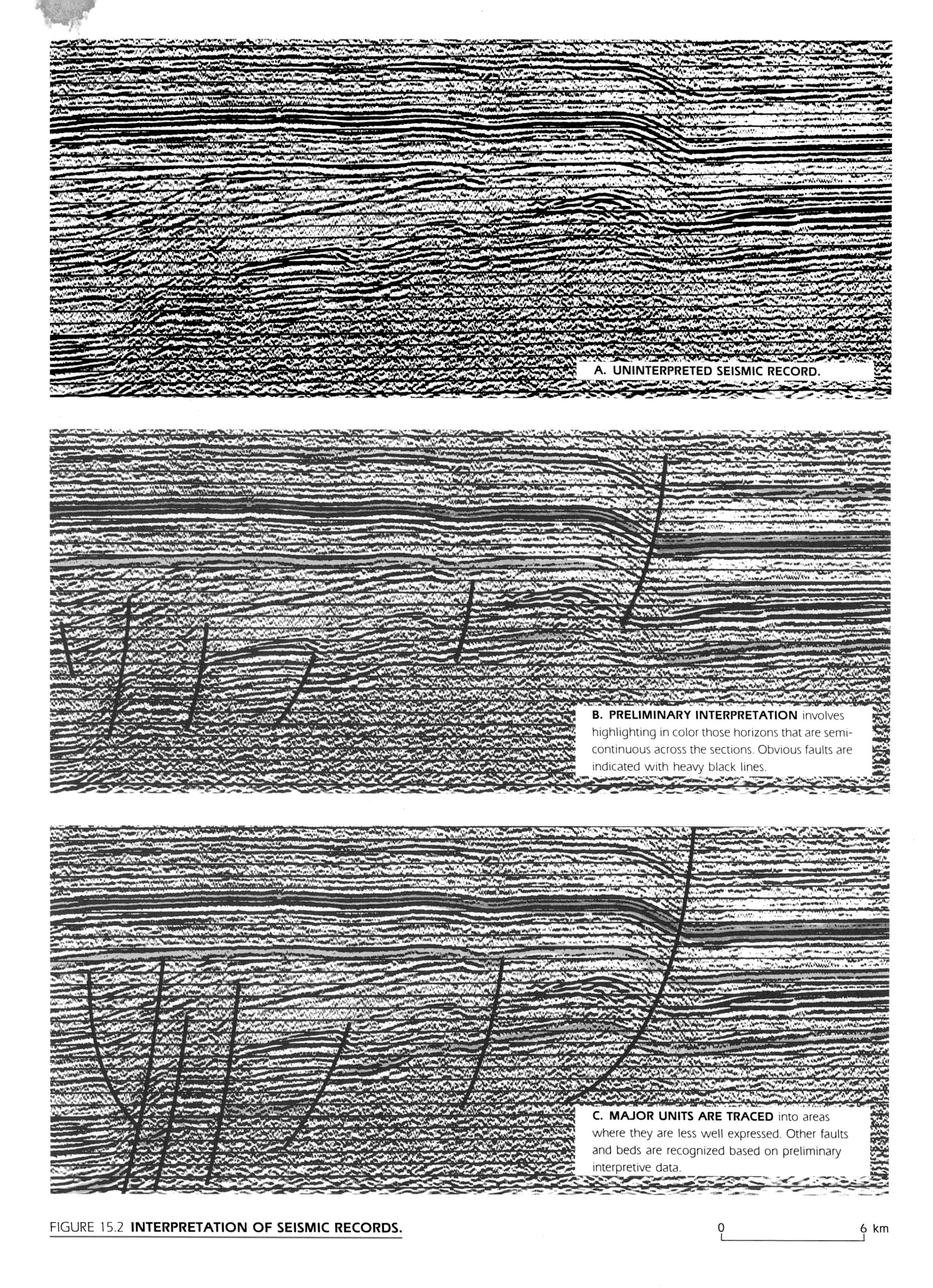

A. UNINTERPRETED SEISMIC RECORD.

B. PRELIMINARY INTERPRETATION involves highlighting in color those horizons that are semi-continuous across the sections. Obvious faults are indicated with heavy black lines.

C. MAJOR UNITS ARE TRACED into areas where they are less well expressed. Other faults and beds are recognized based on preliminary interpretive data.

FIGURE 15.2 **INTERPRETATION OF SEISMIC RECORDS.**

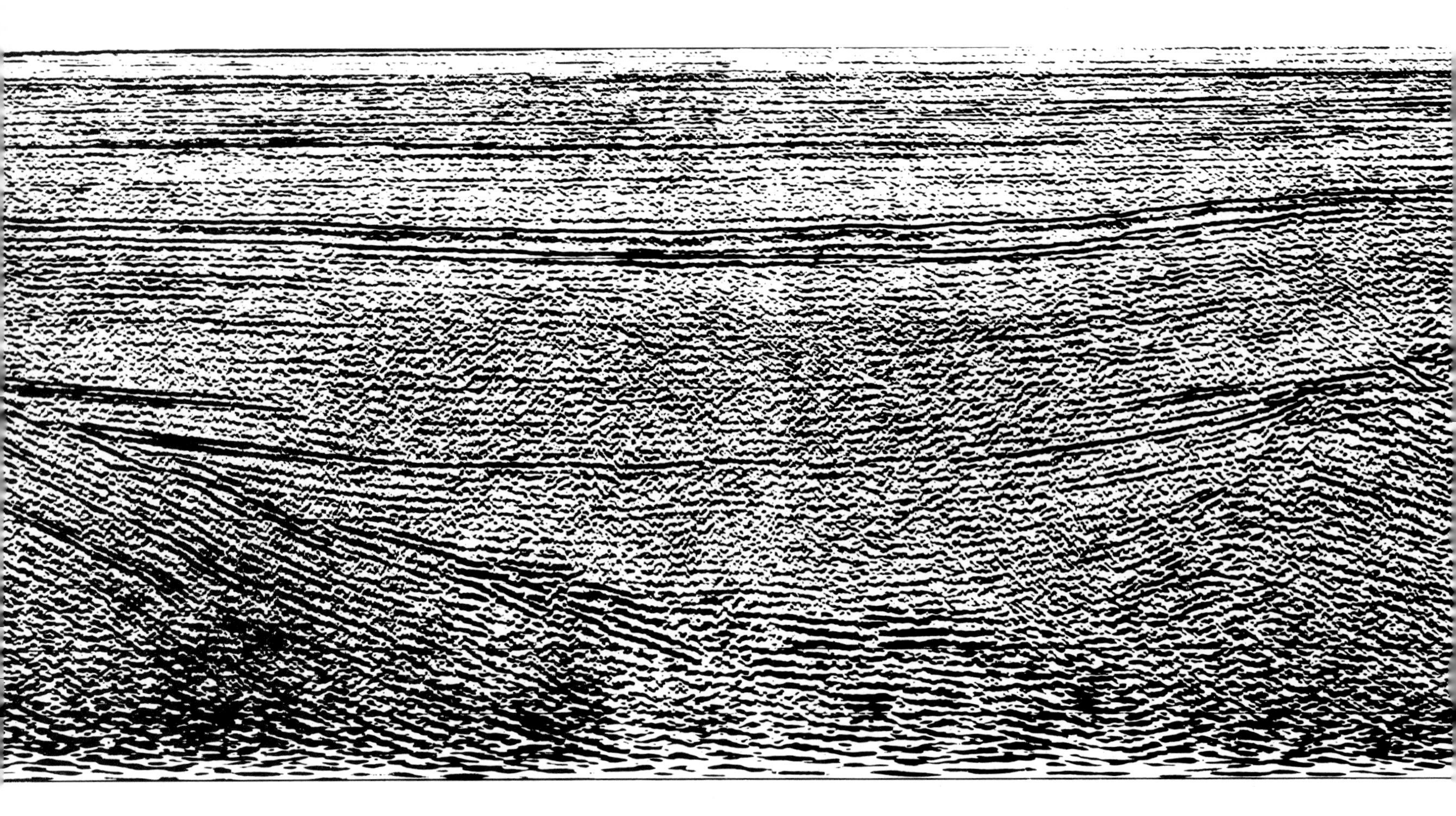

FIGURE 15.3

SEISMIC SECTION FOR INTERPRETATION PROBLEM.

PROBLEM

Study the example of a seismic section shown in Figure 15.2A and its interpretation in Figures 15.2B and 15.2C. Interpret the seismic section in Figure 15.3 by sketching the major boundaries between reflective horizons and by coloring the rock units between boundaries in different colors.

PROCEDURE

1 Lightly sketch with a colored pencil those horizons that are semicontinuous across the section and that separate rock units of similar inclination (see Figure 15.2B).

2 Sketch in the inferred faults, which are expressed by abrupt discontinuities in the major units (Figure 15.2B).

3 Interpolate the extension of the units into areas where they are less well expressed (Figure 15.2C).

4 Outline the major geologic events as interpreted from the data on the seismic section.

THE INTERNAL STRUCTURE OF THE EARTH

Two important facts permit us to determine the radius of the earth's core: (1) The *S*-wave shadow zone begins at 11,500 km (surface distance) from any earthquake epicenter, and (2) the radius of the earth is 6370 km.

PROBLEMS

1 Find the size (radius) of the liquid core by constructing a scale drawing.

2 Compare your result with the accepted value indicated in your textbook. Explain possible reasons for any difference between your result and the accepted value.

PROCEDURE

1 Use a drawing compass to draw a circle (to scale) representing the earth in cross section.

2 Pick any point on the perimeter of the circle and label it "Epicenter."

3 Plot the *S*-wave shadow zone on the circle as follows: The point on the perimeter of the circle where the shadow zone begins may be determined by measuring on the circumference of the circle the surface distance of 11,500 km.

4 Draw a line representing a seismic ray (a line drawn perpendicular to the wave front) between the epicenter and the beginning of the shadow zone. This line in effect defines the outer boundary of the shadow zone that *S* waves cannot enter.

5 Repeat this procedure for several different epicenters.

6 Now use a drawing compass to draw in the earth's core (i.e., a circle that is tangent with the seismic rays that define the shadow zone).

7 Measure the radius of the circle representing the core. (By using simple trigonometry, the size of the core can be obtained by calculation rather than measurement.)

PROBLEM

1 Using the same general procedure, try the following problem: If Mars has a radius of 3400 km and has an *S*-wave shadow zone beginning at 7000 km, what is the size of its core?

Exercise 16

VOLCANISM

OBJECTIVE

To recognize the major features resulting from volcanic activity and to understand how those features were produced by the earth's tectonic system.

MAIN CONCEPT

Volcanism is one of the major products of the tectonic system and is closely associated with plate boundaries.

SUPPORTING IDEAS

1 Landforms developed by volcanism are initially constructional features that build up and add material to the land surface. These landforms, however, are subject to stream erosion, and evolve through a series of stages from fresh cinder cones and lava flows to isolated remnants of volcanic necks, dikes, and lava-capped buttes.

2 The type of volcanic activity depends on the type of plate boundary.

DISCUSSION

Lava, being liquid, will flow down an existing slope and follow previously formed drainage systems. The initial effect of lava flows on the topography is to clog, block, and displace the local drainage system. At first, erosion is concentrated along the margins of the flow, where new drainage channels are established. As erosion proceeds, downcutting and slope retreat produce a lava-capped ridge called an *inverted valley.*

With continued erosion, the inverted valleys become progressively dissected into isolated mesas and buttes, which may rise over 1000 ft above the surrounding area.

Figure 16.1 illustrates the evolution of landforms in an area where local volcanism has occurred. Figure 16.1A shows the early stage of development, in which a lava flow is extruded, moves downslope, and enters a preexisting drainage system. The lava flow follows the river channel and disrupts the drainage. Lakes may be impounded upstream from the flow, and the original drainage is usually displaced so that streams are forced to flow along the margins of the lava. As a result, subsequent stream erosion is concentrated along the flow margins.

With continued erosion (Figure 16.1B), new stream valleys are cut along the lines of the displaced drainage, and grow gradually deeper and wider. The solidified lava occupying the former stream valley is usually far more resistant than the surrounding rock and is thus eroded into a long, sinuous ridge called an *inverted valley.* Cinder cones formed during the initial stage of volcanism are eroded rapidly, leaving in the conduit of the volcano only the resistant lava, which protrudes above the surrounding surface as a *volcanic neck.*

Figure 16.1C illustrates the late stage of landscape development. Erosion has lowered the landscape adjacent to the inverted valleys, so lava-capped mesas and buttes stand high above the surrounding surface as volcanic remnants. Slope retreat along the margins of the inverted valleys reduces the lava flows to small buttes, which in time will be eroded away completely. Volcanic necks and dikes will then remain as the only evidence of volcanic activity.

The surface expressions of the major types of volcanic activity are strikingly different from each other. Volcanism at convergent plate margins is violent and highly explosive. The lava is typically thick and viscous, and the repeated extrusions of ash, andesitic flows, and ash flows build up high, impressive composite volcanoes. In contrast, the fissure eruptions at divergent plate margins are relatively quiet and produce large volumes of fluid basaltic magma. Basaltic volcanism thus forms extensive lava plains and shield volcanoes.

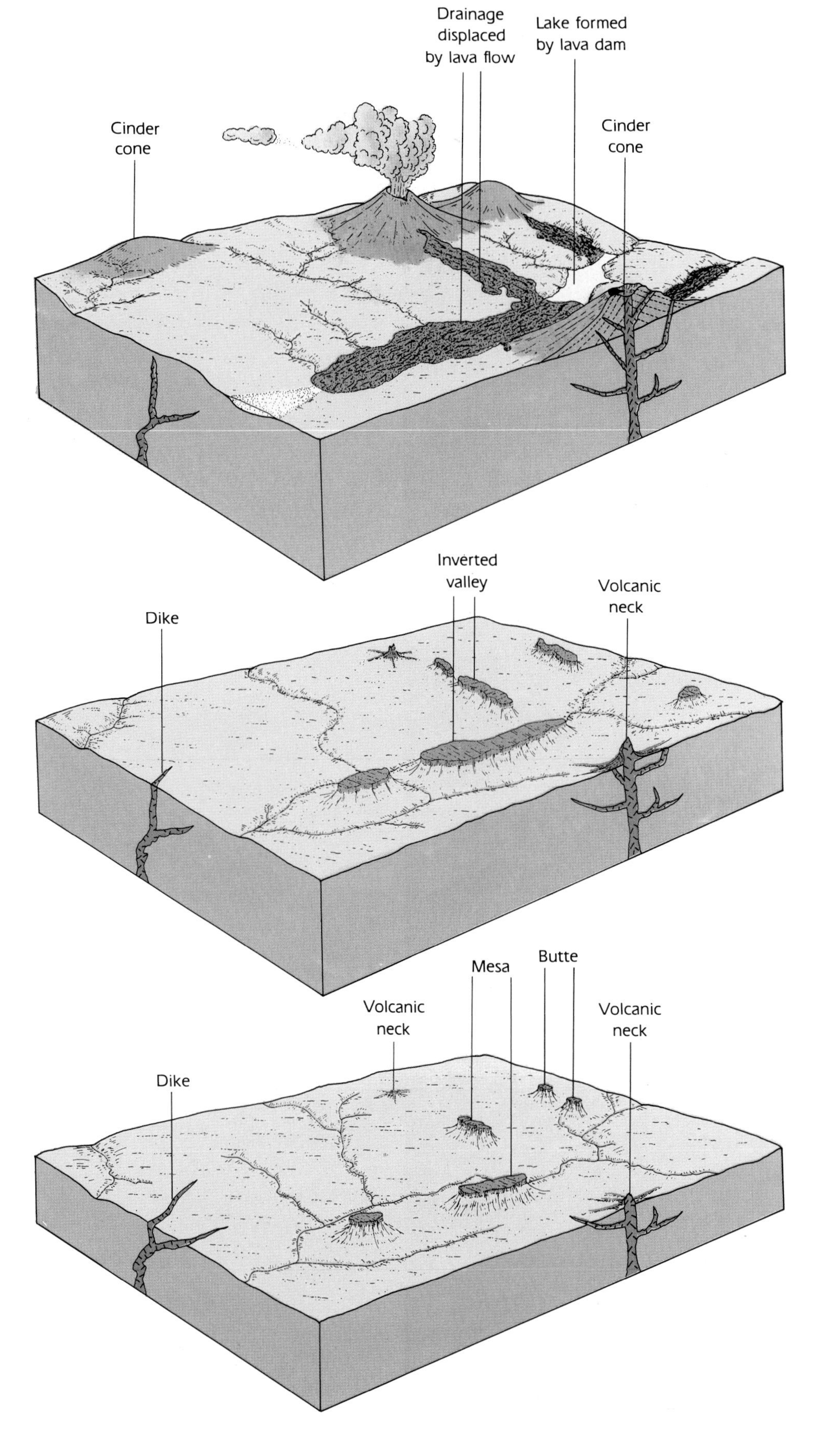

A. EARLY STAGE: Volcanic cones are a conspicuous feature of the landscape. Lava extruded from the vents flows down the regional slope, enters a stream valley, and follows it downgrade. Lakes commonly form behind lava dams and then overflow, forming new stream channels along the margins of the lava flow.

B. MIDDLE STAGE: Volcanic cones are worn down. The solidified lava flow occupying the stream valley is generally more resistant than the surrounding rocks, so erosion is concentrated along the flow margins. The flow develops into low, sinuous mesas called inverted valleys.

C. LATE STAGE: The inverted valleys are dissected into isolated mesas, which may stand over 1000 ft above the surrounding topography. Ultimately, volcanic necks and dikes are the only visible remains of cones and centers of extrusion.

FIGURE 16.1

EVOLUTION OF TOPOGRAPHY ASSOCIATED WITH VOLCANIC ACTIVITY.

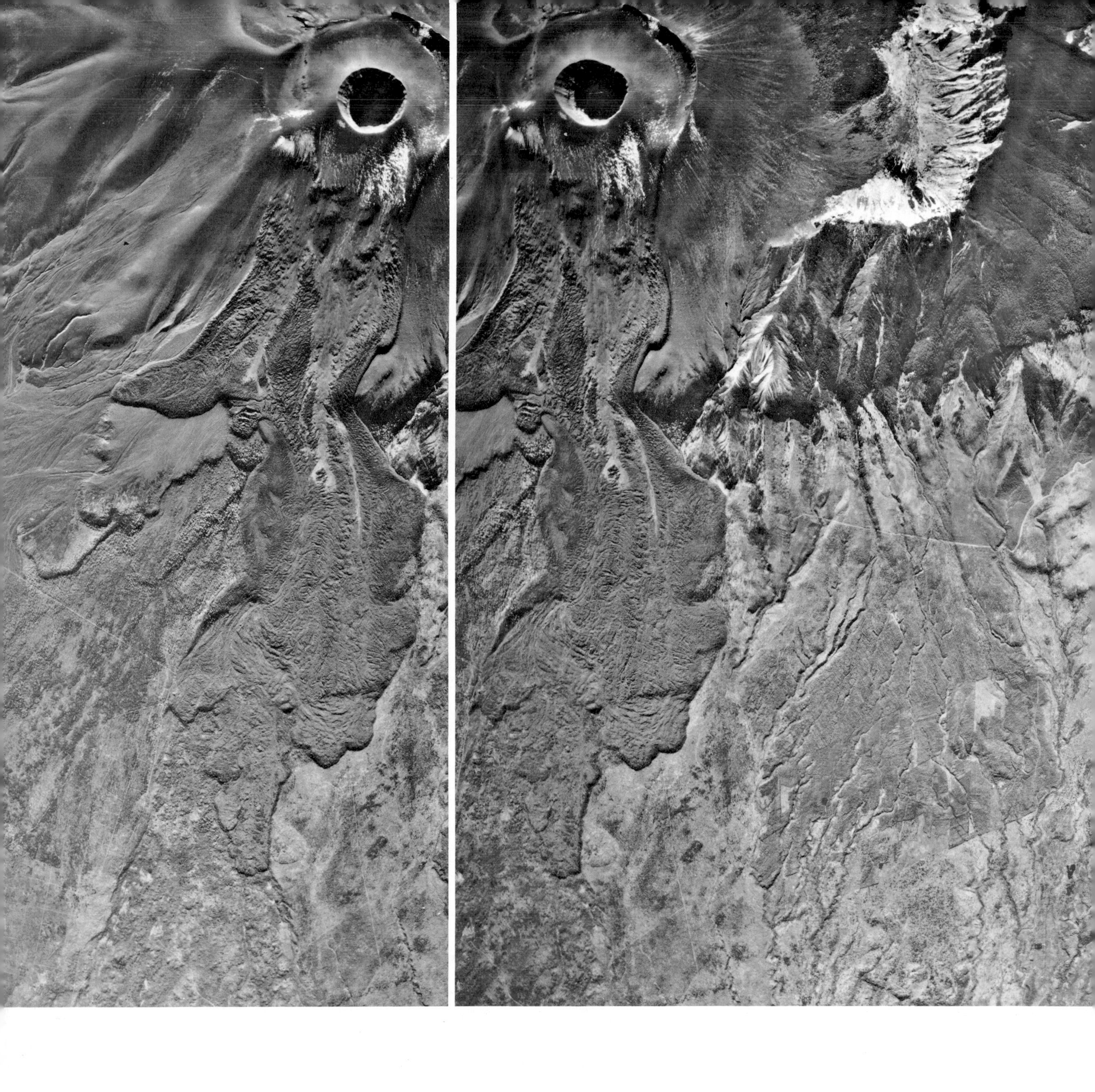

FIGURE 16.2: Asamayama, Japan

0 1 mile

PROBLEMS

1 Trace with a red pencil the outline of the major flow units. Why is new drainage located along the margins of the flow?

2 The most recent flow in this area has a typical rough, blocky surface. Does the flow follow any previous topographic depressions?

3 Has this flow been modified by erosion since it was extruded?

4 How many volcanic cones and cone remnants can you recognize in this area? How are the cones changed with time?

5 Identify and label the oldest lava flow in this area. How do the surface features of the older flows differ from the younger?

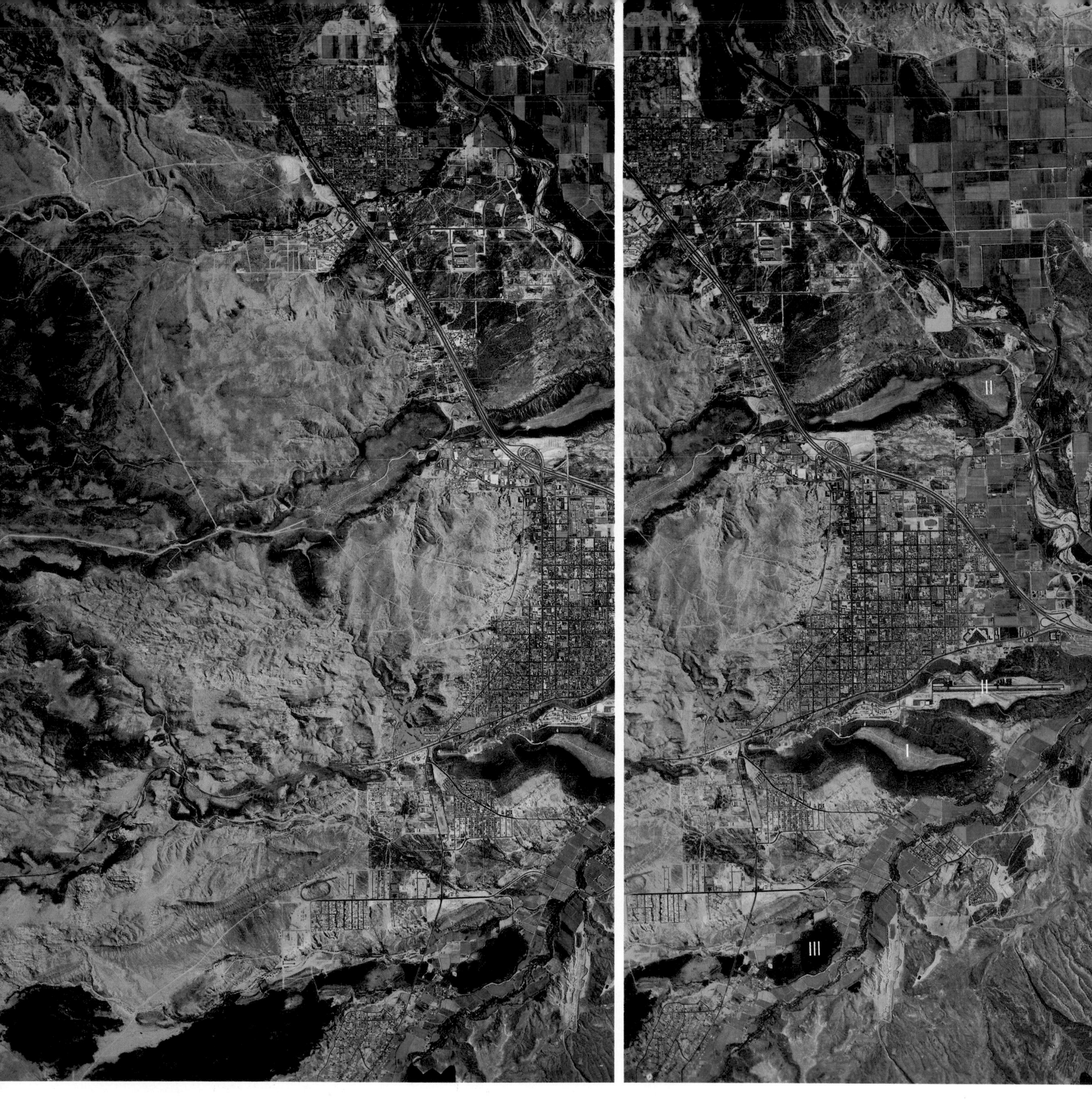

FIGURE 16.3: **St. George, Utah**

Color infrared photo 0 1 mile

PROBLEMS

1 Map the aerial extent of flows I, II, and III. Use a different colored pencil to shade each flow.

2 Note the small creeks on either side of flow III. How will erosion modify this flow with time?

3 Compare and contrast the surface features of flows II and III.

4 Draw an idealized topographic profile across the three major flows.

FIGURE 16.4: **Java**

PROBLEMS

1 Map the location of the volcanoes and remnants of volcanoes in this area. This can be done on a piece of tracing paper by marking the center of eruptions with an asterisk *. How many volcanoes did you map?

2 Identify an example of the younger volcanoes and an example of the older volcanoes. How do the volcanoes in this area change with time?

3 What is the origin of the terrain between the volcanoes?

4 What is the major rock type formed from the volcanic activity in this area?

5 The depth of water to the north is less than 200 meters. Off the south shore the depth is between 6000 and 7000 meters. Two hundred miles to the south the average depth is 5000 meters. Draw a generalized north-south profile across the area. What is the approximate total thickness of volcanic material in this area?

6 What major tectonic features occur on the ocean floor near this area.

7 Why is there such extensive volcanic activity in this area?

8 Compare and contrast the volcanism shown here with that shown in figure 166. Consider such things as volcanic landforms, rock type, nature of flows and cones, type of eruption, etc.

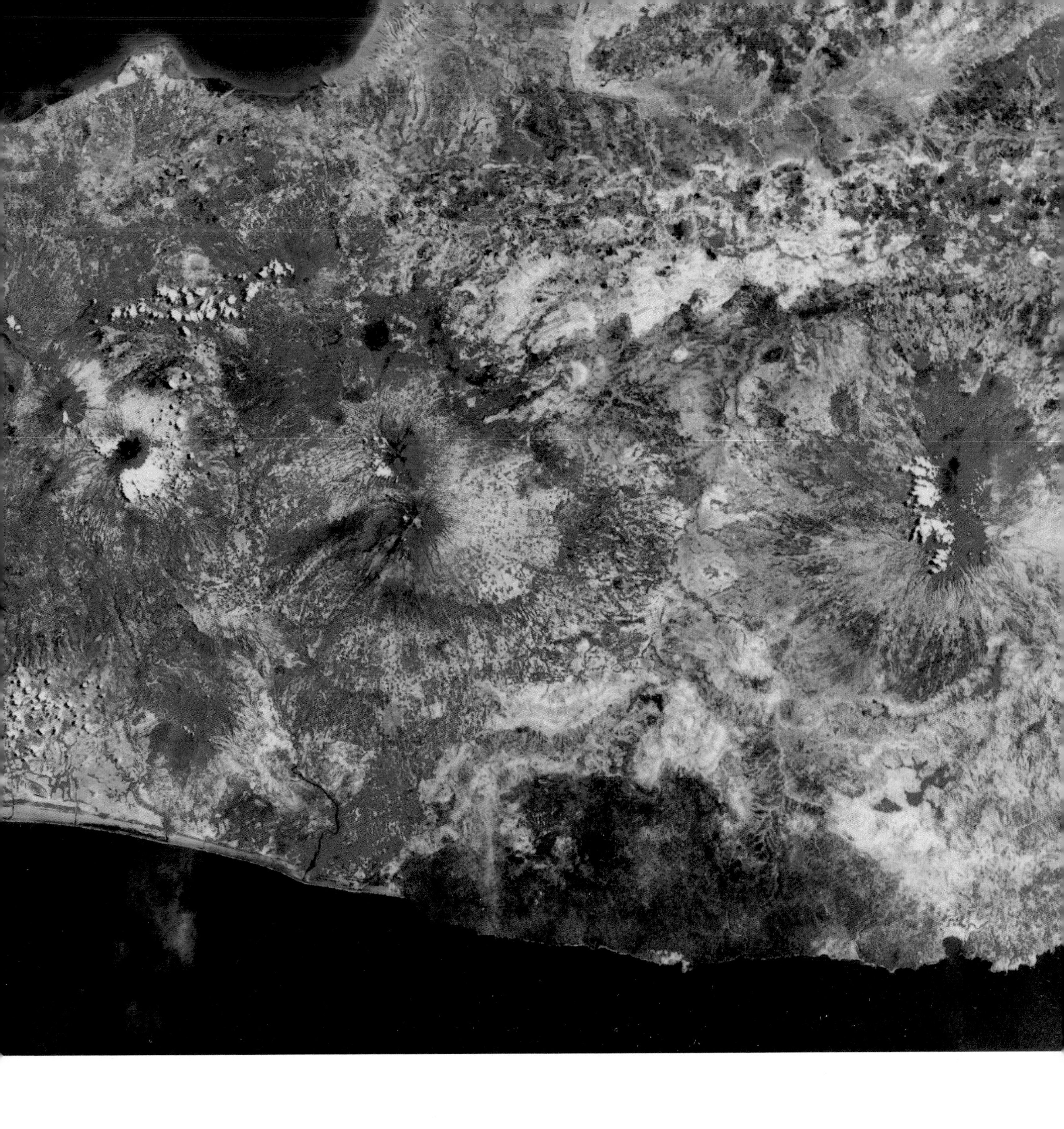

FIGURE 16.5: **Mount Saint Helens**

Color infrared photo 0 0.5 miles

The eruption of Mount Saint Helens on 18 May 1980, is certain to rank among the most significant geologic events occurring in the United States in the 20th century. Some of the major effects evident in the immediate vicinity were (1) ash flows, (2) mudflows, (3) debris avalanches, (4) downed timber from the blast, and (5) flooding.

PROBLEMS

1 Study the stereoscopic photographs (Figure 16.6), and describe briefly the effects of the Mount Saint Helens eruption.

2 Make a sketch showing the tectonic setting for this type of volcanism?

3 Compare and contrast the features shown in Figure 16.6 with those in Figure 16.5.

FIGURE 16.6: **Snake River Plain, Idaho**

Color infrared photo 0 0.5 miles

PROBLEMS

1 What type of volcanic eruptions characterize this area?

2 What are the linear features north of the most recent flows?

3 Why are there no volcanic cones in an area so dominated by the extrusion of lava?

4 Highlight each lava flow with a different color and indicate the relative age of the various flows.

Exercise 17

STRUCTURAL GEOLOGY

OBJECTIVE

To become acquainted with the major structural features of the earth's crust and with the outcrop patterns they produce, and to understand what each major feature implies about crustal mobility.

MAIN CONCEPT

Each major geologic structure (i.e., anticline, syncline, dome, basin, or fault) has a specific geometric form, and when exposed at the earth's surface, produces distinctive outcrop patterns that can be recognized on an aerial photograph or geologic map. These patterns provide a basis for interpreting the geologic history of an area.

SUPPORTING IDEA

1 Geologic maps show the distribution of rock formations, faults, and other structural features as they appear exposed at the earth's surface. Each map is a scale model of the rock bodies of the crust and a fundamental tool for analyzing and interpreting geologic data.

2 Many important aspects of the geologic history of a region can be interpreted from geologic maps, aerial photographs, and remote sensing data.

DISCUSSION

An accurate idea of the size, shape, and extent of rock bodies is essential if we are to interpret the geologic history of a region. Units of sandstone, limestone, or shale, which are called *formations*, may range in thickness from a few feet to several thousand feet, and may cover an area of more than 200,000 mi^2. Formations, therefore, exist as extensive layers of rock that can cover several states. Throughout its extent, a sequence of formations may be warped into folds, displaced by faults, or dissected by erosion.

The problem of studying the distribution and structure of rock bodies is similar to the problem of studying landforms; it is basically a problem of scale. An understanding of the configuration and regional extent of an entire rock body can be obtained only by careful geologic mapping. Geologic maps show the distribution of rock units as they crop out at the surface of the earth. Each rock unit is shown by a specific color or by a tone of gray. A legend accompanies the map to define the symbols and colorations and to indicate the age and stratigraphic relationships of the rock units. The conventional geologic map symbols for common structural features are shown in Figure 17.1

To understand the surface expression of various features, you should first review the basic three-dimensional geometry of the major structures (Figure 17.2) and the discussion in your textbook.

BLOCK DIAGRAMS AND GEOLOGIC MAPS OF MAJOR STRUCTURAL TYPES

The major types of structural features produce distinctive patterns on geologic maps. Examples of various common structural patterns are shown in Figures 17.3–17.22. On each page in this section, the top figure shows a block diagram consisting of a bird's eye view of the structural feature and its topographic expression. In the same figure, the outcrop pattern of each rock unit shown on the block diagram is projected onto a horizontal plane to represent a geologic map in perspective. The bottom figure on each page is a stereoscopic photograph of the structure. This arrangement makes it easy to relate the structure shown on the block diagram to the map patterns. Careful study and understanding of these diagrams and stereoscopic photographs are important prerequisites to the effective interpretation of geologic maps. Continued reference to these figures throughout the exercise is encouraged.

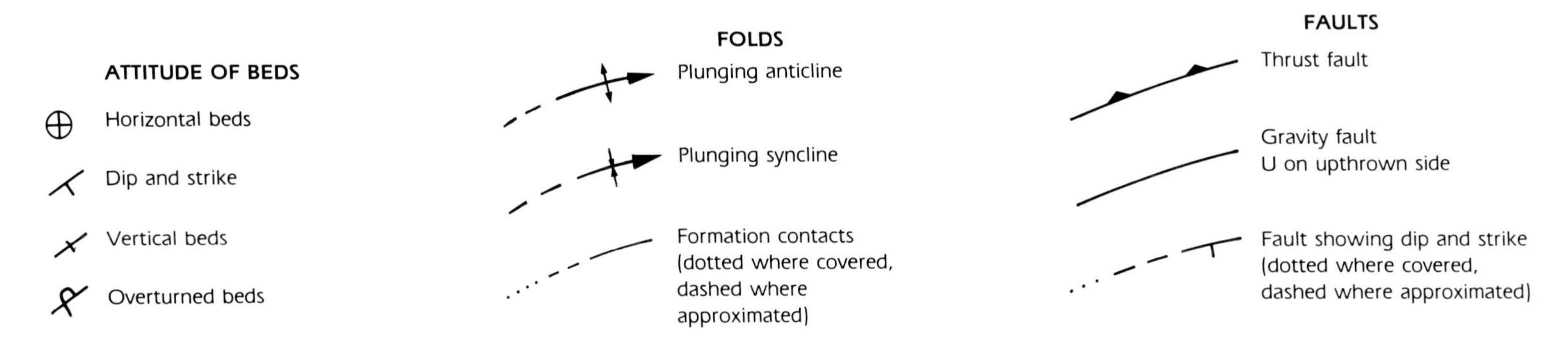

FIGURE 17.1 **GEOLOGIC MAP SYMBOLS.**

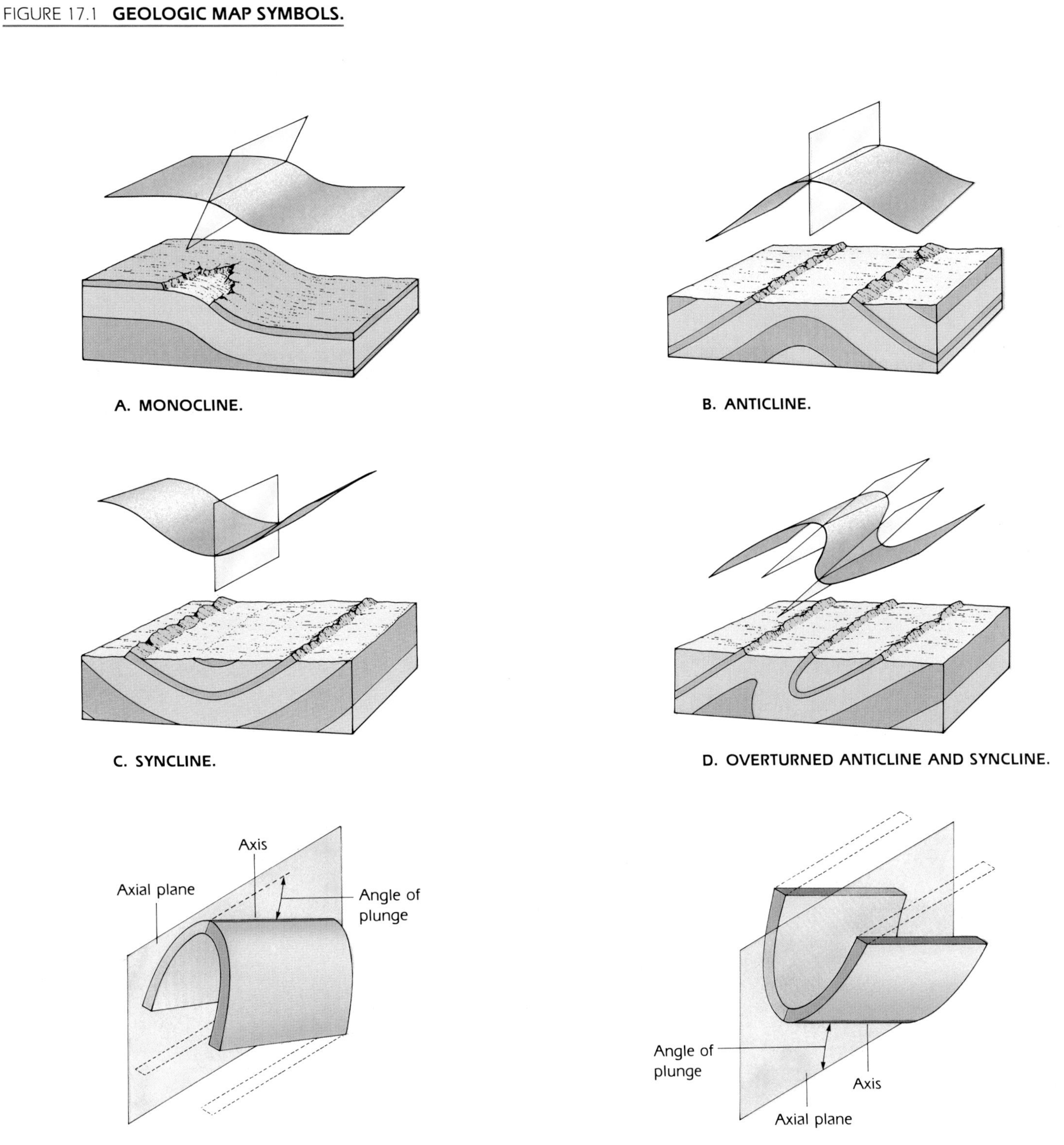

FIGURE 17.2 **THREE-DIMENSIONAL GEOMETRY OF FOLDS.**

FIGURE 17.3: Outcrop Pattern of Horizontal Strata

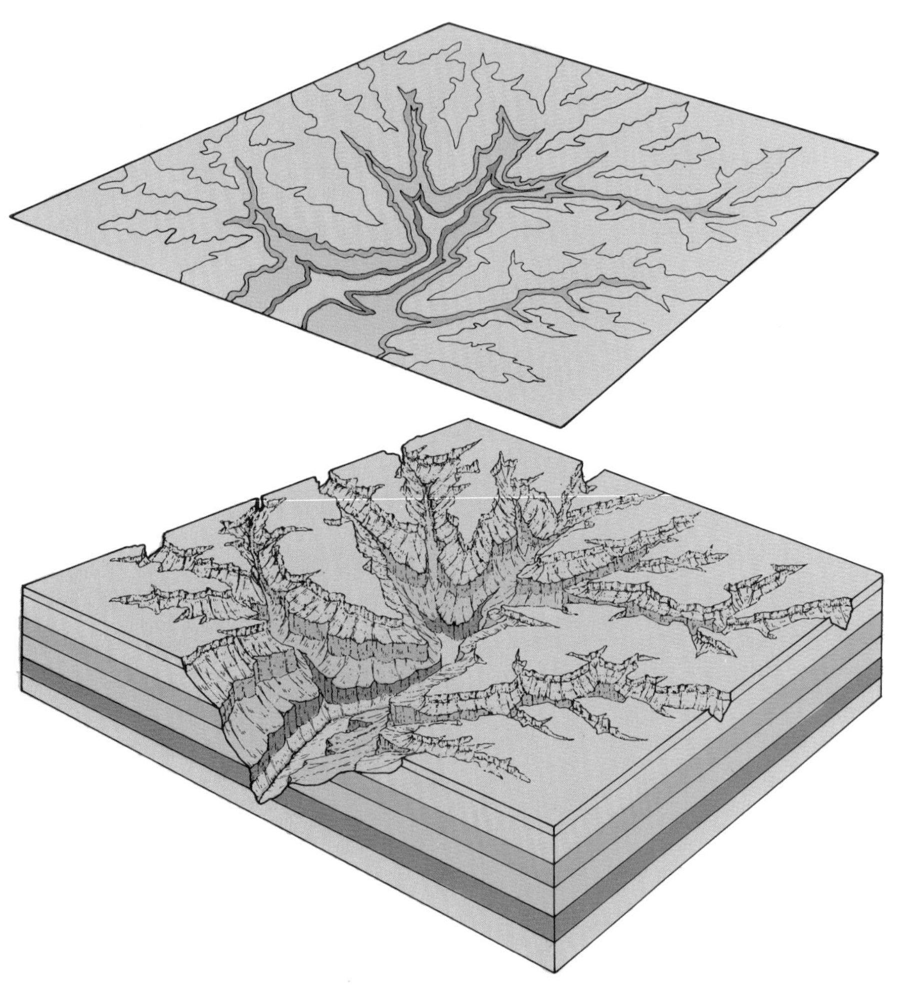

A dendritic drainage pattern characteristically develops on horizontal strata. As erosion proceeds, the river system cuts canyons, or valleys, in which a succession of older rock units is exposed. The outcrop patterns of horizontal strata exposed on the valley walls are therefore parallel to the stream valleys and produce a dendritic pattern on the geologic map. The contacts of the rock units of horizontal strata are essentially parallel to the topographic contour lines. Cliffs and gentle slopes generally develop on resistant and nonresistant beds, respectively, and thus produce variations in the width of the map outcrop pattern of the various rock units. On a steep cliff, the upper and lower contacts (as seen on the map) will appear to be close together, whereas on a gentle slope of the same formation, the contacts will appear to be farther apart. As seen on a map, the width of the outcrop belts of horizontal strata is therefore not dependent solely on the units' thickness, but depends on the topography as well.

0 0.5 miles

FIGURE 17.4: Horizontal Strata—Kansas

PROBLEMS

1 To learn more about geologic mapping, construct a geologic map of the area shown in stereoscopic view by extending the contact between the formations marked by black lines.

2 Number the beds according to age (i.e., 1 = oldest, 2 = next oldest, and so forth).

3 To show the relationship between structure and topography, sketch a geologic cross section through this area.

FIGURE 17.5: Outcrop Pattern of Inclined Strata

When a rock sequence is tilted and truncated by erosion, the outcrop patterns of the rock units (viewed on a regional scale) appear as bands that are roughly parallel. Important variations in the basic map pattern develop in areas dissected by erosion. These variations should be analyzed carefully, for they provide important information about the subsurface structure. When dipping strata are traced across a valley, a V-shaped outcrop pattern that points in the direction of the dip is produced. The size of the V is inversely proportional to the magnitude of the dip, and we can make the following generalizations.

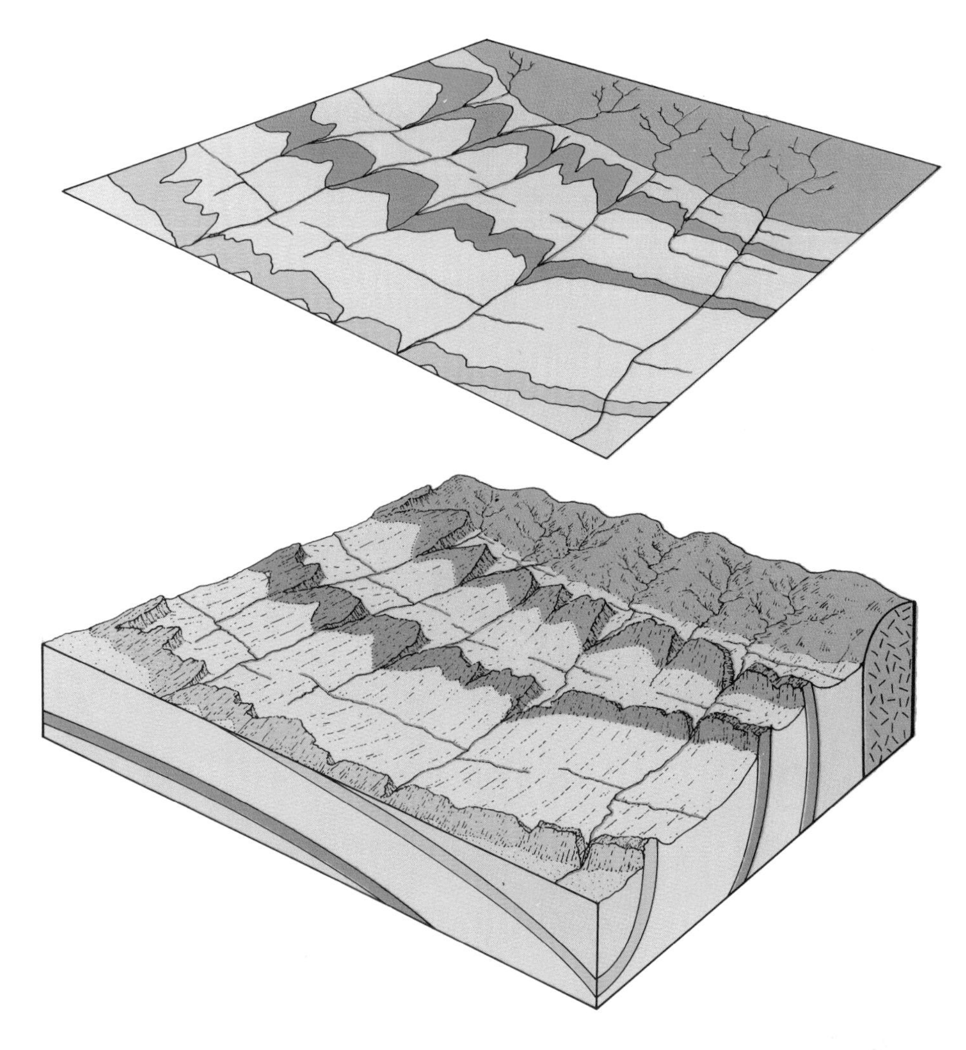

1 Low-angle dip produces a large V (left side of block diagram).

2 High-angle dip produces a small V.

3 Vertical dip produces no V (right side of block diagram).

Careful study of Figure 17.5 will reveal more relationships basic to geologic maps:

1 Older beds dip toward younger beds (unless the sequence is overturned).

2 Outcrop width depends on

- **a** Thickness of the beds
- **b** Dip of the beds (the lower the dip, the wider the outcrop)
- **c** Slope of the topography (the steeper the slope, the narrower the outcrop)

FIGURE 17.6: Colorado Plateau—Utah

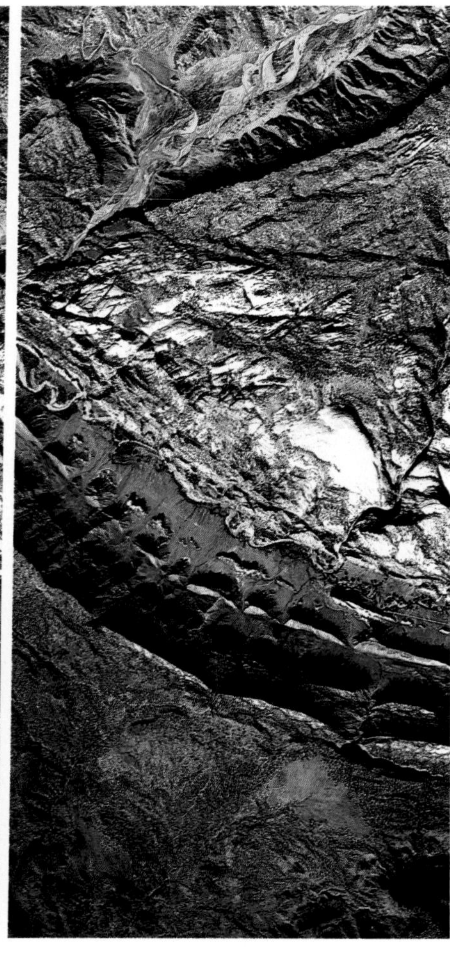

0 0.5 miles

PROBLEMS

1 How can you tell the direction in which the rocks are dipping in this area?

2 Estimate the angle of dip.

3 Show the dip and strike of the rocks in this area by plotting a series of appropriate symbols on the photograph. Show the amount of dip estimated in problem 2.

4 Make a geologic map of the area by extending the contacts of the formations labeled on the photograph.

5 Sketch a geologic cross section perpendicular to the strike of the beds. Show both topography and geologic structure.

FIGURE 17.7: Outcrop Pattern of a Dome

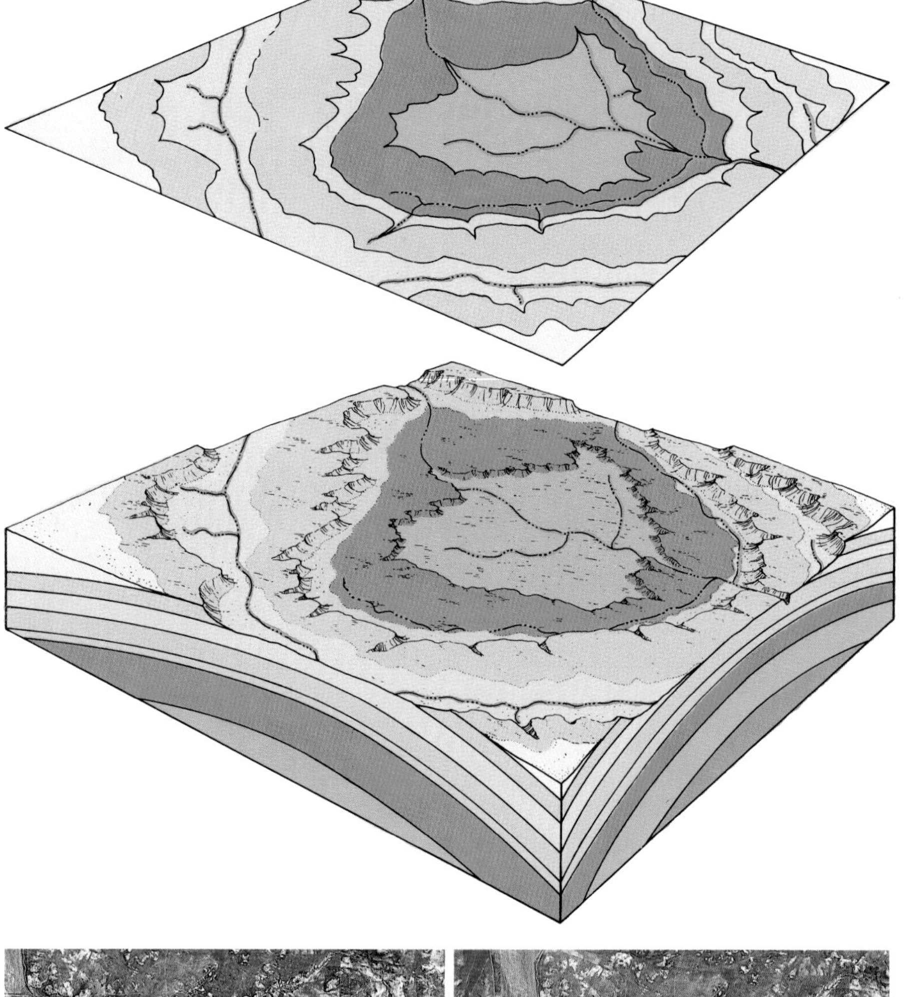

Eroded structural domes form a roughly circular to elliptical outcrop pattern in which the beds dip away from the crest, or central part of the dome. Domes range in size from small upwarps that are a few feet in diameter to regional features that cover hundreds or thousands of square miles. As can be seen in Figure 17.7, the oldest beds are exposed in the central part of an eroded dome, and progressively younger formations are exposed outward from the center of the structure.

Streams cutting across the resistant beds enable us to apply the rule of Vs to determine the direction of dip. If the relative age of each rock unit is shown on the map, then a dome is easily identifiable, because the older rocks are in the center of the structure. Note also that a cliff or scarp formed on the beds of a dome faces inward, and the dip of the slope formed on the top of major units is inclined away from the center of the dome.

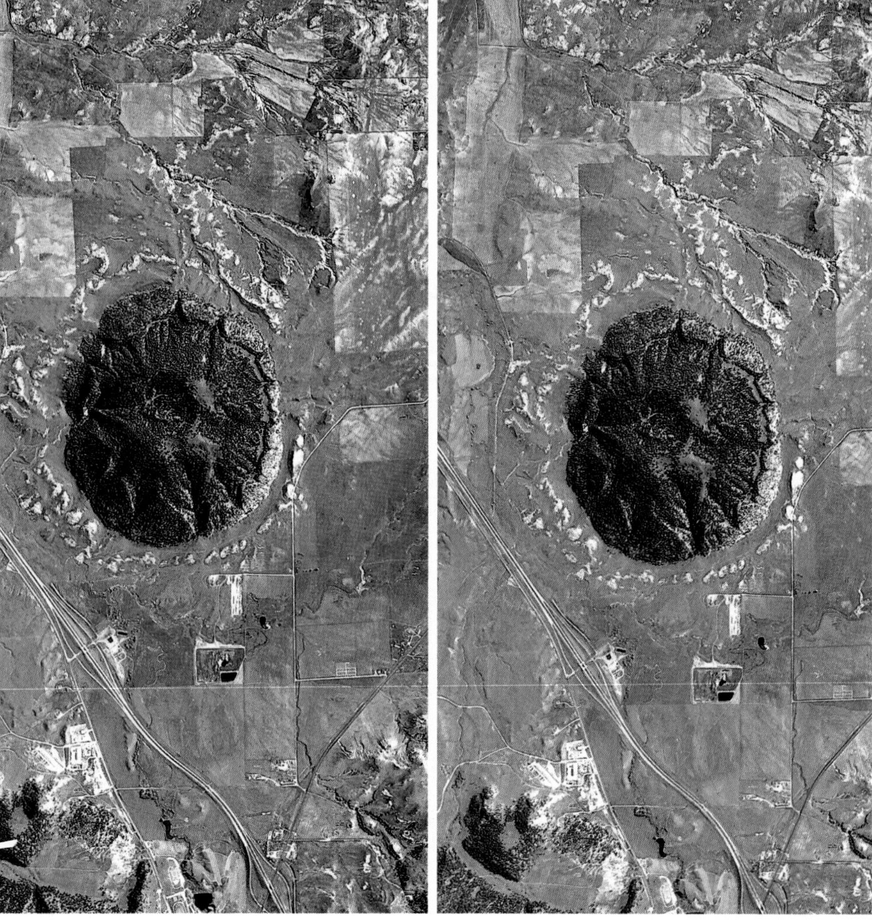

FIGURE 17.8: Domal Structure—Montana

PROBLEMS

1 Make a geologic map of this area by extending the contacts of the formations labeled on the photograph.

2 Estimate the angle of the dips, and plot several dip and strike symbols on your map.

3 Draw a geologic cross section through the structure.

4 Show how the cross section would change if the oldest formation were weak and nonresistant to erosion.

FIGURE 17.9: Outcrop Pattern of a Basin

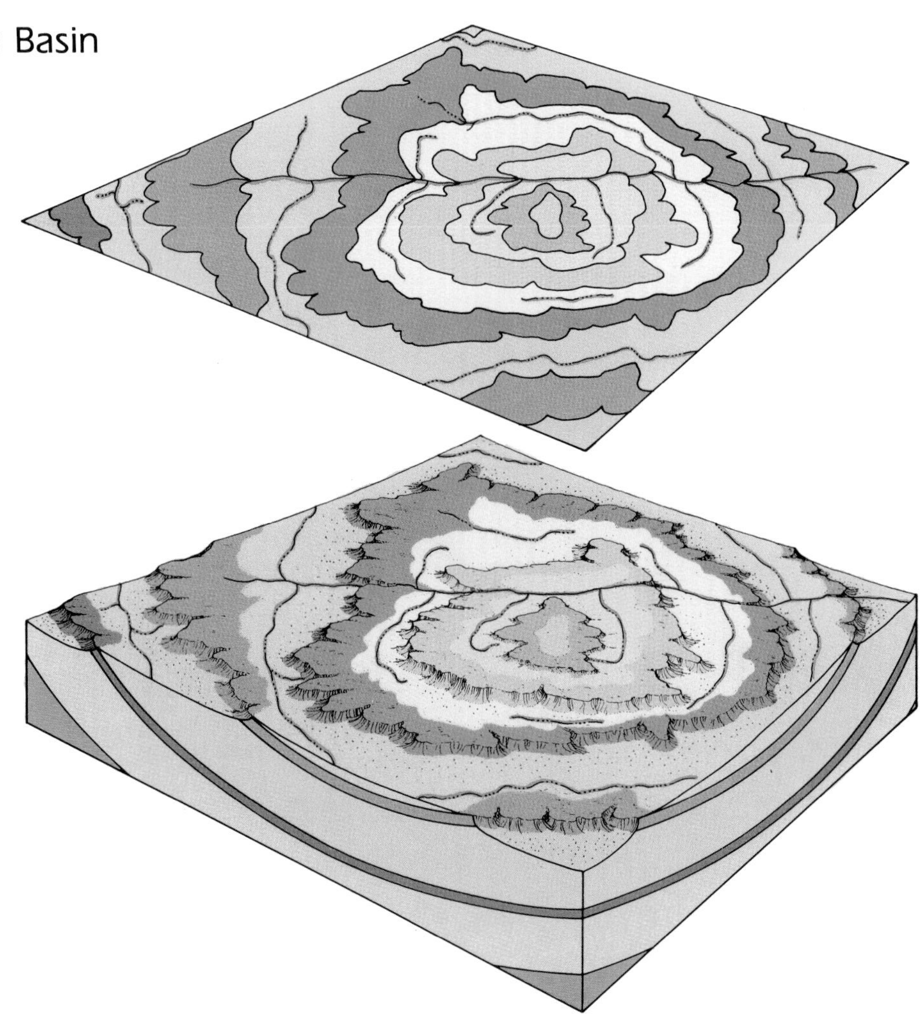

A structural basin, when eroded and exposed at the surface, displays an elliptical or circular outcrop pattern similar to that of an eroded dome. (Compare Figures 17.7 and 17.9.) The general outcrop pattern of both structures is similar, but two major features enable us to distinguish readily a basin from a dome: (1) Younger rocks crop out in the center of a basin, whereas older rocks are exposed in the center of a dome. (2) If the structure has been dissected by stream erosion, the V in the outcrop points toward the center of a basin and away from the center of a dome. In addition, the cliff or scarp formed on the resistant rocks of a basin faces outward, and the dip slope is inclined toward the center of the structure. This is exactly the opposite of the direction in which the slope is inclined in an eroded dome.

FIGURE 17.10: Basin—Alaska

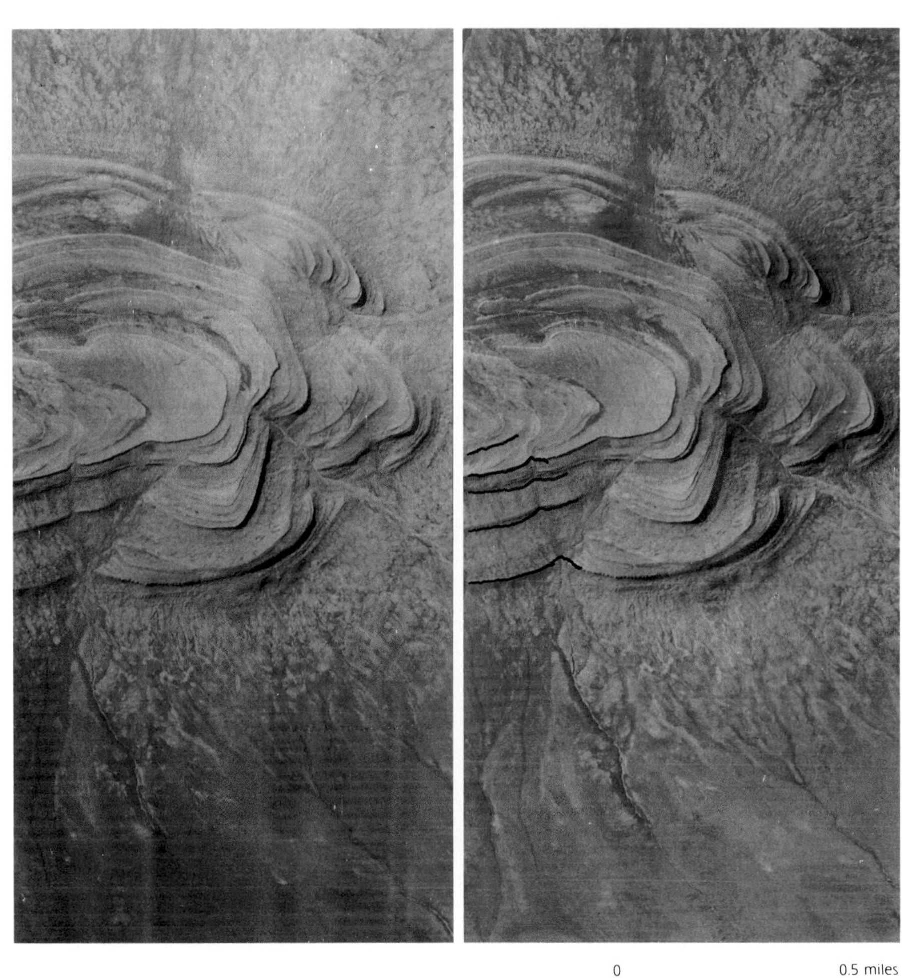

PROBLEMS

1 Make a geologic map of this area by extending the contact of the formations on the photograph.

2 Label the formations 1, 2, 3, and so forth, from the oldest to the youngest formation, respectively.

3 How can you tell the direction of dip?

4 Draw a cross section through the structure.

5 This entire area is part of a basin. Why are there no ridges in the northern and southern part of the area?

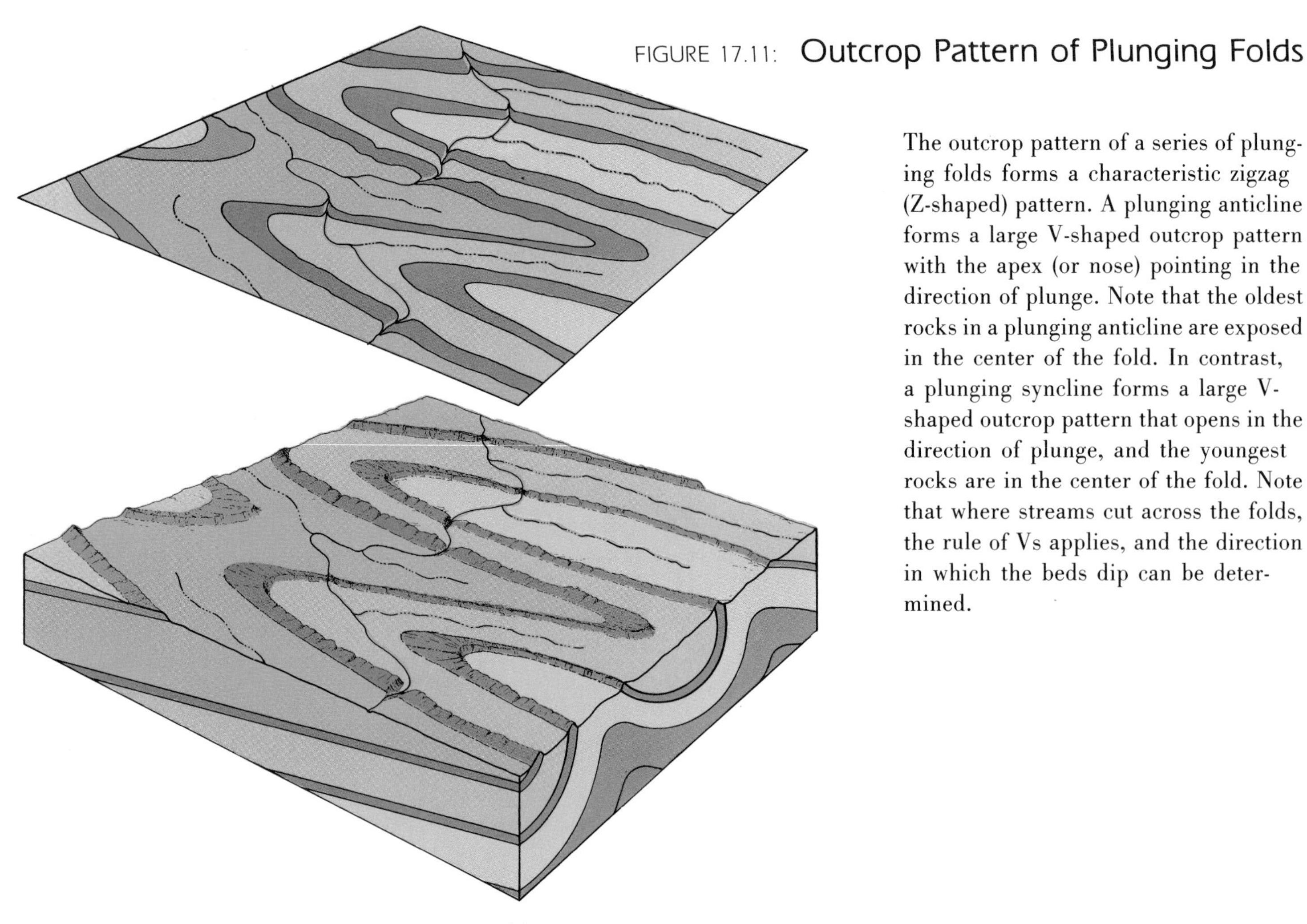

FIGURE 17.11: Outcrop Pattern of Plunging Folds

The outcrop pattern of a series of plunging folds forms a characteristic zigzag (Z-shaped) pattern. A plunging anticline forms a large V-shaped outcrop pattern with the apex (or nose) pointing in the direction of plunge. Note that the oldest rocks in a plunging anticline are exposed in the center of the fold. In contrast, a plunging syncline forms a large V-shaped outcrop pattern that opens in the direction of plunge, and the youngest rocks are in the center of the fold. Note that where streams cut across the folds, the rule of Vs applies, and the direction in which the beds dip can be determined.

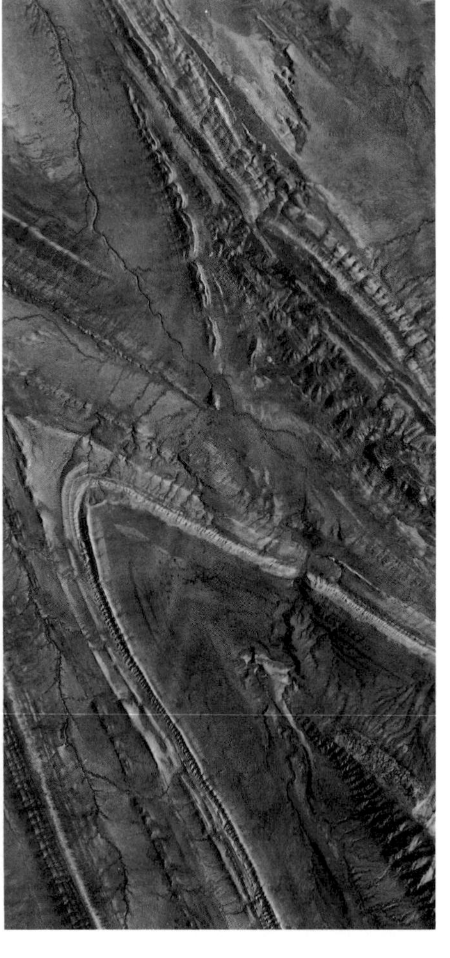

0 0.5 miles

FIGURE 17.12: Plunging Folds—Wyoming

This area illustrates the fine details produced by erosion on folded rocks in an arid region. Resistant formations form ridges, nonresistant units form valleys, and specific textures caused by stream patterns form on the different rock units.

PROBLEMS

1 Make a geologic map of this area by tracing the contacts of the major resistant formations. Show dip and strike symbols and the axis of all folds.

2 Draw a geologic cross section through the structure from the lower left corner to the middle of the photograph.

3 Trace the major drainage pattern with a blue pencil. How has the structure controlled the drainage?

FIGURE 17.13: Outcrop Pattern of an Unconformity

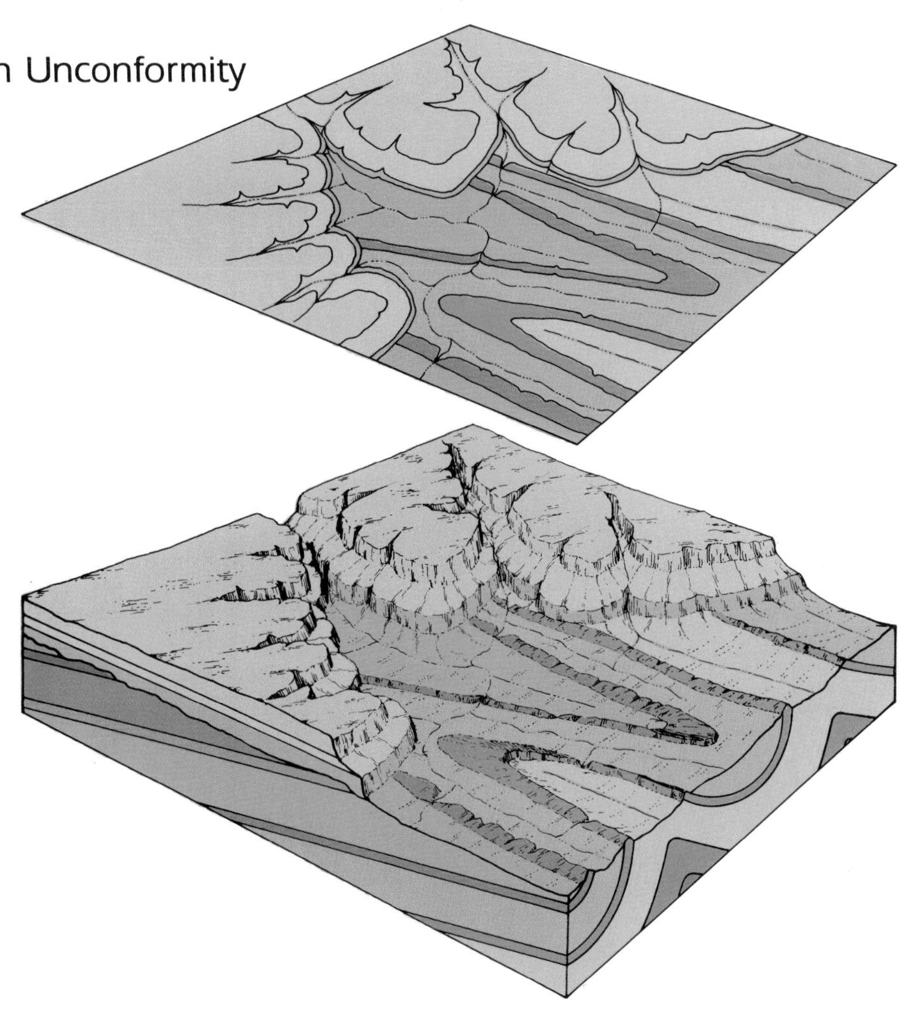

Angular unconformities can be recognized on geologic maps by interruptions, or discontinuities, in the outcrop patterns. The outcrop pattern of older structures is partly covered by younger strata, so on the geologic map, the contacts of the older structures will terminate abruptly against the patterns of the overlying younger beds. In the diagram shown here, the oldest sequence has been warped into plunging folds, eroded, and subsequently covered by a younger sequence of strata. A second period of erosion has partly removed the younger strata and exposed segments of the older folds. The angular unconformity is located at the base of the sequence of younger horizontal strata. All of the map patterns of the older strata terminate against this contact. Vs in the trace of the unconformable surface indicate the direction in which the unconformity dips.

FIGURE 17.14: Angular Unconformity— Montana

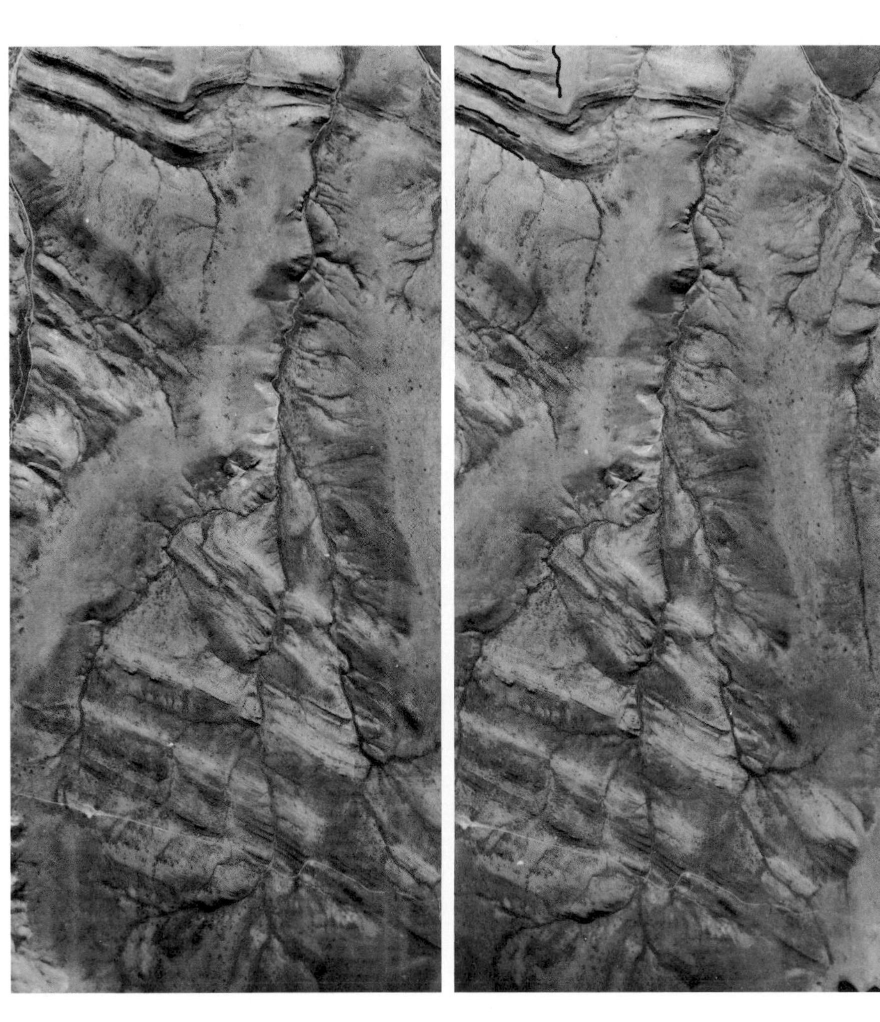

PROBLEMS

1 The unconformity can be recognized by differences in texture, tone, and structural features of the older and younger rocks. Trace the unconformity across the area.

2 What is the structure of the older rocks?

3 What is the structure of the younger sequence of rocks?

4 Make a geologic map of the area and outline the sequence of major geologic events.

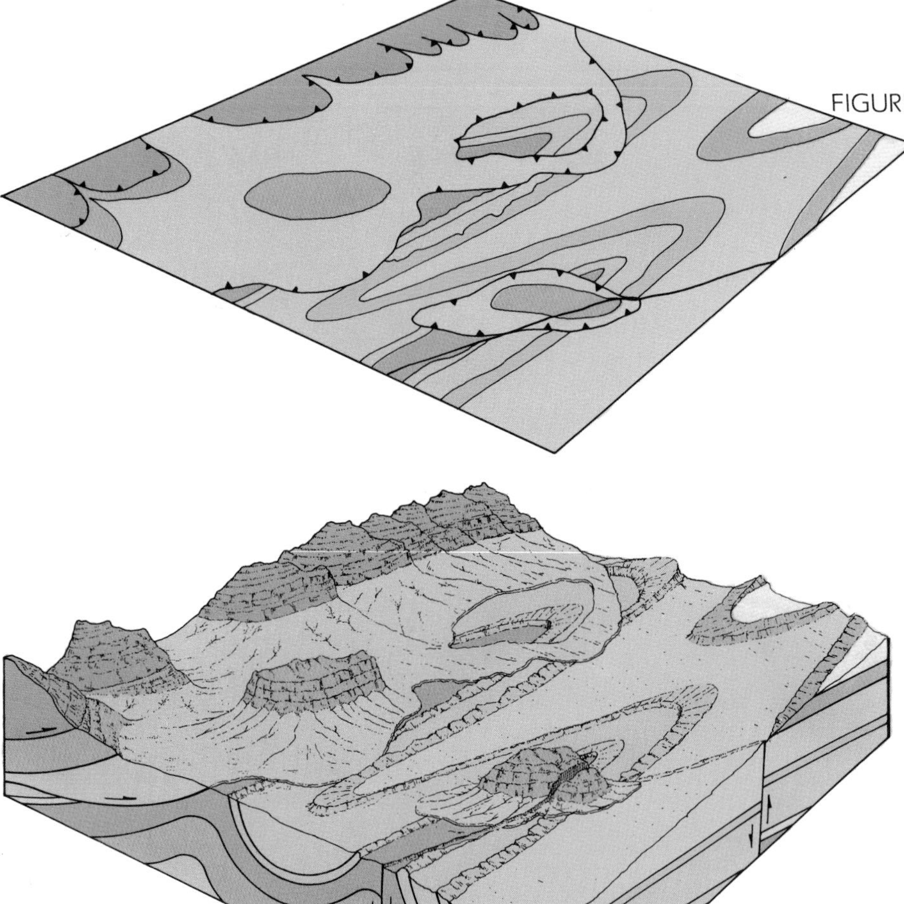

FIGURE 17.15: **Outcrop Pattern of Faults**

Fault patterns on geologic maps are distinctive; they appear as lines or zones of displacement that abruptly offset structures and terminate contacts between rock formations.

Thrust faults generally dip at a low angle. Because of the low-angle dip, the pattern of the fault trace is characteristically irregular and similar in many respects to the trace produced by low-dipping angular unconformities. In Figure 17.15, thrust faults are located at the base of the formations colored purple and blue. The trace of a thrust fault commonly forms a V across valleys, with the V pointing in the direction in which the fault dips.

Normal and reverse faults usually dip at a high angle, so their outcrop patterns are relatively straight. Since older rocks are generally exposed on the upthrown block, the relative movement on most high-angle faults can be determined from the map relations alone.

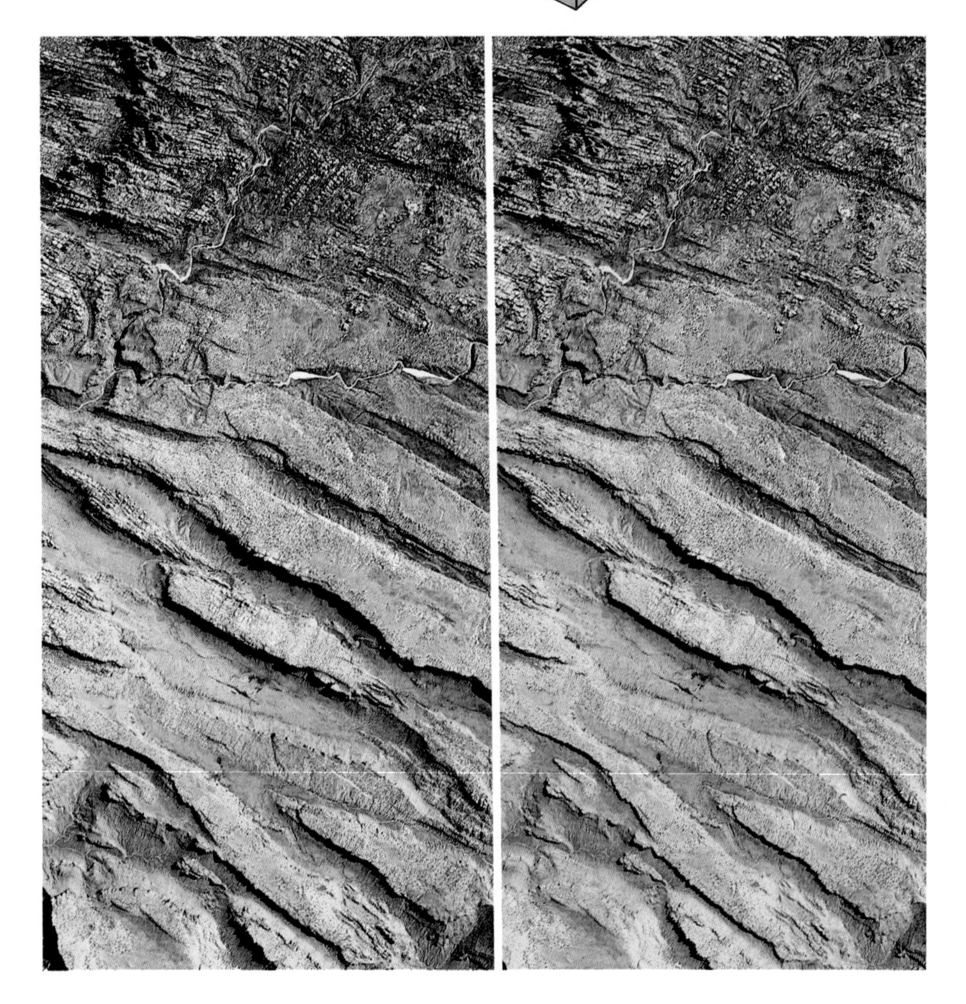

FIGURE 17.16: **Faults—Colorado Plateau**

Thick, resistant sandstone formations in the Colorado Plateau commonly show the characteristics of fracture systems in remarkable detail. The sandstone formation in this area is nearly horizontal. Normal faults are shown where the surface is displaced vertically. Joint systems are accentuated by weathering and are expressed as cracks.

PROBLEMS

1 Map the normal faults in this area, and label the relative movement *D* on the downthrown block and *U* on the upthrown block.

2 Compare and contrast the cliff or scarp formed by normal faulting (Figure 17.16) with that produced by strike-slip faulting (see Figure 17.18).

FIGURE 17.17: Outcrop Pattern of Strike-Slip Faults

In a strike-slip fault, the displacement is parallel to the strike of the fault plane. Displacement on strike-slip faults may reach several hundred miles, so rock types of very different structure and geologic characteristics may be placed side by side after prolonged periods of movement. The trend of strike-slip faults is typically straight, in contrast to the irregular trace of thrust faults and the zigzag trace of normal faults. Small slivers or slices of foreign rock bodies may be caught in the strike-slip fault zone and are commonly expressed either as elongate troughs or ridges.

The lateral displacement of the crust in strike-slip faults does not produce high scarps. The fault line is, however, commonly marked by structural and topographic discontinuities, linear ridges and valleys, and offset drainage patterns. The offset drainage is usually very significant, because it indicates the direction of displacement.

FIGURE 17.18: Strike-Slip Faults— California

PROBLEMS

1 Trace the fault line with a red pencil and the drainage lines in blue. Is the fault in a single plane or in a zone of several parallel planes?

2 Study the stream pattern on both sides of the fault. How does the stream pattern show the relative movement along the fault?

3 Draw arrows to show the relative movement on either side of the fault plane and determine the amount of recent displacement?

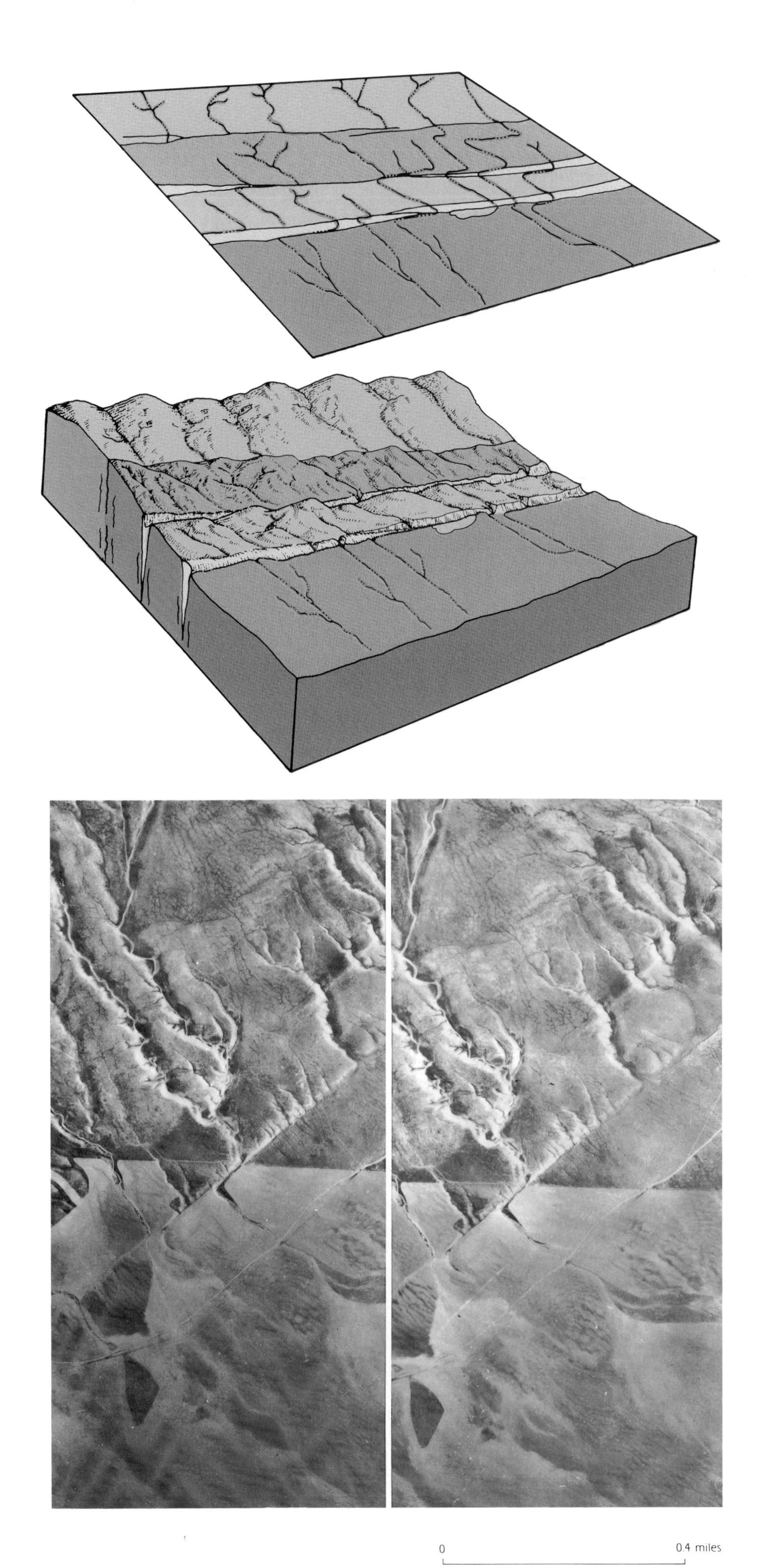

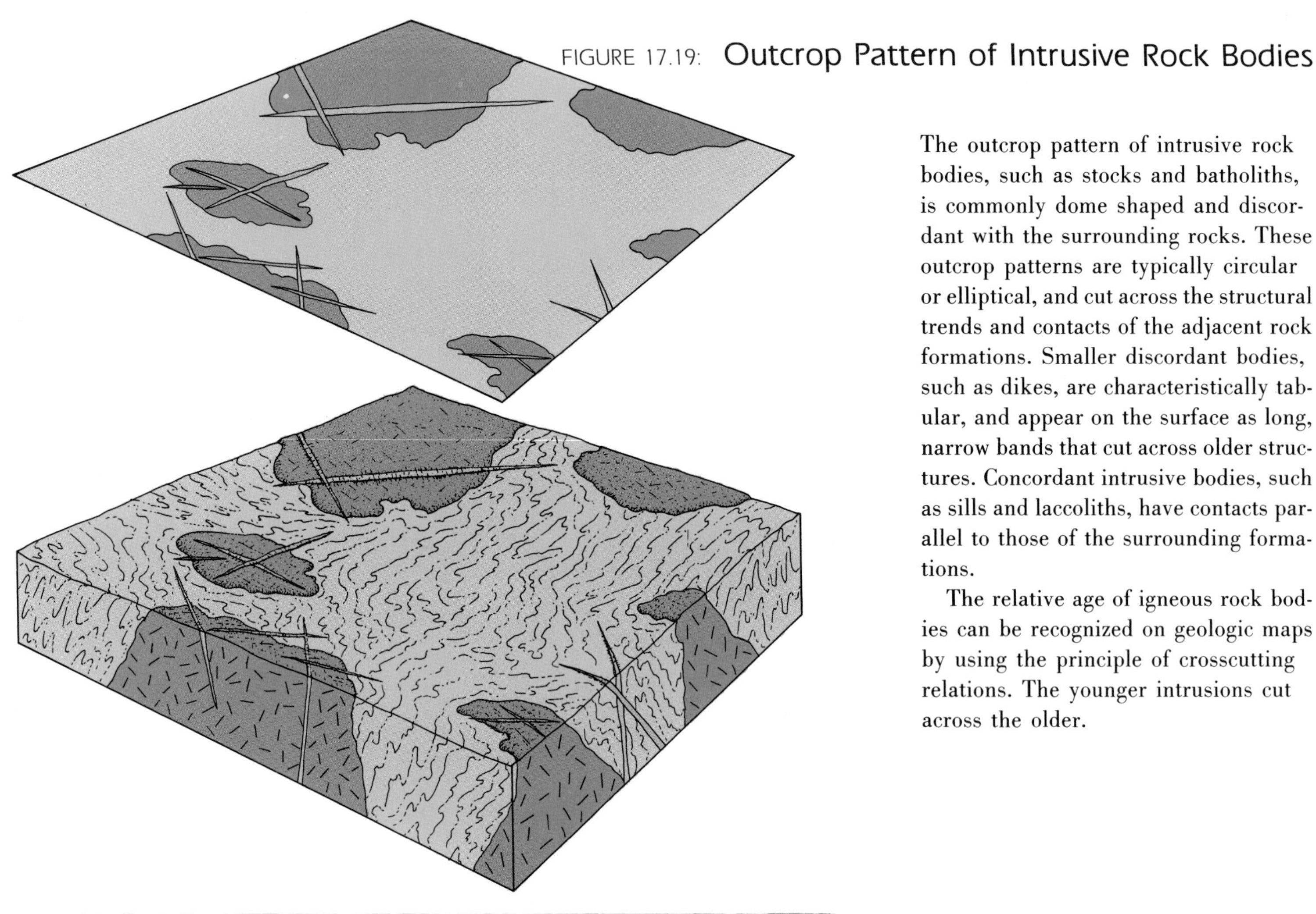

FIGURE 17.19: **Outcrop Pattern of Intrusive Rock Bodies**

The outcrop pattern of intrusive rock bodies, such as stocks and batholiths, is commonly dome shaped and discordant with the surrounding rocks. These outcrop patterns are typically circular or elliptical, and cut across the structural trends and contacts of the adjacent rock formations. Smaller discordant bodies, such as dikes, are characteristically tabular, and appear on the surface as long, narrow bands that cut across older structures. Concordant intrusive bodies, such as sills and laccoliths, have contacts parallel to those of the surrounding formations.

The relative age of igneous rock bodies can be recognized on geologic maps by using the principle of crosscutting relations. The younger intrusions cut across the older.

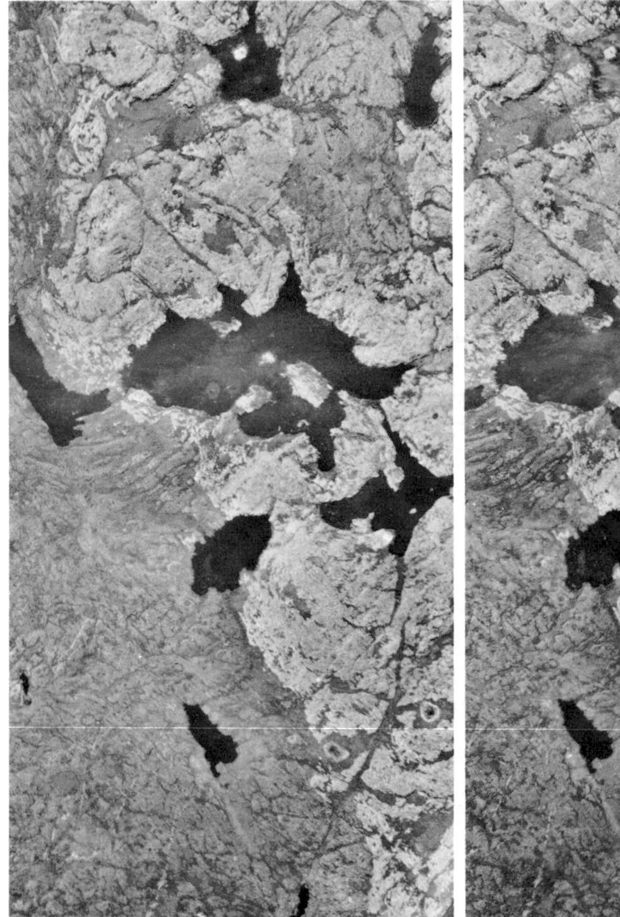

FIGURE 17.20: **Igneous Intrusions— Canadian Shield**

Three major rock bodies are exposed in this area: (1) metamorphosed sediments (dark gray), (2) granite (light tones), and (3) dikes (long, narrow bands of dark gray tones).

PROBLEMS

1 Make a geologic map of the area shown in stereoscopic view. What types of intrusive bodies are formed by the granite?

2 Study the contacts between the granite and the dikes, and determine the relative age of these rock bodies.

3 Which rock types are most resistant to weathering? Which weather most rapidly?

FIGURE 17.21: Outcrop Pattern of Surficial Deposits

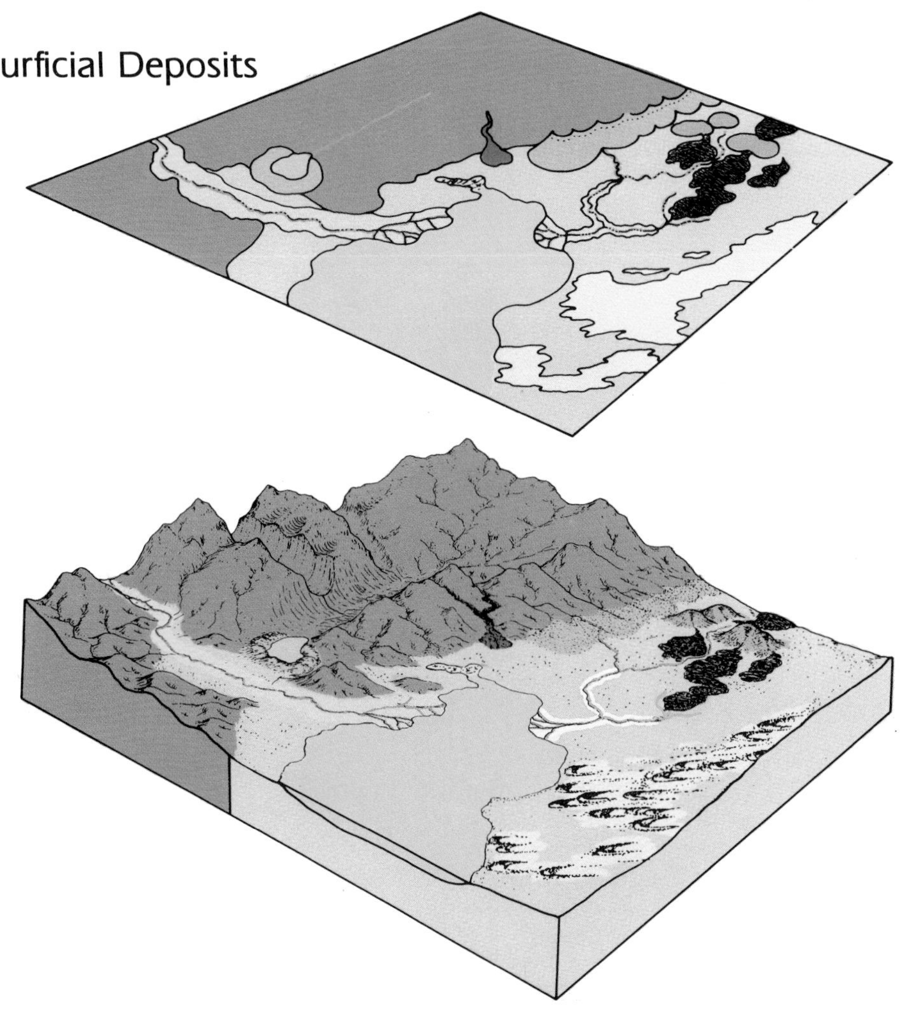

The principal types of surficial deposits are stream deposits, windblown sand and loess, landslide deposits, glacial deposits, beaches and other shoreline sediments, and recent volcanic deposits. The accumulation of materials in these deposits is characteristically thin, and generally masks the underlying rock units. Surficial deposits may or may not be shown on a geologic map, depending on the objectives of those who made the map. Some geologic maps do, however, emphasize surficial deposits when the deposits are important to engineering projects or have significant environmental implications.

On a geologic map, recent surficial deposits typically form an elongate, irregular pattern, and are commonly associated with the major processes actively operating on the landscape. In the diagram, note the relationship between glacial deposits and the glacial valley, between stream deposits and the present stream valleys, and between landslide deposits and the mountain front. Surficial deposits rest unconformably on the underlying bedrock and do not extend very far into the subsurface.

FIGURE 17.22: Surficial Deposits—Arizona

0 0.5 miles

PROBLEMS

1 Map and label the major surficial deposits in this area.

2 How are the surficial deposits related to the geologic processes presently operating in the area?

FIGURE 17.23: Geologic Map of Part of the Grand Canyon, Arizona

PROBLEMS

1 How do the outcrop patterns indicate that most of the Paleozoic strata in this area are horizontal?

2 What is the approximate thickness of the total sequence of Paleozoic rocks in the area? How did you arrive at your answer?

3 Draw a geologic cross section along line **A-A′**.

4 Note the relationship of rock type to cliff and slope topography on the map and on your geologic cross section. Which rock types develop cliffs and which develop slopes?

5 Note the outcrop pattern of the Temple Butte Limestone. Why is this pattern discontinuous?

6 How many unconformities are shown on the map? Between which units are they located?

7 What type of faulting is shown on the map?

8 What effect has faulting had on the course of the Colorado River and on the development of tributary valleys?

9 Two sets of faults are shown, one trending northwest and the other northeast. Which is the older?

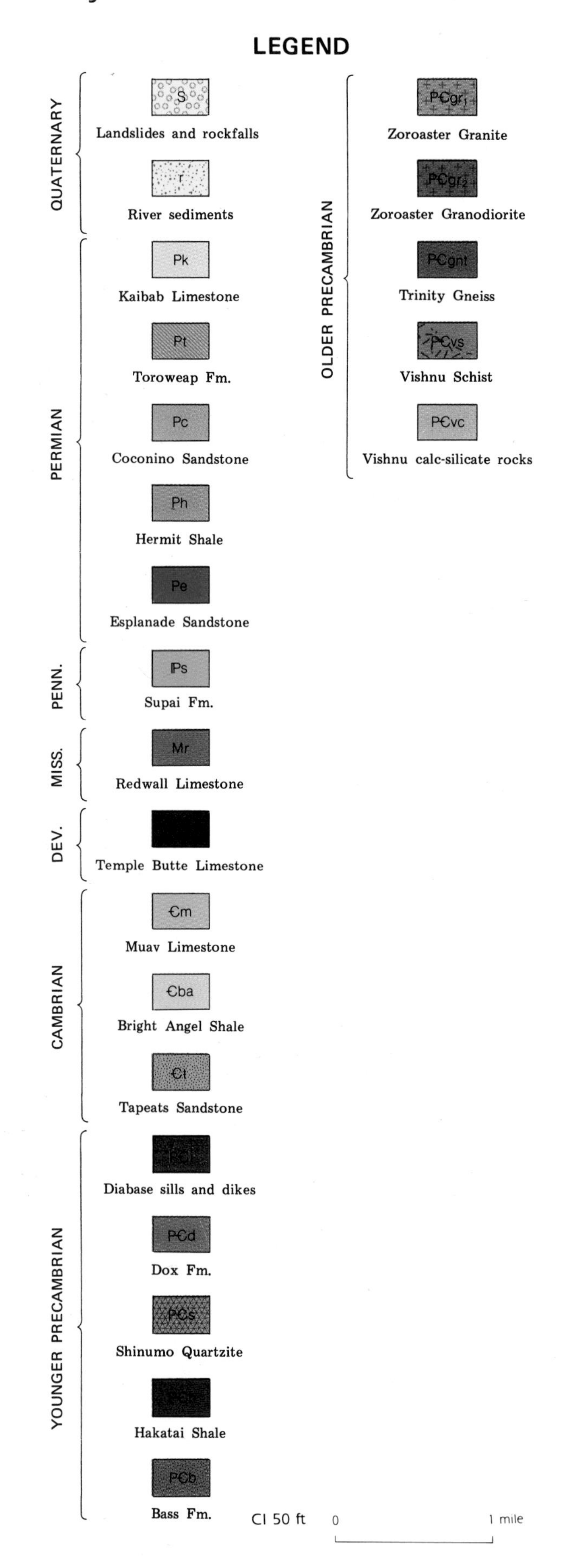

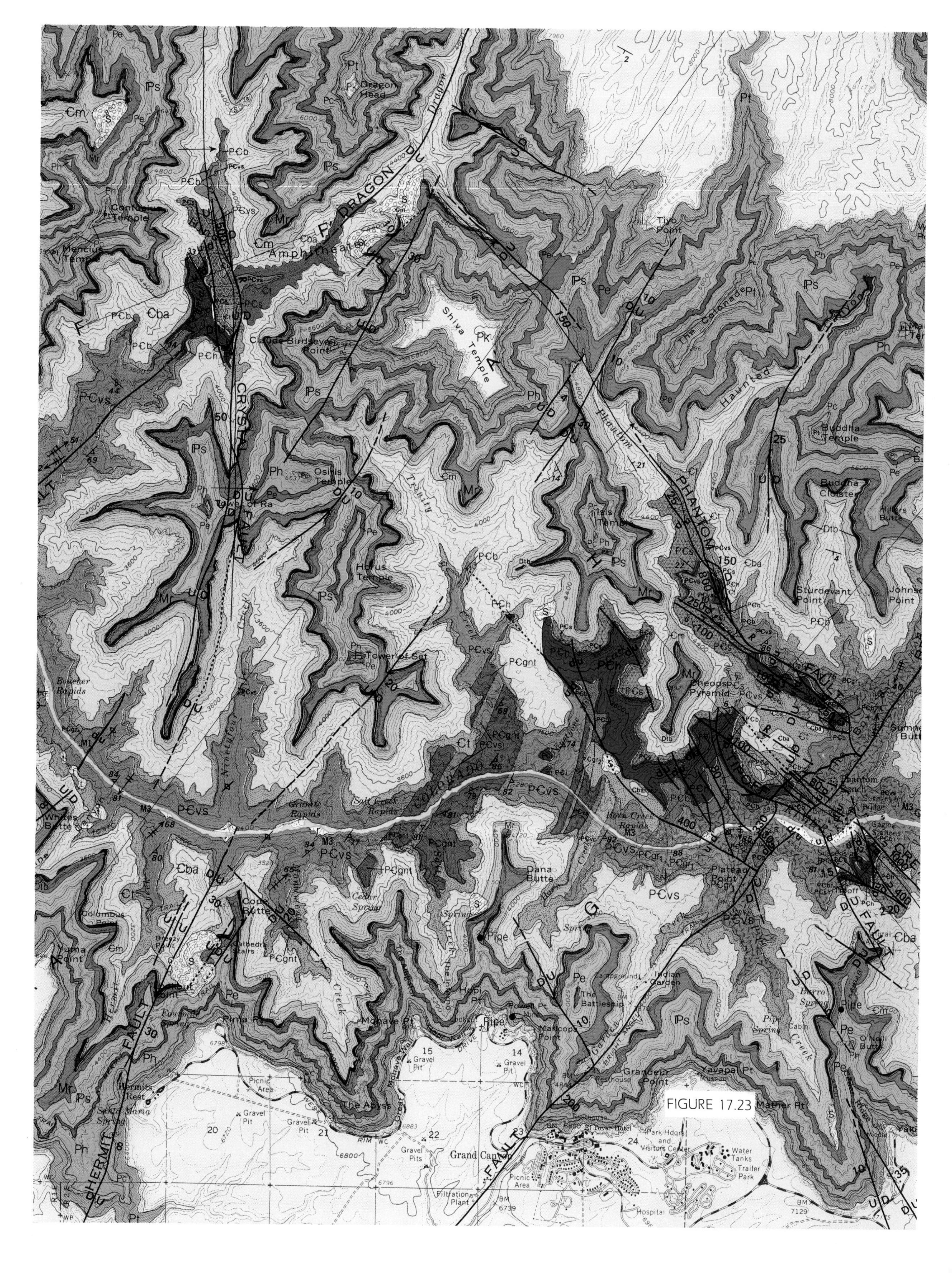

Dragon Head
DRAGON
Amphitheater
Shiva Temple
Claude Birdseye Point
Confucius Temple
Mencius Temple
CRYSTAL
FAULT
Osiris Temple
Tower of Ra
Horus Temple
Tower of Set
Trinity
Creek
Tiyo Point
The Colonade
Haunted
Phantom
PHANTOM
Buddha Temple
Buddha Cloister
Hillers Butte
Isis Temple
Sturdevant Point
Johnson Point
Cheops Pyramid
Phantom Ranch
Suspension Bridge
Boucher Rapids
Granite Rapids
Salt Creek Rapids
COLORADO
Horn Creek Rapids
Whites Butte
Columbus Point
Yuma Point
Cope Butte
Cathedral Stairs
Dana Butte
Plateau Point
Pipe
Indian Garden
Campground
The Battleship
Hopi Pt
Powell Pt
Maricopa Point
Mohave Pt
Pima Pt
The Abyss
Hermits Rest
Picnic Area
Gravel Pit
Gravel Pits
Grand Canyon
Grandeur Point
Yavapai Pt
Mather Pt
Park Hdqrs and Visitors Center
Water Tanks
Trailer Park
Filtration Plant
Hospital
Burro Spring
O'Neill Butte
Cedar Spring
Pipe Spring
HERMIT
FAULT
BRIGHT ANGEL

FIGURE 17.23

FIGURE 17.24: Geologic Map of Michigan and Surrounding Area

PROBLEMS

1 Where are the youngest rocks exposed in this area?

2 Study the patterns made by the rock units in Michigan and in the adjacent areas. What major structural feature is shown?

3 Note the age of the deformed units in this structure. What is the oldest age assignable to the deformation?

4 Assuming that the units on this map dip at a low angle, sketch an east-west geologic cross section through the center of Michigan.

5 Note the outcrop pattern of the lower Paleozoic rocks (Ordovician and Devonian) and that of the upper Paleozoic rocks (Carboniferous) in the area around the Illinois-Indiana state line. What major structural feature is present in this area?

6 What type of fault is evident south of Saginaw Bay? What was the relative movement—north side up or down?

7 What major unconformities occur in the rock sequence in this area?

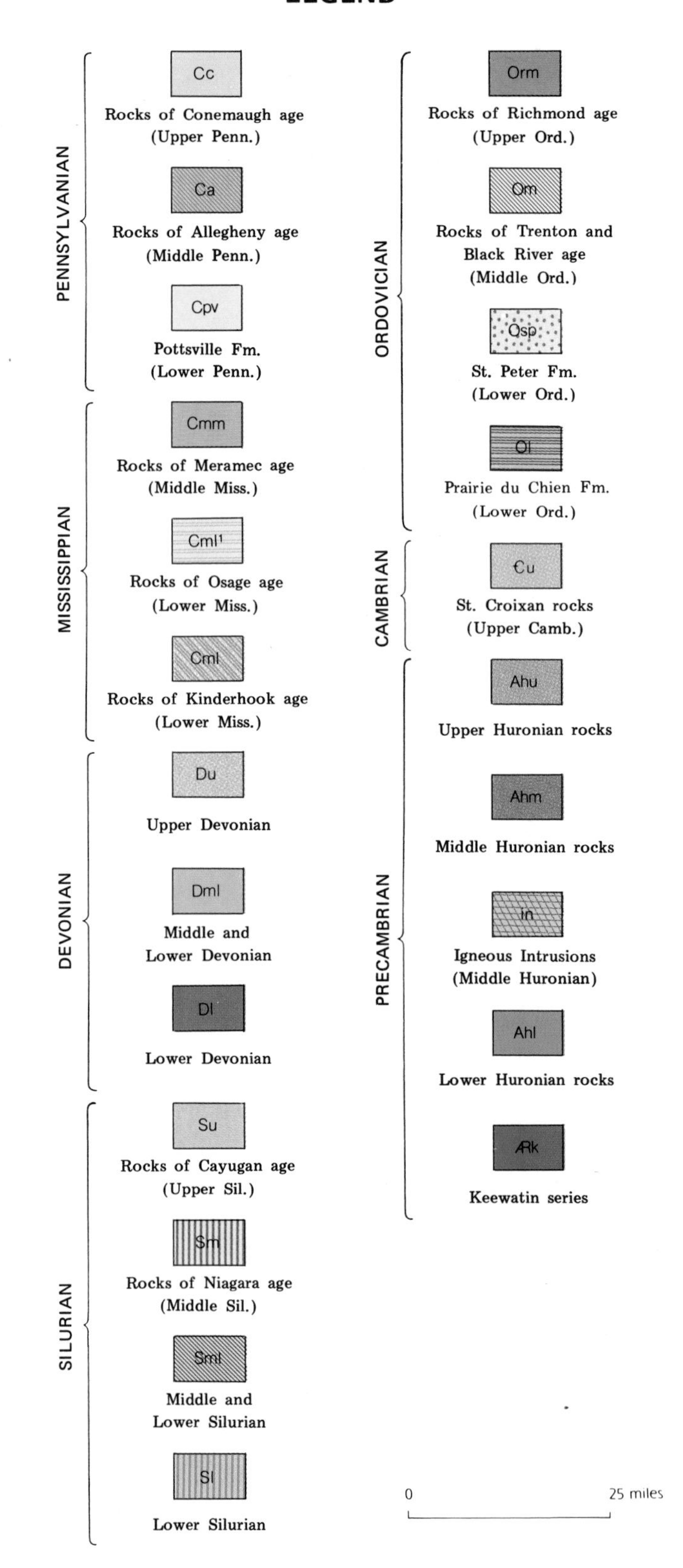

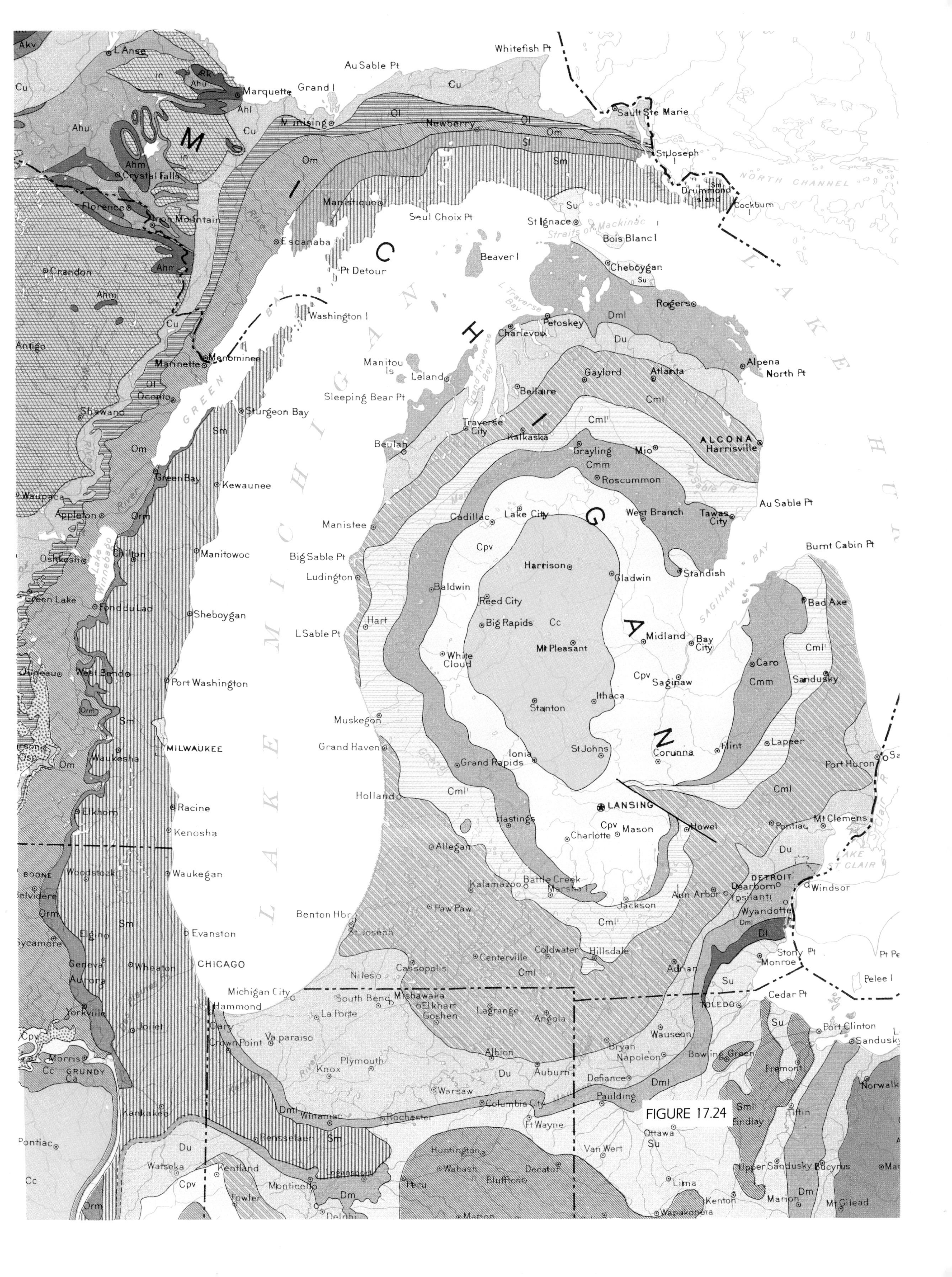

FIGURE 17.24

FIGURE 17.25: Geologic Map of the Wyoming Area

PROBLEMS

1 Study the outcrop pattern in the northeast quarter of the map. This area encompasses the Black Hills of South Dakota and Wyoming. Describe the geologic structure.

2 What is the age of the oldest rocks in the central part of the Black Hills?

3 Is this structure of the Black Hills symmetrical or asymmetrical? On what facts did you base your answer?

4 Draw a geologic cross section along an east-west line through the center of the structure, and show all beds at the surface and in the subsurface.

5 Judging from the outcrop pattern of the White River Group (Φw) on the east flank of the structure, what would be the age of the deformation in this area?

6 Are the igneous intrusions older or younger than the deformation of the sedimentary rocks in this area? What map patterns support your answer?

7 What major structure is located between the Black Hills and the town of Buffalo, Wyoming?

8 Study the outcrops of Precambrian rock. What structures are associated with these outcrops?

9 Study the outcrops of Tertiary rocks in Wyoming, especially the Wasatch (Ews), Fort Union (Efu), and Lance (El) formations. What structures are indicated by the outcrop of these formations?

10 Using a red pencil, complete the geologic map by drawing the axes of the major domes and basins.

11 Study the outcrop patterns of the White River formation (Φw) throughout Wyoming. What do these patterns tell you about the age of the deformation of the Rocky Mountains?

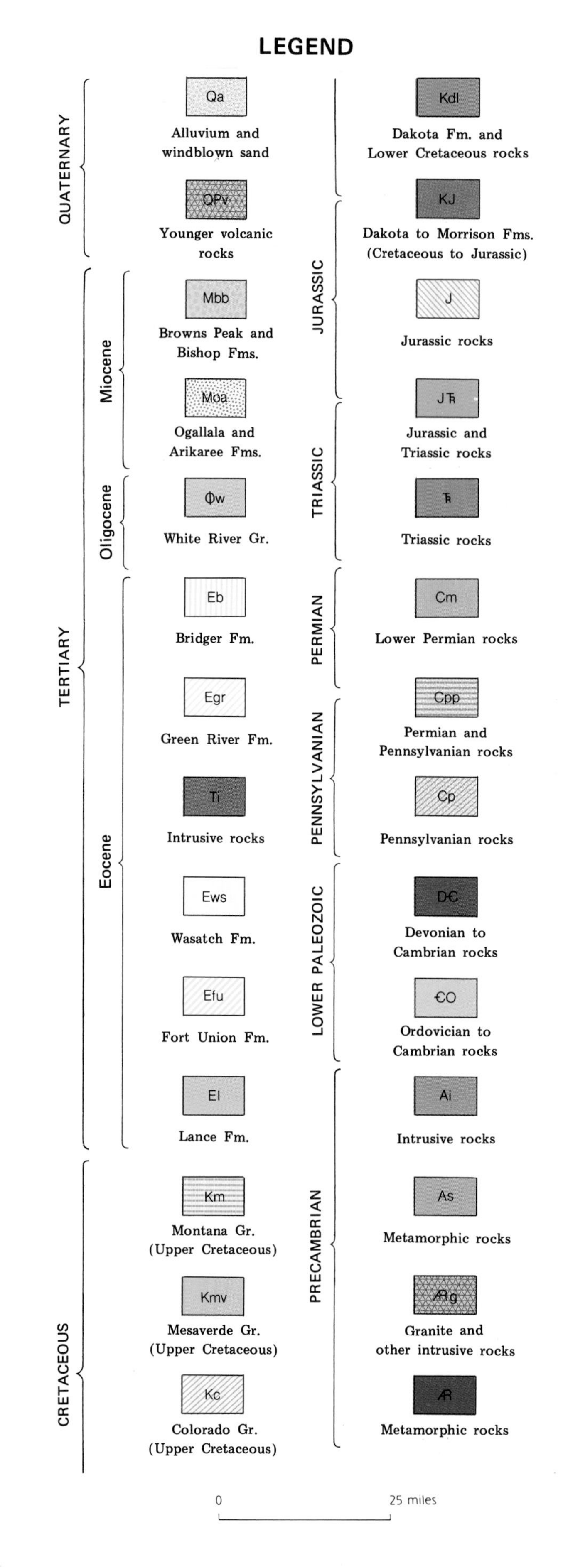

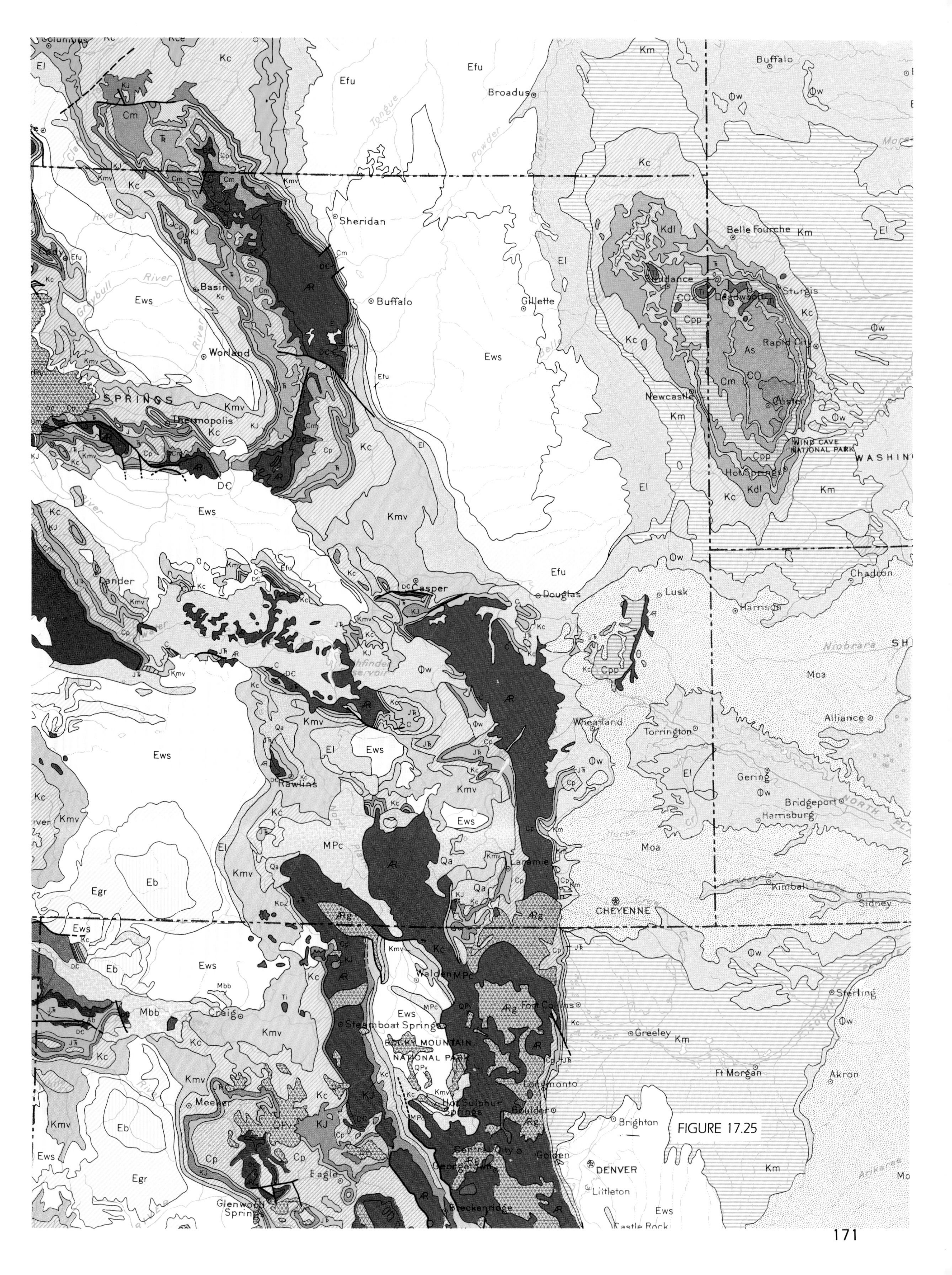
Buffalo
Broadus
Sheridan
Buffalo
Gillette
Belle Fourche
Sundance
Deadwood
Sturgis
Rapid City
Newcastle
Custer
Hot Springs
WIND CAVE NATIONAL PARK
Basin
Worland
Thermopolis
Lander
Casper
Douglas
Lusk
Harrison
Chadron
Wheatland
Torrington
Alliance
Gering
Bridgeport
Harrisburg
Rawlins
Laramie
Kimball
Sidney
CHEYENNE
Walden
Fort Collins
Greeley
Sterling
Craig
Steamboat Springs
ROCKY MOUNTAIN NATIONAL PARK
Ft Morgan
Akron
Meeker
Hot Sulphur Springs
Longmont
Boulder
Brighton
Golden
DENVER
Littleton
Castle Rock
Eagle
Glenwood Springs
Breckenridge
Central City
Georgetown
Ews
Kmv
Kc
Km
El
Efu
Eb
Egr
Moa
MPc
Qa
Cpp
Kdl
Niobrara
Tongue
Powder

FIGURE 17.25

FIGURE 17.26: Geologic Map of the Southern States

PROBLEMS

1 What is the regional strike of the Cretaceous and Tertiary rocks in southern Alabama?

2 In what direction do these rocks dip?

3 What is the regional strike of the Paleozoic rocks in Tennessee, Georgia, and northeastern Alabama?

4 What structural features are indicated by the outcrop patterns of these Paleozoic rocks?

5 What is the structural relationship of the Cretaceous rocks and the underlying Paleozoic rocks?

6 If you were to divide the area into geologic regions, where would you draw the boundaries between major provinces? On what factors did you base your decision?

7 Outline a sequence of events to account for the differences in outcrop patterns and age relationships in the area.

8 Are the faults in eastern Tennessee normal or thrust faults?

9 From the structural trend of the Paleozoic rocks, determine the orientation of the forces responsible for pre-Cretaceous deformation.

10 What structural feature dominates the northwest quarter of the map?

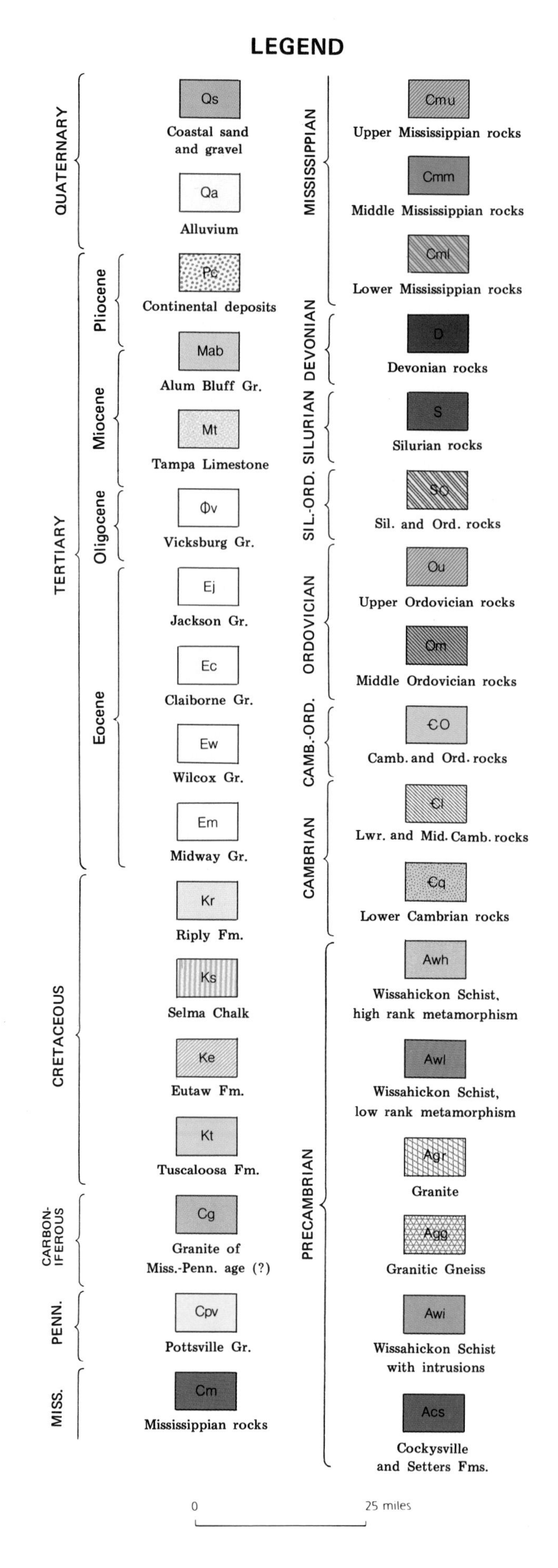

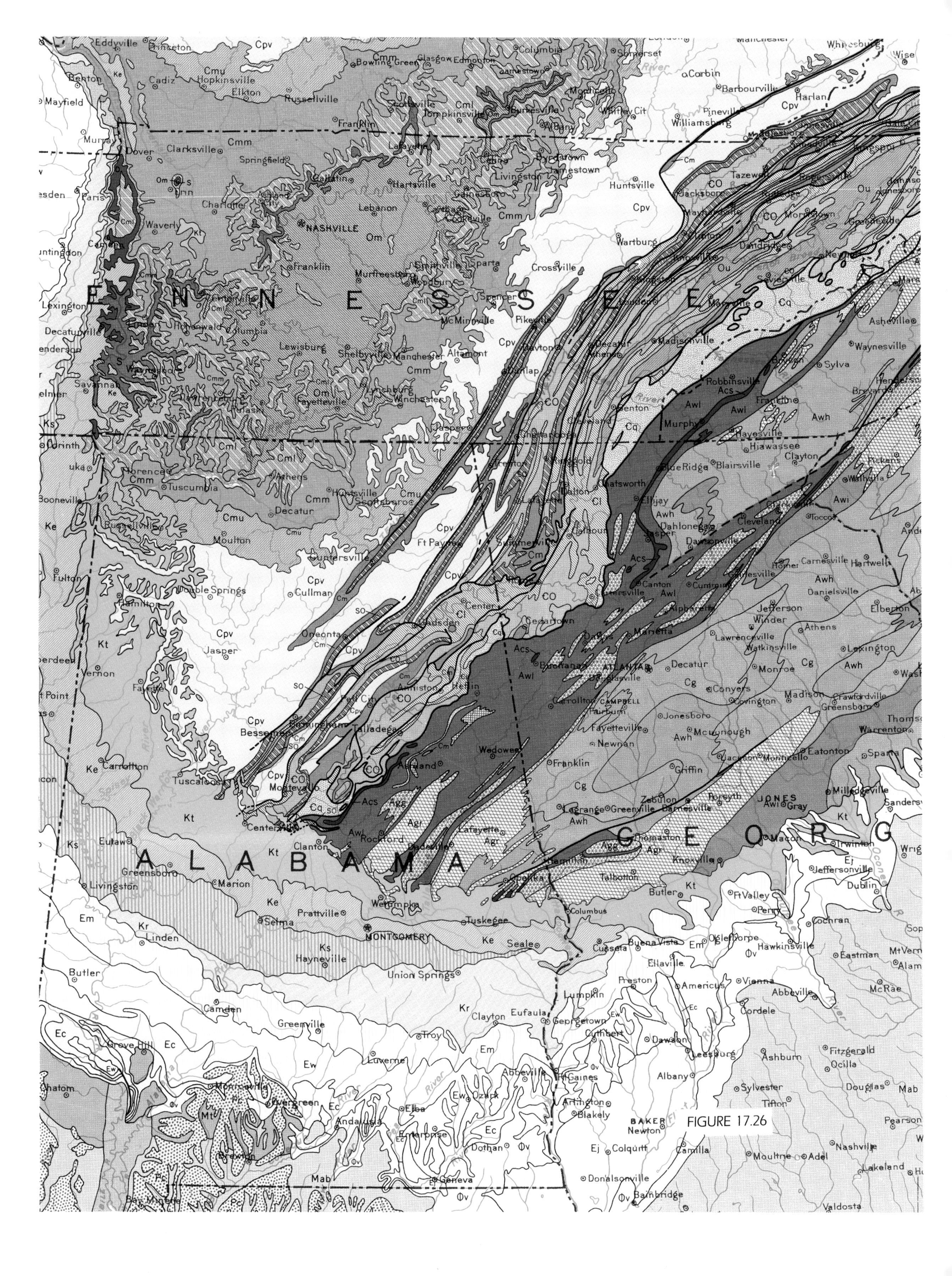

FIGURE 17.26

FIGURE 17.27: Geologic map of Chico, California

PROBLEMS

1 What map evidence indicates that the plutonic rocks are younger than the metamorphic rocks?

2 How many different periods of intrusion are shown on this map?

3 What types of faults occurred in this area?

4 List the major types of surficial deposits.

5 What is the relationship between the Lovejoy basalt and the present drainage system.

6 What map evidence indicates the relative age of the Lovejoy basalt and the pyroclastic andesites?

7 How do the outcrop patterns of the major igneous intrusions compare with those shown in Figure 17.19?

SURFICIAL DEPOSITS

Qgr Gravel deposits

EXTRUSIVE ROCKS

Tbo Olivine basalt

Ta Pyroxene andesite

Tpa Pyroclastic andesite

Tl Lovejoy basalt

PLUTONIC ROCKS

K1tr Trondhjemite

KJqd Quartz diorite/tonalite

KJhd Hornblende-biotite quartz diorite

KJpd Pyroxene diorite

Jgb Gabbro

METAMORPHIC INTRUSIVES

sp Serpentinite-talc schist

mtr Metatrondhjemite

mdi Metadiorite

mgb Metagabbro

mqp Metaquartz porphyry

hgn Hornblende gneiss

METAMORPHIC ROCKS

mb Metabasalt

ma Meta-andesite

md Metadacite

mr Metarhyolite

mt Metatuff

mm Marble

mq Quartzite-metachert

mp Phyllite

FORMATIONS

FRANKLIN CANYON FORMATION

BLOOMER HILL FORMATION

DUFFEY DOME FORMATION

FAULT

CI 50 m 0 2 miles

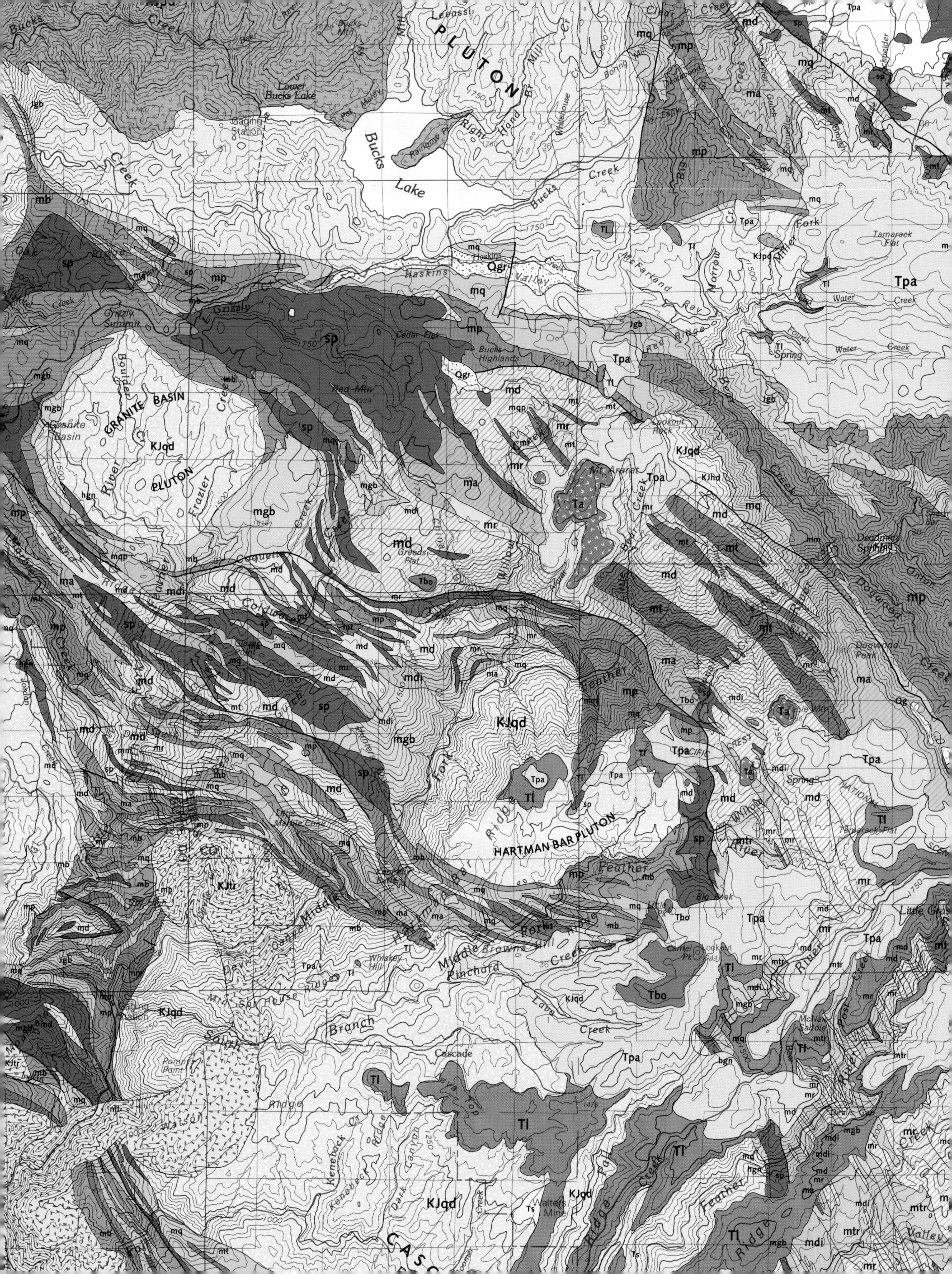
PLUTON
Bucks Lake
Lower Bucks Lake
Gaging Station
Bucks Creek
Rainbow Pt
Right Hand Br
Mill Cr
Haskins Valley
McFarland Ravine
Oak Ridge
Grizzly Summit
L. Grizzly
Cedar Flat
Bucks Highlands
Red Mtn
GRANITE BASIN
Granite Basin
KJqd
PLUTON
Frazier
Boulder Cr
Red Ridge
Lookout Rock
Mt. Ararat
Greens Flat
Coquette
Corduroy
Tamarack Flat
Fourth Water Creek
Spring
Dogwood Peak
Deadman Spring
Onion
HARTMAN BAR PLUTON
Feather River
Middle Fork
Little Marble Cone
Big Peak
Camel Pk
Lookout
Little Grass
Whiskey Hill
Pinchard
Middle Browns Hill
Cascade
Rompts Point
Kenebeck Cr
Dark Canyon
Walters Mine
McNair Saddle
PACIFIC CREST
NATIONAL
South Branch
Watson Ridge
Lava Top
Valley
CASC
sp
mp
mq
md
mt
ma
mb
mr
mgb
mdi
mtr
Tpa
Tl
Tbo
Ta
Ts
Ogr
Og
Jgb
hgn
KJtr
KJhd
KJpd

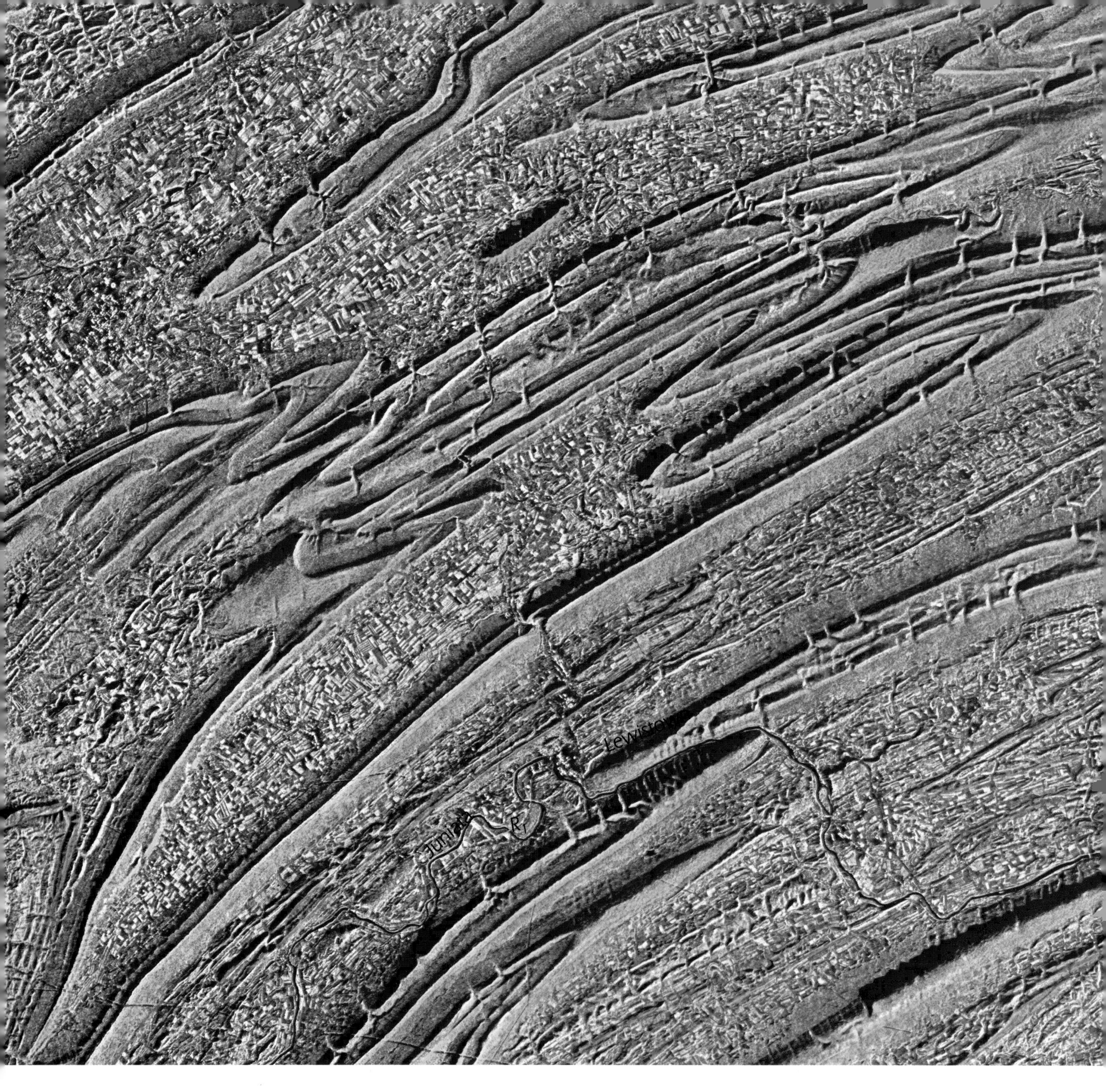

FIGURE 17.28: **Central Pennsylvania**

Radar image 0 5 miles

PROBLEMS

1 Complete the geologic map by extending the colored formations into the unmapped area. Identify the major anticlines and synclines, and complete the geologic map by adding the proper symbols for fold axes. Show direction of plunge.

2 Identify the minor folds (both anticlines and synclines). How are the minor folds related to the major folds with respect to their orientation (trend of axial trace), and direction of plunge?

3 Study the large anticlines shown on the southern part of the map. How can you identify these folds as anticlines?

4 What was the orientation of the force that deformed this sequence of rocks?

5 Where are the oldest rocks exposed in the mapped area?

6 Draw an idealized north-south cross section through the map.

7 Make an explanation for this map similar to those in Figures 17.23–17.26.

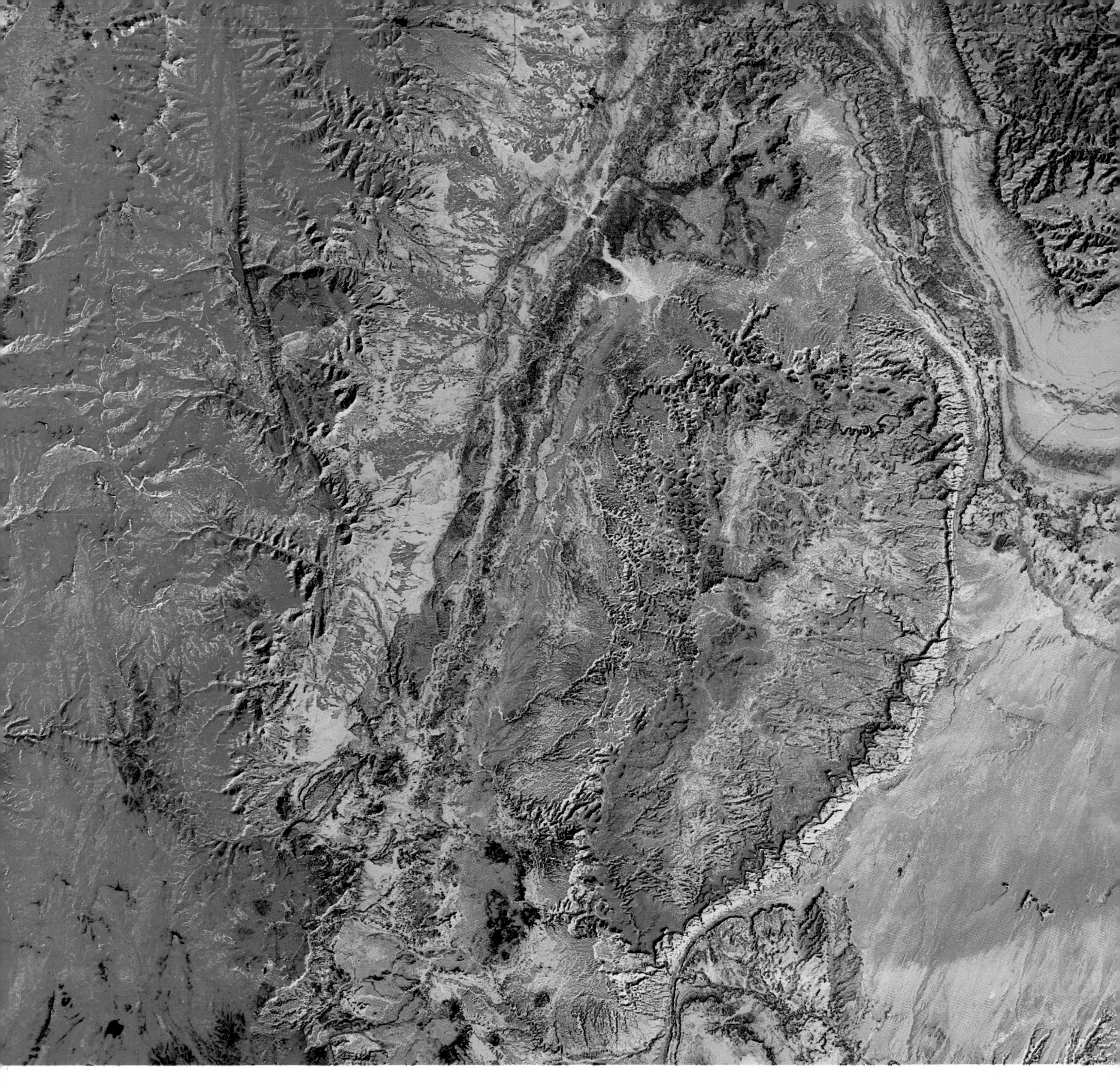

FIGURE 17.29: San Rafael Swell—Central Utah

Enhanced Landsat image 0 10 miles

The desert regions of the Colorado Plateau have little soil and vegetation, so the color and tone of a Landsat image of the area are produced largely by the coloration of the exposed rock. This image has been enhanced by the computer. The false colors are exaggerated to emphasize the contrasts, so the image becomes almost a geologic map. The brilliant pink indicates areas covered with vegetation.

PROBLEMS

1 Make a geologic map of this area by tracing the contacts of the major rock units indicated by different colors and by different topographic expressions.

2 What geologic structure dominates this area?

3 How does this structure compare with that of the Black Hills area shown on the map on page 170?

4 Where are the oldest rocks exposed?

5 Where are the youngest rocks exposed?

6 What features produce the linear texture (colored yellow) in the lower right corner of the image?

7 If you were exploring for petroleum and natural gas, where in this region would you recommend drilling?

8 What types of surficial deposits occur in this area?

Exercise 18

THE OCEAN FLOOR

OBJECTIVES

To become familiar with the geology of the ocean floor, and to understand how the major structural and topographic features of the ocean floor are explained by the plate tectonic theory.

MAIN CONCEPT

The topography of the ocean floor has been mapped by means of seismic reflection profiles. These profiles are of great value because they show both the topographic form and the structure of shallow features on the oceanic crust. The major topographic features of the ocean floor are (1) the oceanic ridge, (2) the abyssal floor, (3) fracture zones, (4) seamounts, (5) deep-sea trenches, and (6) continental margins.

SUPPORTING IDEAS

1 The oceanic ridge is a broad, fractured swell, or arch. Sediment is thin or absent on the crest and thickest on the flanks of the ridge.

2 The abyssal floor consists of abyssal hills and abyssal plains.

3 Fracture zones cut the oceanic ridge, and show strong evidence of strike-slip movement and horizontal displacement of the sea floor.

4 The size, shape, composition, and structure of seamounts indicate that they are submarine shield volcanoes.

5 Deep-sea trenches are long, linear depressions of the sea floor adjacent to the most active seismic and volcanic zones in the world.

6 The continental slopes are the margins of the continents. They are commonly cut by submarine canyons.

DISCUSSION

MAPPING THE OCEAN FLOOR

Significant advances in our understanding of the earth have been made possible by new techniques for mapping the sea floor. The seismic profiler is an especially effective instrument, for it makes a continuous topographic profile and it also is able to detect the interfaces of rock layers. These data are plotted automatically as the ship moves, thus creating a topographic and structural profile of the ocean floor. The information obtained from thousands of seismic profiles has been used to map the relief of the sea floor with an accuracy comparable to that of preliminary topographic maps of the land surface. Maps and charts compiled from recent oceanographic studies show that the sea floor has many unique and spectacular features.

PROFILE A

PROFILE B

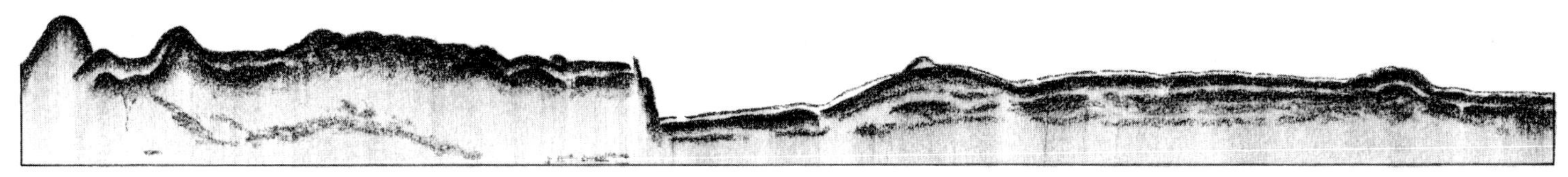
PROFILE C

FIGURE 18.1 Seismic Profiles of the Atlantic Ocean

PROBLEMS

1 What physiographic feature is shown in profile **A?**

2 Sediment accumulating on the ocean floor is derived largely from the shells of floating organisms. Explain why the sediment is much thicker on the flanks than on the crest of the structure in profile **A.**

3 Is the crust in profile **A** being compressed or pulled apart? Cite evidence to support your answer.

4 What physiographic feature is shown in profile **B?**

5 Explain the origin of the topography of the sea floor shown in profile **B.**

6 Study profiles **A** and **B.** Why do geologists conclude that the oceanic crust has not been subjected to strong compressive forces?

7 The oldest sediment in profile **B** is 30 million years old. What major geologic events have occurred in this part of the ocean during the last 30 million years?

8 Profile **C** was made across the trend of a fracture zone. What evidence on the profile indicates that the "fractures" are faults?

9 Draw the faults on profile **C.** Do they occur along a single fracture or along a fracture zone?

10 Maps of the ocean floor show great horizontal displacement along fractures. Is there apparent vertical displacement as well? Explain.

PROFILE A

PROFILE B

FIGURE 18.2 Seismic Profiles—Atlantic Ocean Basin

PROBLEMS

1 What major physiographic feature of the ocean floor is shown in profile **A?**

2 Explain the absence of thick sediment deposits in profile **A.**

3 Describe briefly the topography shown in profile **A,** and explain its origin.

4 The ocean is only 300 ft deep on the left-hand side of profile **B,** but is nearly 17,000 ft deep on the right-hand side. What material makes up the ocean floor in each area?

5 What are the channels cut in the surface of profile **B?** How did they form?

6 Has this part of the oceanic crust been deformed? Cite evidence to support your answer.

7 What geologic processes have been most active in this area (i.e., folding, faulting, deposition of sediment, or erosion).

8 Compare and contrast the type of sediment that is likely to be deposited near the left and right margins of profile **B.**

FIGURE 18.3 Seismic Profiles—Pacific Ocean Basin

PROBLEMS

1 What evidence can you find on profile **A** that the "mountains" are of volcanic origin?

2 What type of volcanic would these features be?

3 What type of sea-floor topography is adjacent to the "mountains"?

4 Explain the difference in elevation of the sea floor shown in profile **B.**

5 Sediments commonly accumulate in the lowest part of an area. Explain the absence of sediments in the depression shown in profile **B.**

6 Note the difference in elevation of the sea floor on either side of the depression in profile **B.** Explain this difference in light of the plate tectonic theory.

7 Draw the probable faults on profile **B** and explain how they may have originated.

8 Explain the origin of the "mountainous" topography just left of center in profile **B.** What rock types would you expect to compose these features?

MAJOR TOPOGRAPHIC AND STRUCTURAL FEATURES OF THE OCEAN FLOOR

The physiographic map accompanying this manual shows, in perspective, the regional features on the ocean floor. Although there is considerable vertical exaggeration and generalization, the essential elements shown on the map have been firmly established by seismic profiles.

PROBLEMS

1 Label on the physiographic map a specific example of each of the following ocean-basin features:

a Fracture zone
b Guyot
c Abyssal plain
d Chain of composite volcanoes
e Chain of shield volcanoes
f Broad continental shelf
g Seamount
h Trench
i Abyssal hills
j Deep-sea fan

2 What topographic evidence is there that the continental crust is different from the oceanic crust?

3 Are most submarine canyons located on the continental shelf, the continental slope, the oceanic ridge, or the abyssal plains?

4 What is the probable origin of submarine canyons?

5 Color the rift valley at the crest of the mid-Atlantic Ridge red. What types of stress are implied by the way the ridge has been displaced along the fracture zone?

6 What is the orientation of fracture zones with respect to the continental margins on either side?

7 Explain why Japan has numerous severe earthquakes and why central Australia has few.

8 Why are there only a few abyssal plains in the Pacific Ocean, whereas abyssal hills cover over 80% of the Pacific Ocean floor?

9 Why are there no abyssal hills adjacent to the continental slope in the Atlantic Ocean basin?

10 If you were to test the theory of plate tectonics by a deep-sea drilling program and by geophysical studies, where would you plan to drill to find the following:

a Oldest sediments on the sea floor
b Thickest sediments on the sea floor
c Youngest rocks of the sea floor
d Area of greatest seismic activity
e Area of maximum heat flow

11 Many seamounts have flat tops. If the moving seamounts rest on plates, how would you expect the flat surface of the seamounts to be tilted in relation to an approaching trench?

12 Note the linear trend of many of the seamounts. If the theory is correct that these volcanoes develop from a lithospheric plate passing over a "hot spot" in the mantle, where would you expect to find the oldest island (or seamount) in the Pacific Ocean?

13 Color the deep-sea trenches to emphasize their geographic distribution. How are the trenches related to island arcs?

14 How are trenches and island arcs explained by the theory of plate tectonics?

15 Study the structure and landforms associated with the Puerto Rico Trench just north of South America. Indicate with arrows the direction of plate movement in this area. What tectonic processes occur in this area?

16 Explain the origin of the islands in the Caribbean Sea.

17 Recent drilling in the northwestern Pacific has uncovered deposits of the shells of organisms that today thrive only in warm water near the equator. How do you explain this anomaly according to the plate tectonic theory?

18 Several types of islands and small, submerged landmasses rise above the ocean floor. These can be categorized as follows:

a Basaltic islands and seamounts (shield volcanoes)

b Simple island arcs composed of andesitic volcanic products

c Complex island arcs composed of granite, metamorphic rock, and andesitic volcanic products

d Microcontinents consisting of small fragments of granitic continental crust

e Islands formed when shallow seas spread over the continental platform

Locate the following features on the physiographic map and categorize each island (or submerged landmass) according to the preceding descriptions.

a New Zealand
b Philippines
c Borneo
d Java
e Madagascar
f Seychelles
g Azores
h Bermuda
i Baffin Island
j Sri Lanka
k British Isles
l Aleutian Islands

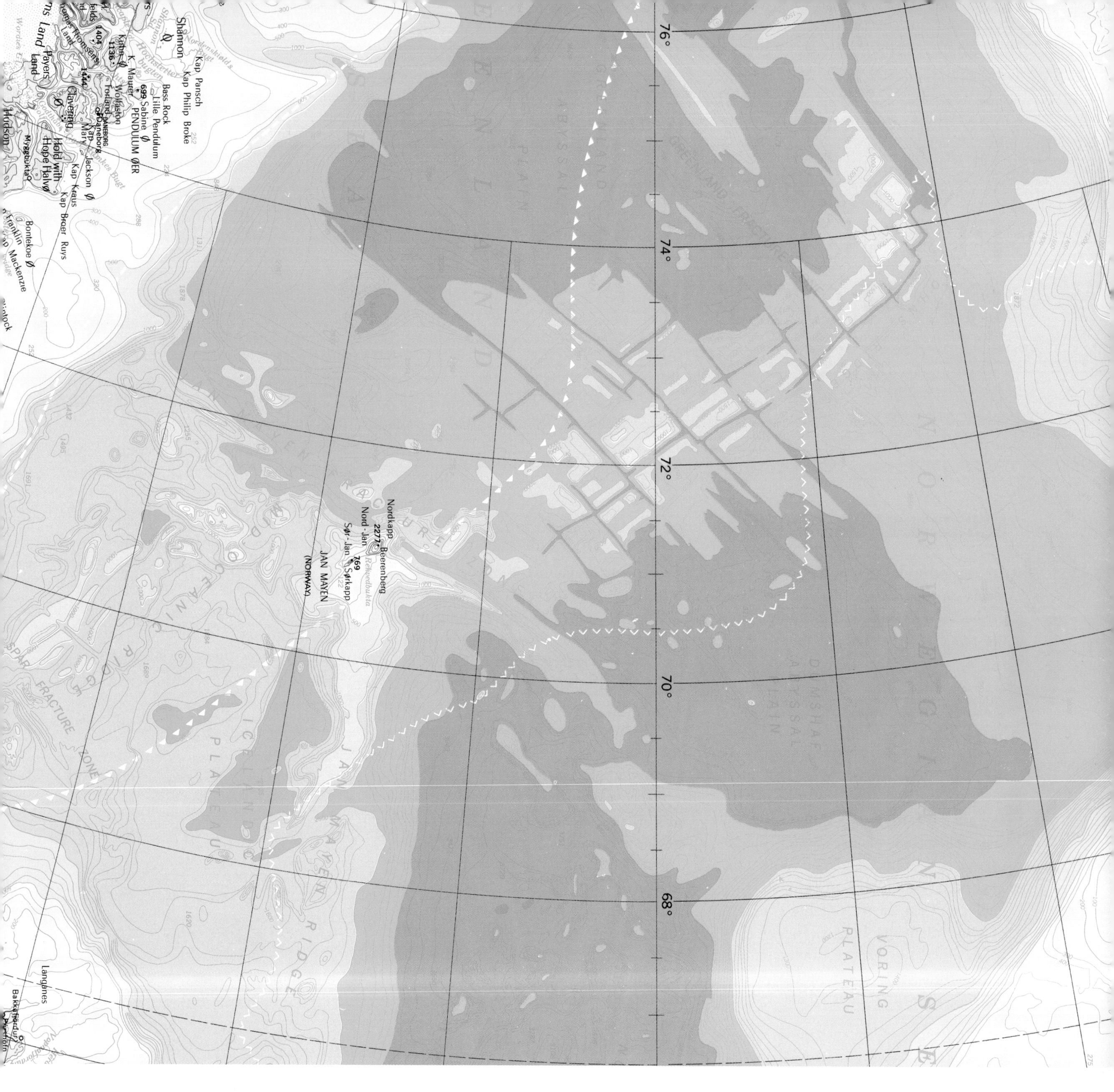

FIGURE 18.4: **Bathymetric Map of a Portion of the North Atlantic Ocean**

The bathymetric data obtained from seismic profiles and from sonar soundings is commonly summarized on bathymetric maps like the one shown in Figure 18.4. The elevation of the surface of the ocean floor is indicated by contour lines. Blue contour lines show depth in meters.

PROBLEMS

1 Outline and highlight with a colored pencil the following physiographic provinces and features of the ocean floor: (a) continental shelf, (b) continental slope, (c) abyssal plains, (d) abyssal hills, (e) oceanic ridge, (f) crest of oceanic ridge, (g) fracture zones, (h) submarine canyons on continental slope, and (i) seamounts.

2 How high do the abyssal hills rise above the surrounding area?

3 How high does the crest of the oceanic ridge rise above the abyssal plains?

4 How deep is the rift valley?

Exercise 19

PLANETARY GEOLOGY

OBJECTIVE

To recognize the major types of landforms on the various planetary bodies in the solar system and to understand the processes that formed them.

MAIN CONCEPT

Space photographs of the planetary bodies in the solar system show that the rocks and surface features of a planet contain a record of the events that result from the specific energy systems operating on or within the planet. With this new information, we can compare the geology of the earth with the geology of other planetary bodies, and identify those processes that are fundamental to all and those that are unique or of special importance to each planet.

SUPPORTING IDEAS

1 The sequence of events in a planet's geologic history can be determined by using the principle of superposition.

2 The impact of a meteorite produces a new landform called a *crater*, and a new rock body consisting of the blanket of *ejecta* that surrounds the crater.

DISCUSSION

Most scientists believe that the planets were formed by the accretion of matter from a cold, dry, dust-laden cloud of gas that had some degree of internal motion. The small particles collected by gravitational attraction to form increasingly larger, solid bodies, called *planetoids*, which were up to several miles in diameter. Accretion of the planetoids by gravitational attraction eventually formed a planet, which then swept up most of the remaining planetoids and meteorites in its orbital path.

The dominant surface features of the Moon, Mercury, and Mars are impact craters, which were probably produced during the last phases of planetary accretion. In simple terms, the impact of a meteorite produces two geologic features: (1) an impact crater, and (2) a body of fragmented rock (an ejecta blanket) thrown out of the crater by the impact. The smaller particles are thrown out great distances from the point of impact and form huge "splash marks" called *rays*.

On the basis of the superposition of ejecta or volcanic material or both, the relative sequence of many events in the geologic history of a planetary body can be determined.

To solve the following problems, remember that the age of a planetary surface is reflected by the relative number of impact structures. Old surfaces have been subjected to long periods of bombardment and are saturated with craters. Younger surfaces have fewer craters.

FIGURE 19.1: Mare Imbrium Area of the Moon

Telescopic photo

PROBLEMS

1 What evidence is there in the photograph that the craters Archimedes **(A)** and Plato **(B)**, together with their ejecta blankets, are older than Aristillus **(C)**, Autolycus **(D)**, Timocharis **(E)**, and Eratosthenes **(F)**?

2 Note that the Imbrium Basin **(G)** is considered to be a multiringed crater. What evidence in the photograph supports this description?

3 What photographic evidence can you find that the crater Sinus Iridum **(H)** and its ejecta blanket are younger than the ejecta of the multiringed crater that forms the Imbrium Basin?

4 What photographic evidence suggests that Copernicus **(I)** and its ejecta blanket are the youngest features in the area?

5 Arrange the following Moon features in proper chronologic sequence:

- **a** Eratosthenes (ejecta blanket)
- **b** Mare Imbrium (basalt flows)
- **c** Archimedes (ejecta)
- **d** Copernicus (ejecta)

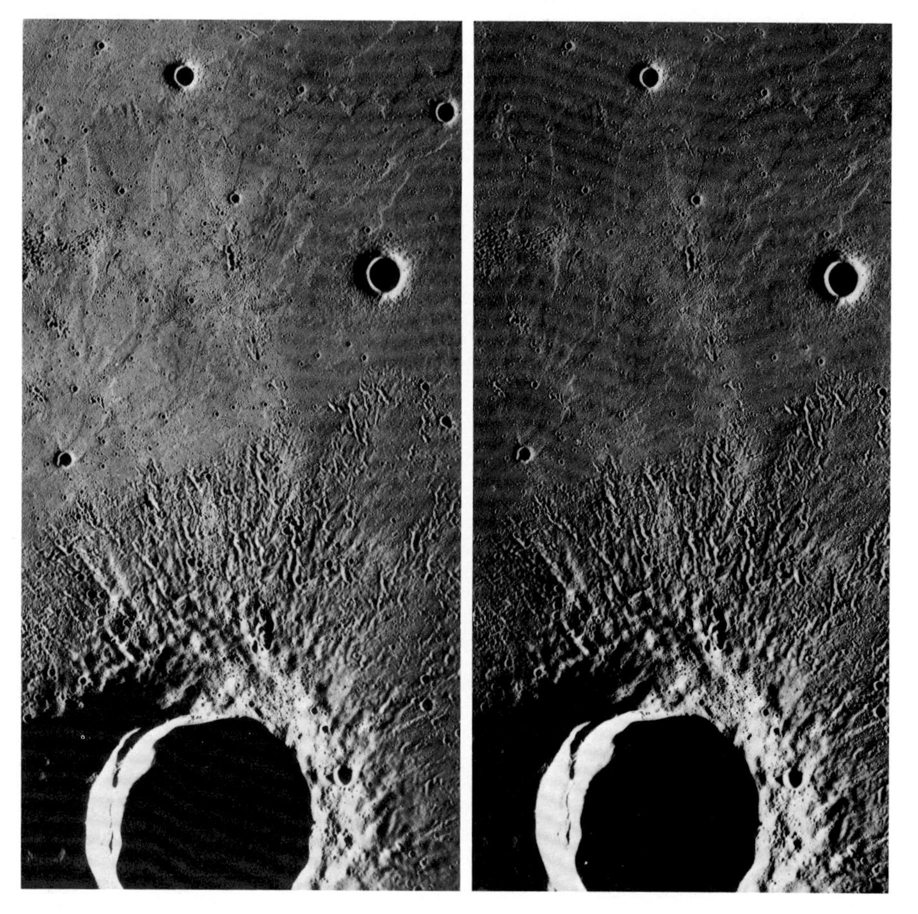

Orbiting satellite photo 0 10 miles

FIGURE 19.2: Mare Imbrium

Parts of the lunar surface have been photographed stereoscopically and can be studied in the same way that we study stereoscopic photographs of the earth.

PROBLEMS

1 Study the ejecta blanket of the large crater in the lower part of the photograph. Is this body of rock older or younger than the basalts of the surrounding maria? Explain.

2 A number of individual lava flows can be seen in this photograph. Trace their margins and try to determine their relative age. On the basis of your tracings, where do you think the center of extrusion is?

3 What type of volcanic eruptions have occurred in this area?

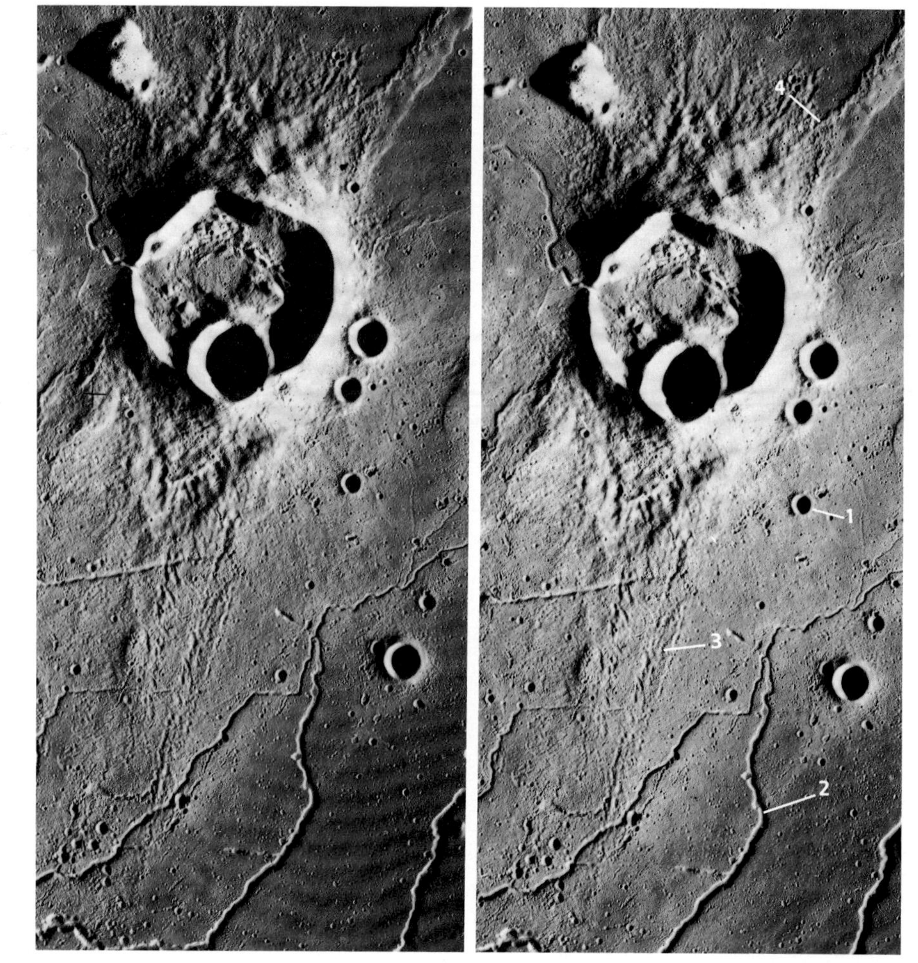

Orbiting satellite photo 0 10 miles

FIGURE 19.3: Mare Imbrium

PROBLEMS

1 Name the features identified on the photograph by the numbers 1, 2, 3, and 4.

2 Are the craters older or younger than the lava flows of Mare Imbrium?

3 Map the extent of the ejecta blanket associated with the largest crater.

4 What was the most recent geologic event in this area?

5 Where are the oldest rocks exposed in this area?

6 How do the younger craters in this area and in Figure 19.2 differ from the older craters?

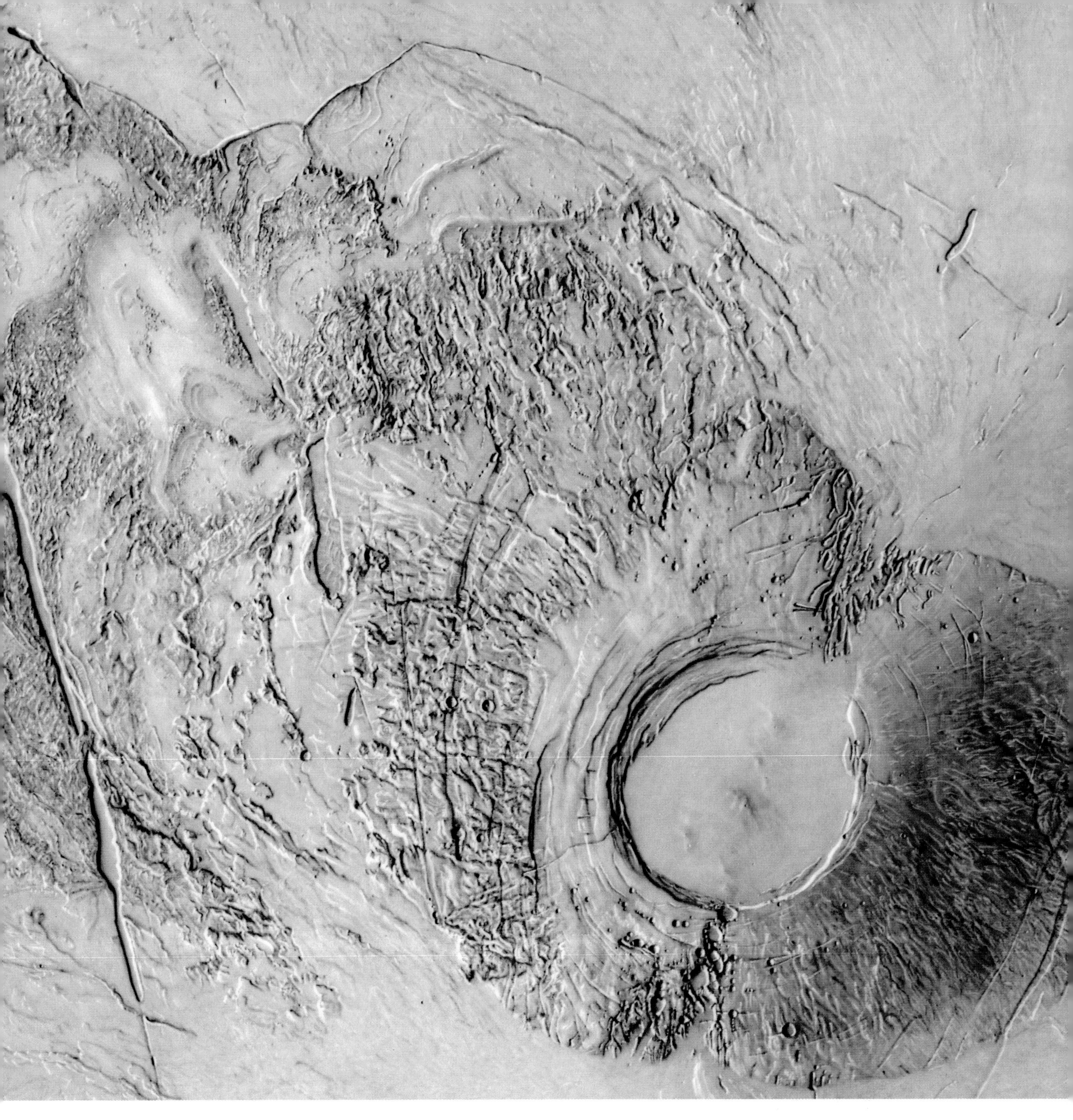

FIGURE 19.4: Phoenicis Lacus Region, Mars

Shaded relief map 0 60 miles

PROBLEMS

1 What features indicate that the large structure is a volcano and not an impact crater?

2 How has the volcano been modified by tectonic processes?

3 How has the volcano been modified by erosional processes?

4 What evidence indicates that the entire surface shown on this map is relatively young?

5 What evidence indicates that a period of fissure eruptions of flood basalts occurred before the formation of the large volcano?

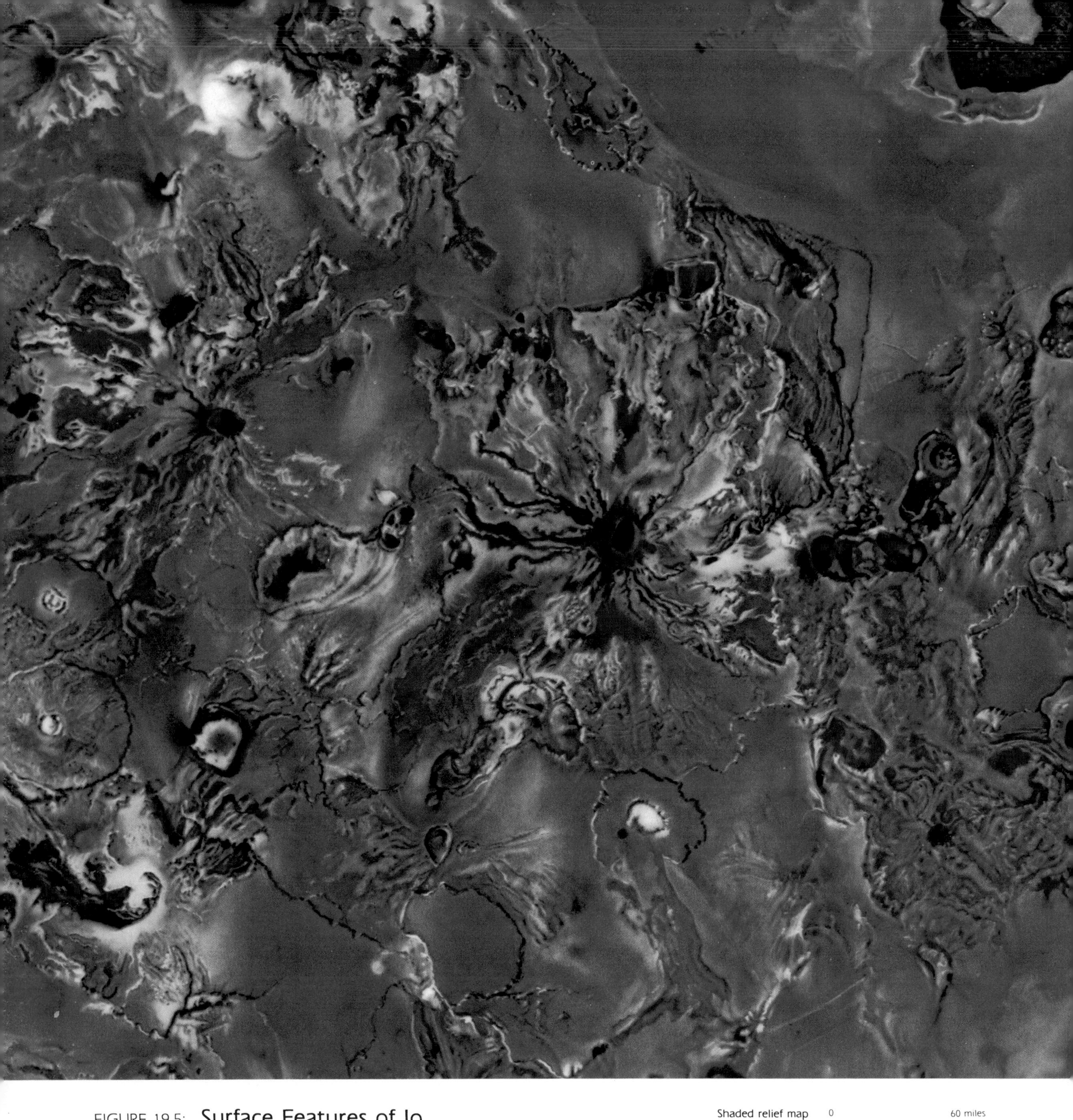

FIGURE 19.5: Surface Features of Io

Shaded relief map 0 60 miles

PROBLEMS

1 Identify and label the following features: (a) major centers of volcanic eruption, (b) lava flows, (c) plains covered with sulfur, and (d) active volcanic eruptions.

2 Why do geologists believe that the surface of Io is much younger than the surface of the other moons of Jupiter?

3 What is the origin of the volcanic activity on Io?

4 Is the surface of Io shown here older, younger, or about the same age as the surface of Ganymede shown on the facing page?

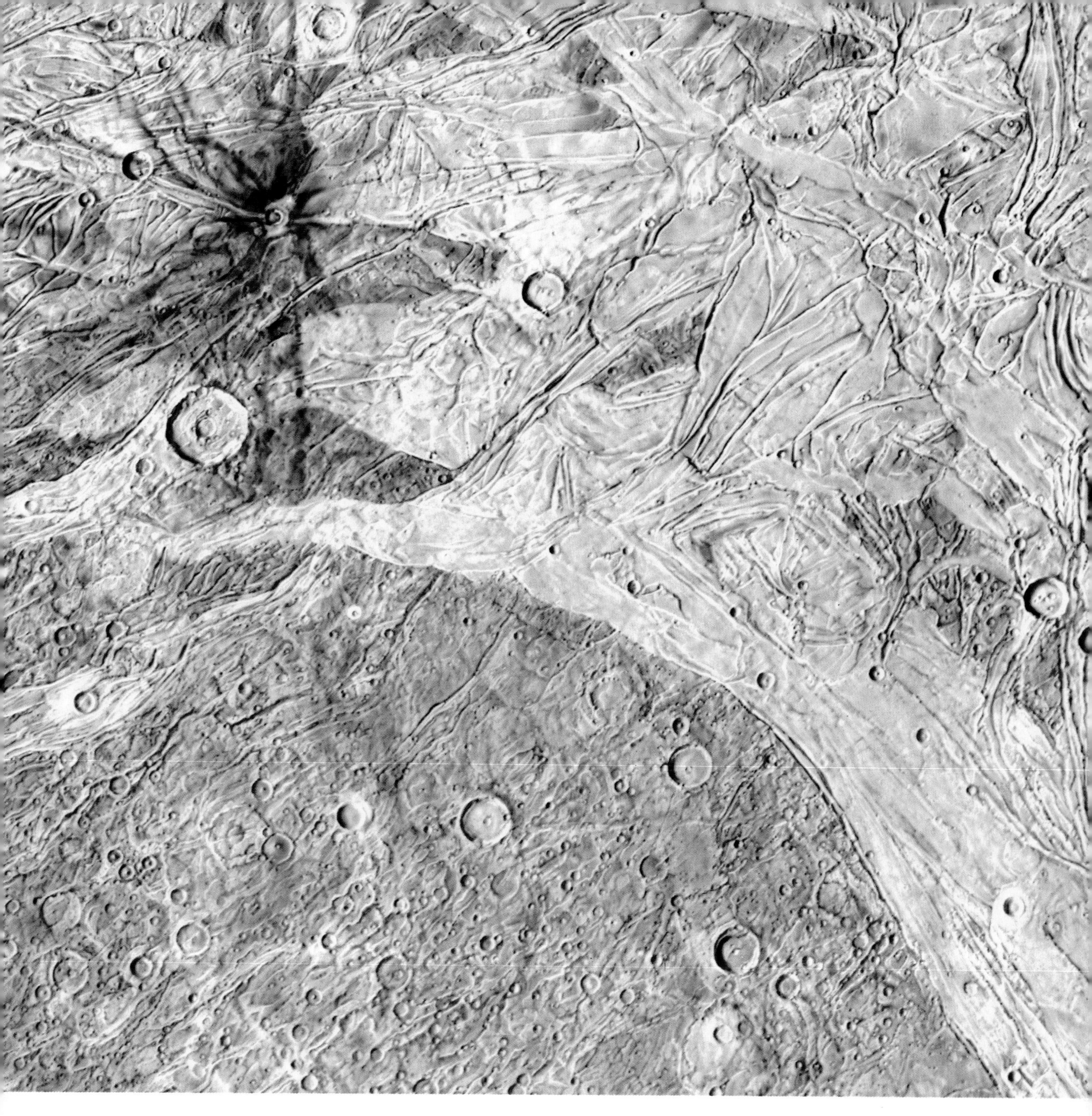

FIGURE 19.6: Ganymede

Shaded relief map 0 60 miles

PROBLEMS

1 Based on the relative number of craters, which terrain type would you judge to be older: (a) the dark, fragmented terrain, or (b) the light, grooved terrain?

2 What are the youngest features on Ganymede?

3 What evidence indicates that the crust of Ganymede has moved and shifted horizontally?

4 Explain the nature and origin of the volcanism on Ganymede.

FIGURE 20.1: Colorado Plateau—Utah and Colorado

This image is a Landsat mosaic that has been manipulated by computer to obtain a stereoscopic pair. The image is unique in that it covers an exceptionally large area (most of eastern Utah and western Colorado). **Pages 190 and 191 must be removed from the manual in order to be used as a stereoscopic pair.**

PROBLEMS

1 What is the regional structure of the sedimentary rocks?

2 Why are there deep canyons in this part of the United States and no canyons in the midcontinent states such as Kansas and Ohio.

3 What evidence can you find that indicates recent uplift on a regional scale?

Enhanced Landsat mosaic 0 50 miles

4 In light of the regional structure of the sedimentary strata, explain the origin of the high, isolated, conical peaks. (They are composed of the rock type illustrated in Figure 3.14.)

5 Study the area closely and map the major faults and folds. What type of faulting occurs in this area?

6 What single geologic process is responsible for the development of most of the landforms in this area?